Genders and Sexualities in History

Series Editors: **John H. Arnold**, **Joanna Bourke** and **Sean Br**

Palgrave Macmillan's *Genders and Sexualities in History* series
ter new approaches to historical research in the fields of gen.
will promote world-class scholarship that concentrates upon the interconnected themes of
genders, sexualities, religions/religiosity, civil society, class formations, politics and war.

Historical studies of gender and sexuality have often been treated as disconnected fields,
while in recent years historical analyses in these two areas have synthesized, creating new
departures in historiography. By linking genders and sexualities with questions of religion,
civil society, politics and the contexts of war and conflict, this series will reflect recent developments in scholarship, moving away from the previously dominant and narrow histories
of science, scientific thought and legal processes. The result brings together scholarship from
contemporary, modern, early modern, medieval, classical and non-Western history to provide a diachronic forum for scholarship that incorporates new approaches to genders and
sexualities in history.

Wolfenden's Witnesses: Homosexuality in Postwar Britain is a landmark critical source edition.
The Wolfenden Report of 1957 has long been recognized as a watershed in gay law reform,
in the United Kingdom and throughout the anglophone world. Brian Lewis' superbly edited
source edition is an invaluable contribution to scholarship. The testimonials and evidence
given to the Wolfenden Committee and published here in this volume represent a fascinating and broad insight into how male homosexuality was understood in Britain in the 1950s.
The testimonials give account not only of official attitudes towards male homosexuality – by
the judiciary, police, politicians, and a range of other professionals and 'experts' – but also
of the self-fashioning of the mostly anonymous homosexual men who gave evidence to the
committee. In common with all volumes in the Genders and Sexualities in History series,
the source edition *Wolfenden's Witnesses: Homosexuality in Postwar Britain* presents a multi-
faceted and meticulously researched scholarly study, and is a sophisticated contribution to
our understanding of the past.

Titles include

John H. Arnold and Sean Brady (*editors*)
WHAT IS MASCULINITY?
Historical Dynamics from Antiquity to the Contemporary World

Valeria Babini, Chiara Beccalossi and Lucy Riall (*editors*)
ITALIAN SEXUALITIES UNCOVERED, 1789–1914

Victoria Bates
SEXUAL FORENSICS IN VICTORIAN AND EDWARDIAN ENGLAND
Age, Crime and Consent in the Courts

Heike Bauer and Matthew Cook (*editors*)
QUEER 1950s

Cordelia Beattie and Kirsten A. Fenton (*editors*)
INTERSECTIONS OF GENDER, RELIGION AND ETHNICITY IN THE MIDDLE AGES

Chiara Beccalossi
FEMALE SEXUAL INVERSION
Same-Sex Desires in Italian and British Sexology, c. 1870–1920

Gillian Williamson
BRITISH MASCULINITY IN THE *GENTLEMAN'S MAGAZINE*, 1731–1815

Midori Yamaguchi
DAUGHTERS OF THE ANGLICAN CLERGY
Religion, Gender and Identity in Victorian England

Genders and Sexualities in History Series
Series Standing Order 978–0–230–55185–5 Hardback
978–0–230–55186–2 Paperback
(*outside North America only*)

You can receive future titles in this series as they are published by placing a standing order. Please contact your bookseller or, in case of difficulty, write to us at the address below with your name and address, the title of the series and the ISBN quoted above.

Customer Services Department, Macmillan Distribution Ltd, Houndmills, Basingstoke, Hampshire RG21 6XS, England

Wolfenden's Witnesses

Homosexuality in Postwar Britain

Brian Lewis
McGill University, Montreal, Canada

First published 2016 by
PALGRAVE MACMILLAN

The author has asserted his right to be identified as the author of this work in accordance with the Copyright, Designs and Patents Act 1988.

Palgrave Macmillan in the UK is an imprint of Macmillan Publishers Limited, registered in England, company number 785998, of Houndmills, Basingstoke, Hampshire RG21 6XS.

Palgrave Macmillan in the US is a division of Nature America, Inc., One New York Plaza, Suite 4500, New York, NY 10004-1562.

Palgrave Macmillan is the global academic imprint of the above companies and has companies and representatives throughout the world.

Hardback ISBN: 978–1–137–32149–7
Paperback ISBN: 978–1–137–32148–0
E-PUB ISBN: 978–1–137–32151–0
E-PDF ISBN: 978–1–137–32150–3
DOI: 10.1057/9781137321503

Distribution in the UK, Europe and the rest of the world is by Palgrave Macmillan®, a division of Macmillan Publishers Limited, registered in England, company number 785998, of Houndmills, Basingstoke, Hampshire RG21 6XS.

A catalog record for this book is available from the Library of Congress.

A catalogue record for the book is available from the British Library.

For Edgar

Contents

Acknowledgements

I would like to thank the Social Sciences and Humanities Research Council of Canada, which helped fund this project. For information regarding copyright, I am indebted to Tim Padfield of the National Archives at Kew. For myriad suggestions, clarifications and help with sources, my thanks to Lisa Jorgensen, Roger Davidson, Chris Waters, Josh Mentanko, Julia Laite, Laura Doan, Elsbeth Heaman, Allan Hepburn, Justin Bengry, Jeff Meek, John Hall, Elizabeth Elbourne, Nancy Partner, Stephen Brooke, Frank Mort, Emma Vickers, Guy Baxter (University Archivist, the University of Reading) and Erin O'Neill (BBC Written Archives Centre, Caversham). The staff at the National Archives, the British Library and the McLennan Library at McGill have been models of courtesy and efficiency, as have the editorial staff at Palgrave Macmillan. Last but by no means least, my thanks to family, friends and colleagues, who kept me (relatively) sane through their support and encouragement, and to my partner, Edgar Navarro, who has lived with Wolfenden as long as I have known him. The book is dedicated to him, with love.

Note on Style

The memoranda and transcripts of interviews in the Wolfenden Papers consist of typed pages, all typed up by a team of stenographers, each deploying slightly different standards. Apart from the correction of an occasional obvious typographical error, I have retained the original spelling, grammar, punctuation, capitalization, underlining and (so far as possible) spacing. I have used ellipses (...) to indicate where text has been omitted.

Part I
Introduction

Introduction

In May 1954 Gilbert Andrew Nixon, 37, a company director of a firm of manufacturing chemists from West Kirby, Cheshire, killed himself with cyanide in a gaol cell. He had just been sentenced to 12 months' imprisonment at Somerset Assizes at Wells on a charge of gross indecency. Fourteen other men, some with Taunton addresses, had also pleaded guilty to committing or attempting to commit unnatural acts and acts of gross indecency; nine of them received prison sentences, ranging from one to four years. During the Second World War Nixon had received the Military Cross in Sicily and had recently retired as a lieutenant-colonel from the Territorial Army. 'It is a terrible thing', said Mr Justice Oliver in passing sentence, 'to see a man like you with a gallant military record standing as you are.' As recorded in *The Times*,

> The Judge said that it was an appalling thing for him that an ancient, historic, not very large town like Taunton should at one single Assize exhibit as many cases of homosexual crime as in the ordinary way he met with in a whole year. The answer, as he saw it, was not that the population of Taunton was more debased than other groups of the community, but that once that vice got established in any community it spread like a pestilence and unless held in check threatened to spread indefinitely.[1]

Gilbert Nixon's tragic end caused only a ripple in the national press. But it came amidst a period of intense introspection about homosexuality and the law among those hankering for reform and those, like Mr Justice Oliver, who wanted the 'pestilence' eliminated. (The two groups were not mutually exclusive.) On 24 August 1954 the Conservative government of Winston Churchill appointed a Departmental Committee on Homosexual Offences and Prostitution. Its remit was to consider and recommend changes to the laws relating to homosexual offences and prostitution, and to consider the treatment of those convicted of homosexual offences. This would involve and necessitate a thorough investigation into the causes and consequences of sexual deviancy. Chaired by John Wolfenden, the Vice-Chancellor of the University of Reading, the committee met on 62 days over the next three years; 32 of those days were devoted to the 'oral examination' of 'witnesses', mostly at the Home Office in Whitehall, partly at the Scottish Home Department, St Andrew's House, Edinburgh.[2] This volume provides a selection from the memoranda submitted by those witnesses and from the transcripts of the interviews themselves, all housed at the National Archives at Kew. It covers solely the homosexual concerns of the committee's remit; Julia Laite's companion volume deals with the prostitution.

The Wolfenden Report came out in 1957. It has long been recognized as a landmark in moves towards gay law reform. What is less well known is that the testimonials and written statements of the witnesses provide by far the most complete and extensive array of perspectives we have on how homosexuality was understood in Britain in the middle decades of the twentieth century. Those giving evidence, individually or through their professional associations, included a broad cross-section of official, professional and bureaucratic Britain: police chiefs, policemen, magistrates, judges, lawyers and Home Office civil servants; doctors, biologists (including Alfred Kinsey), psychiatrists, psychoanalysts and psychotherapists; prison governors, medical officers and probation officers; representatives of the churches, morality councils and progressive and ethical societies; approved school headteachers and youth organization leaders; representatives of the army, navy and air force; and a small handful of self-described but largely anonymous homosexuals.

This introduction gives a concise overview of the history of the committee and of the report, and maps out the major debates surrounding homosexuality in the 1950s. Part II, the meat of the collection, includes a representative range of the differing perspectives before the committee, contextualized and annotated. Part III highlights excerpts from the Wolfenden Report itself as a logical culmination of the committee's three years of information-gathering and deliberation.

*

When Sir David Maxwell Fyfe, the Home Secretary, brought the twin evils of prostitution and homosexuality before the cabinet table for discussion in February 1954, he thought that he had a growing and increasingly visible problem on his hands. The Conservative government craved an ordered society of gendered conformity, enhanced fecundity and contented domesticity, but the postwar reality appeared instead to have thrown up a host of social problems requiring urgent solutions.[3] For example, as John Wolfenden later reminisced, there was increasing alarm in official circles about the 'shamelessness' of prostitutes on the streets of London (their numbers and visibility especially embarrassing during Queen Elizabeth II's coronation in 1953, when the city was on show) and about an apparent increase in homosexual behaviour. The policing of both of these problems was no longer fit for purpose: the police had resorted to pulling in working girls in rotation (the prostitutes would be fined—in effect, taxed—and then return to the streets), and the policing of homosexual offences varied wildly across the country and had included some recent high-profile cases. All of this was serving to bring the law into disrepute.[4]

The greater visibility of homosexuality was critical in generating what a number of scholars have characterized as a moral panic.[5] There had been a considerable spike in England and Wales over the previous quarter of a century in cases known to the police of buggery, gross indecency and indecent assault—from 622 in 1931 to 6,644 in 1955—and in prosecutions for the same offences: from 390 in 1931 to 2,504 in 1955. The more than threefold rise since the end of the war was especially alarming.[6] A number of prominent individuals had found themselves before the courts in 1953 on homosexual charges, including the actor Sir John Gielgud (fined for persistently importuning in a public lavatory in Chelsea),[7] the Labour MP for Paddington North, William Field (ditto, in public lavatories in Piccadilly Circus and Leicester Square)[8] and the writer Rupert Croft-Cooke (sentenced to nine months for indecent acts with sailors picked up in the Fitzroy Tavern, near Tottenham Court Road).[9] The conviction in March 1954 of Lord Montagu of Beaulieu, his fellow landowner Michael Pitt-Rivers and Peter Wildeblood, the diplomatic correspondent of the *Daily Mail*, for private and consensual offences, and after highly questionable police methods, caused particular disquiet.[10] Croft-Cooke wrote darkly of a sexual McCarthyism, and the story became ingrained of a witch-hunt against homosexuals orchestrated in high places.[11] As Patrick Higgins, Matt Houlbrook and others have pointed out, the reality is more prosaic: lower-level decision-making in a small number of Metropolitan Police districts and among certain provincial forces accounts for most of the rise in statistics.[12] But the bulk of commentators at the time discounted the potential impact of more vigorous policing, preferring to believe that more prosecutions meant more offences committed—that homosexual practices were on the rise. It was difficult to pinpoint any one culprit, but the dislocation of families and the alleged breakdown of communal moral values because of the war featured prominently in most explanations.[13]

Just as worrisome for the government was the impression that more people were *talking* about homosexuality. The popular press in the late 1940s and early 1950s was jettisoning its former reticence about reporting in any but condensed and opaque terms on acts of sexual deviancy. The *Sunday Pictorial* featured a three-part series on 'Evil Men' in 1952, making it clear that these sick individuals were organizing in a degenerate sexual underworld and had designs upon *your* children.[14] More soberly, a serious newspaper such as the *Sunday Times* was calling for reform, picking up not only on the unsatisfactory spectacle in the courts but also on many decades of medical, scientific and religious questioning about sexual inversion. Its leader of 1 November 1953 argued, on the one hand, that 'the law that makes intercourse between males as such an indictable offence is neither enforceable nor consonant with current ethical standards', but also, on the other, that, 'It is not, in the long run an uncontrollable phenomenon. If, for some, perversion is an inherent and deep-rooted psychopathic state, for a far greater number it is a tendency which can be resisted, sublimated, or never awakened.'[15]

One of the most significant of these recent intellectual sallies upon which the editorial built was an article in 1952 by a Church of England clergyman, Derrick

Sherwin Bailey, followed by a report in 1954 commissioned by the Church of England Moral Welfare Council, both of which concluded that at least some homosexuals were born that way and—so long as they left the children alone and didn't frighten the horses—they should not find themselves up before the magistrate for whatever they chose to do behind closed doors.[16] Another was the 1952 study *Society and the Homosexual*, by the homosexual psychologist and sociologist Michael Schofield, writing under the pseudonym of Gordon Westwood, which sought to recast homosexuality as a psychological condition largely determined in early childhood rather than a deliberate, morally perverse choice.[17] A third was the 1953 novel *The Heart in Exile*, by a Hungarian expatriate, Adam de Hegedus, writing under the pseudonym of Rodney Garland, which pleaded for tolerance for the middle-class homosexual.[18]

Maxwell Fyfe did not favour a relaxation in the law for homosexuals. Yet, in presenting to his cabinet colleagues the arguments for an official inquiry, he recognized the 'considerable body of opinion which regards the existing law as antiquated and out of harmony with modern ideas'.[19] His first concern was clearly the prostitution problem, but he believed that the government could not strengthen the law and penalties against streetwalkers without the backing of a Royal Commission's authoritative findings. And, given the noise surrounding homosexuality, he reasoned that launching a thorough inquiry into prostitutes while ignoring the other sexual deviants would be scarcely credible. Churchill's preferred method of dealing with the unwelcome chatter was to curtail press freedom to publish the details of criminal prosecutions for homosexual offences, to prevent a repetition of the sensational coverage of the Montagu trial, but Maxwell Fyfe was able to persuade the cabinet that a dispassionate inquiry, which might educate the public, was preferable to censorship.[20]

In the event, the cabinet agreed to a Home Office departmental committee rather than a full-scale Royal Commission. This would have the advantage of allowing possibly reticent witnesses to speak off the record and in private.[21] With the exception of a small minority of MPs, such as Sir Robert Boothby, there was no strong parliamentary pressure for the decriminalization of homosexual offences, and the public remained largely hostile, so Maxwell Fyfe and the cabinet were not pushing for a progressive agenda here.[22] Their aim was to control the threat that marginal sexual figures posed to public morals and decent family values, however that might best be done. And this opened up the prospect that a more liberal variation on the state regulation of sexuality might prevail if this seemed to be the optimal way to achieve these goals.

*

The Home Office brought together a committee of 15 members. John ('Jack') Wolfenden (1906–85) was a safe pair of hands to act as chairman. A grammar school boy who won a scholarship to Oxford, started his career as an Oxford philosophy don, then became the headmaster of Uppingham and Shrewsbury Schools, he had been appointed to the vice-chancellorship at Reading in 1950. He was establishing his reputation as a diligent public servant and chair of councils and committees; a knighthood duly followed in 1956.[23] In common with the rest of the committee, throughout the proceedings and in his memoirs Wolfenden was adept at preserving a façade of impartiality and innocence on the question of homosexuality.[24] His knowledge—their knowledge—was ostensibly based purely upon professional experience (in his case predominantly as a headmaster in boarding schools[25]); family secrets or personal desires remained necessarily hidden. But if, as seems likely, he knew about the ostentatious homosexuality of his brilliant son Jeremy, this was disingenuous.

As an astonishingly self-aware 18-year-old, fresh from Eton and living in London, Jeremy Wolfenden wrote in 1952, in a statement anticipating many of the themes that the committee was going to encounter,

> I am a queer; so much is physically evident. But I have a lot more important things to do than waste my time hunting young men... I may end up with an undemanding and unsensational ménage with a single boy-friend; I may end up unsatisfied except for an occasional Sloane Street tart... I may, I suppose, turn to heterosexuality; but if by a pretty mature (physically) eighteen, I am not attracted by girls either physically or emotionally or aesthetically it seems unlikely.[26]

Jeremy's biographer, Sebastian Faulks, claims that Jack Wolfenden must have known about his son's sexual tastes for at least two years before he accepted the chairmanship of the committee, which invites speculation about an unacknowledged motivation for taking on the challenge. According to Jeremy, his father wrote to him, 'I have only two requests to make of you at the moment. (1) That we stay out of each other's way for the time being; (2) That you wear rather less make-up.'[27]

Home Office civil servants and Wolfenden selected a cross-section of the Establishment to make up the rest of the committee: representatives of the legal profession, medicine, government, education and religion; attempts at some kind of balance from England, Scotland and Wales; and three token women, mainly to contribute to the prostitution side of the mandate.[28] James Adair (1886–1982), OBE, was a solicitor, a former procurator fiscal (public prosecutor) for Glasgow and Chairman of the Scottish Council of the YMCA.[29] Mary Cohen (1893–1962), OBE, was vice-president of the Scottish Association of Mixed Clubs and Girls' Clubs.[30] Dr Desmond Curran (1903–85) was consultant psychiatrist at St George's Hospital, London.[31] The Rev. Canon Vigo Auguste Demant (1893–1983), theologian and social commentator, was Regius Professor of Moral and Pastoral Theology at Oxford.[32] Kenneth Diplock (1907–85), QC, the Recorder of Oxford, was appointed a judge of the Queen's Bench Division, with the customary knighthood, in 1956.[33] Sir Hugh Linstead (1901–1987) was a barrister, a pharmaceutical chemist and the Conservative MP for Putney.[34] Peter Francis Walter Kerr, 12th Marquess of Lothian (1922–2004), farmed estates in the Borders and subsequently held junior positions in Conservative administrations.[35] Kathleen Lovibond (1893–1976) chaired the Uxbridge Juvenile Magistrates' Court, was appointed CBE in 1955 and became Mayor of Uxbridge in 1956.[36] Victor Mishcon (1915–2006), the son of a rabbi, was a Brixton solicitor and a Labour member (and chairman) of the London County Council.[37] Goronwy Rees (1909–79) was the Principal of the University College of Wales, Aberystwyth.[38] The Rev. R. F. V. Scott (1897–1975) was the minister of St Columba's Church of Scotland, Pont Street, London.[39] Lady (Lily) Stopford (1890–1978) was an ophthalmologist and magistrate; she was married to Professor Sir John Stopford, Vice-Chancellor of the University of Manchester.[40] William Wells (1908–90), barrister and Labour MP for Walsall and then Walsall North, was appointed QC in 1955.[41] And Dr Joseph Whitby (1900–60), a former national bridge champion, was a general practitioner in north-west London with wartime psychiatric experience.[42]

Two members of the committee did not see it through to the end. The Presbyterian minister, the Rev. Scott, resigned in March 1956 on his appointment as Moderator of the General Assembly of the Church of Scotland; he could no longer spend so much time in London.[43] Goronwy Rees was forced out a month later. He was the most colourful and curious appointee to the committee. Born into a Welsh-speaking family, the son of a Calvinist Methodist minister in Aberystwyth, he flourished as a student at Oxford.[44] Here he gravitated towards the 'aesthetes' among the undergraduates: those who devoted themselves to poetry and the arts, spent their vacations soaking up the supposed decadence and sexual hedonism of Weimar Germany and who were—or affected to be—homosexual.[45] In spite of falling in love 'with a very beautiful and wild young man' in his second year,[46] he claimed that he himself was heterosexual, and he went on to marry and have a family. (Given his involvement with homosexual coteries at university and also during the war years, he at least was no innocent about gay lifestyles and identities at meetings of the Wolfenden Committee—and it was he who arranged for some homosexuals to appear as witnesses.) It was also at Oxford that he was seduced by Marxism and met the Cambridge Apostle and spy Guy Burgess, who became a close friend. Burgess recruited him for some low-level intelligence-gathering for the Comintern—spying that probably did not extend beyond the Nazi-Soviet Pact of August 1939. During the years when he was nurturing his family and making his living variously as a journalist, novelist, wartime officer, industrialist and college bursar, he remained close to Burgess—and when Burgess and Donald Maclean defected to the Soviet Union in 1951, Rees was interviewed by MI5. None of this prevented his appointment at Aberystwyth in 1953 and to the Wolfenden Committee a year later. Only when he recklessly provided material for a sensational series of articles on Burgess in the *People* in 1956 (anonymously, although his identity was swiftly revealed by the *Daily Telegraph*)—partly for the money, partly to exorcise his guilt for his previous support of communism—did his position become untenable. Wolfenden and Sir Frank Newsam, Permanent Under-Secretary at the Home Office, had no doubt that—since Rees had made it clear that he had been best friends with such a notorious and promiscuous homosexual and traitor as Burgess—any report that advocated homosexual law reform with his signature attached would be irreparably compromised. He was compelled to step down.[47]

*

More than 200 witnesses and organizations presented written evidence to and/or appeared before the committee.[48] The majority, reflecting a mid-century sense of what constituted expert opinion and which voices should be privileged, were invited to contribute at the suggestion of committee members or of the Home Office, though some submitted statements or other written materials unsolicited.[49] The usual, but not invariable, procedure was for a memorandum to be circulated around the committee and then for a follow-up interview with the author(s) to take place. As the excerpts below reveal, the range of opinions varied between those (mainly in the law enforcement community) who favoured continued criminalization of gay sex and those (mainly in medical, religious and ethical ranks) who tended to believe that homosexuality was a psycho-medical, not a legal, problem, that it was either innate (as sexologists such as Havelock Ellis would have it) or acquired at some stage during childhood development (possibly through early seduction) or a form of arrested development (the sub-Freudian perspective). Biology, hormones, dysfunctional families, degenerate societies: all might be to blame.[50] In addition to the multifarious attempts at discovering aetiologies, there is much fascinating information in the Wolfenden archive about possible treatments and about the policing of public sex and of cottaging (gay sex in public lavatories)—even a set of instructions on how to conduct physical examinations for sodomy in the Royal Navy. There are also many case studies of homosexuals and how they lived their lives. The focus is on men since the law was silent on consensual sex between women, but occasionally some of these expert commentators shared their thoughts on lesbians as well.

The final report recommended that homosexual sex between consenting males over the age of 21 in private be decriminalized, drawing a very firm public/private distinction, and that street prostitution be more strictly regulated. The latter was acted upon swiftly, but it took another decade for the gay sex suggestions to be enacted, in the Sexual Offences Act of 1967.[51] Despite this stuttering start, those who favour a liberal, progressive narrative have seen in Wolfenden a step in the right direction—the acknowledgment that homosexuals existed for reasons quite often beyond their control, that they deserved sympathy rather than censure and that they had the right to a private sexual life beyond the reach of the law. The Wolfenden Report may have been patronizing, condescending and limited (in that it recommended an unequal age of consent and the stricter policing of public sex), but it was, for its time, as enlightened and rational a pronouncement as one could expect from Official Britain, and it paved the way for all that followed: decriminalization, gay liberation and the raft of twenty-first-century reforms culminating in gay marriage. Critics of Wolfenden see something quite different: an attempt, in Foucauldian terms, to call 'the homosexual' into being in order better to control him. They allege that Wolfenden allowed only a very limited tolerance of a certain type of respectable, domesticated, masculine homosexual at the expense of all other varieties of sexual or gender expression, that the report crystallized a strict hetero/homo binary that ill favoured sexual and gender fluidity. They assert that the committee aimed to perpetuate the moral stigma against homosexual conduct and to eradicate it as far as possible beyond the hopeless class of congenital inverts, and that this identification or invention of a corralled, essential homosexual minority entailed the systematic denial of the universalizing reality of same-sex desire running through everyone.[52]

There is much to be said for all these variations on a critical theme, even if the normalizing, assimilationist intent behind Wolfendenian discourse could be and was subverted in practice. But this is to anticipate. In the early months of the committee's deliberations there was a widespread perception that the problem of homosexuality needed to be squared up to in a manly fashion. 'Homosexuality is no new phenomenon, but it is still not openly discussed', wrote J. Tudor Rees and Harley V. Usill in their introduction in 1955 to *They Stand Apart: A Critical Survey of the Problems of Homosexuality*. 'Is it not time that it was brought into the light of day for investigation with a view to its eradication? ... Are those who indulge in these corroding practices to be pitied as the victims of a disease or punished as criminals who have broken both the legal and the moral codes of law?'[53] The Wolfenden Committee's *raison d'être* was to provide an answer.

Part II
The Witnesses

1
Law Enforcers

Introduction

Since the Wolfenden Committee's mandate was to consider whether legal changes were necessary, the first document outlines the state of the law in 1954. This Home Office memorandum [i(a)] describes the three principal statutes that could ensnare men committing homosexual acts in England and Wales: sections 61 and 62 of the Offences Against the Person Act of 1861 (sodomy and attempted sodomy); section 11 (the Labouchère Amendment) of the Criminal Law Amendment Act of 1885 (gross indecency); and section 1 of the Vagrancy Act of 1898 (persistent soliciting and importuning). (By the time that the Committee reported in 1957, the Sexual Offences Act of 1956 had consolidated and replaced much of this legislation.) The second memorandum, from the Scottish Home Department [i(b)], points out that only the 1885 statute applied to Scotland as well. Instead, unlucky males committing sexual acts with other males might fall foul of the common law or of provisions in the Immoral Traffic (Scotland) Act of 1902 and the Criminal Law Amendment Act of 1912.[1] The third document, courtesy of Sir John Nott-Bower, Commissioner of the Metropolitan Police [i(c)], provides more detail about the offences, about the potential penalties and about the statistics of arrests, charges and sentences for 1953 in London—including additional ways of targeting homosexuals: municipal by-laws for gross indecency and the use of the Metropolitan Police Act of 1839 to clean up indecent or disorderly conduct in pubs.

The statements from Maxwell Fyfe's Home Office, from Nott-Bower and from the Director of Public Prosecutions (DPP), Sir Theobald Mathew [i(e)], all clearly called for the laws against homosexual acts to remain.[2] They pointed to the significant increase in numbers of offences in recent years and—although Nott-Bower gave some credence to the notion that greater police activity in certain districts was part of the explanation[3]—tended to believe that the growth in homosexual practices was real. If a glance at the Scottish figures, which had changed little over the same period, gave them pause, they failed to register it.[4] Instead, they argued that the deterrent against a great and growing threat to individual and community values needed to be maintained, in spite of the acknowledged pressure for reform. Wolfenden's witnesses were unanimous in declaring a need to protect the young from the depredations of homosexual corrupters and seducers and to preserve public order and decency from revolting displays; but the reformers argued that the law should not interfere with moral issues and that sex between consenting adult men in private should be allowed. The Home Office and the DPP insisted that only rarely did 'genuine' homosexuals, who conducted their relations in private and discreetly, end up before the courts, so no change in the law was necessary. It was to be a repeated refrain: the curious argument that the law as it stood might be rather harsh to 'respectable', monogamous homosexuals, but that these people need not worry because it was rarely applied to them. Any relaxation of the law to take account of this, on the other hand, would only encourage the expansion of the deplorably corrupting and promiscuous varieties of homosexual practice.

One such variety was importuning and/or sex between men in public lavatories (cottaging). Nott-Bower provided some information about this in his report, but it was Police Constables Darlington and Butcher of the Metropolitan Police who really went into detail in their interview before the committee [i(d)], describing how they and other plain-clothes officers working in tandem observed cottaging over prolonged periods.[5] Their accounts were notable for descriptions of the different types who populated the lavatories (from office workers to 'mincing' male prostitutes) and for contrasting sexual geographies (lunchtime importuning was popular in the old-fashioned, dimly lit urinals of Mayfair and Soho, evening cottaging in the newer, white-tile, well-lit conveniences of Chelsea). Their appraisals of conventional thinking regarding homosexuality are striking: 'to accuse a man of importuning male persons is very nearly as serious as accusing him of murder, and it is the most awful thing that could happen to a man.' But PC Butcher was at pains to stress that no innocent man need ever fear using a public lavatory lest his actions be misconstrued ('To prove a charge of persistently importuning it must mean persistent importuning'), and this was backed up by Nott-Bower in his interview.[6] But a number of Wolfenden's witnesses provided evidence or hearsay suggesting that the police often took short cuts because they knew their word would be accepted by the magistrate, or they accepted bribes, or—as the memorandum from the Council of the Bar put it [ii(b)]—the police and magistrates made mistakes that could ruin a man. A stench of police corruption hung over the proceedings, and even Jack Wolfenden 'thought it prudent to avoid public lavatories in the West End' while they lasted.[7]

If homosexuals could expect little sympathy from police and prosecutors, the opinions of lawyers, judges and prison officials were more mixed, even if a conventional language of horror and disgust about the moral evil of homosexual conduct permeated nearly all the discourse, pro- or anti-reform. The Law Society was against reform [ii(a)], but the Bar Council was divided as to whether consenting, adult, private, homosexual acts should be decriminalized. R. E. Seaton, speaking against change [ii(c)], posited the widely held 'seduction thesis', that every homosexual was a potential danger to youth. Confirmed homosexuals were very often attracted to boys, he continued rather more eccentrically, since boys were the nearest thing to women. Deterring the threat of homosexual contagion, he argued, was enough to justify the preservation of existing legislation. R. Ormrod countered by pleading for the separation of law and private morality and brought in examples from other times (ancient Greece) and places (continental Europe) where the toleration of homosexual practices did not produce a deluge. He, like many others, thought that the current laws incited blackmail.[8]

Ormrod recommended an age of consent of 21; the Magistrates' Association preferred 30 [iii(a)]. A narrow majority of its council (but not the membership) favoured reform; many other practices were, after all, also dangerous and undesirable evils, they maintained, and yet these escaped criminal sanction. The magistrates' recommendation for such a high age of consent rested on the belief that homosexuals could be divided into true inverts (whose chances of cure were slight) and those suffering from arrested development (who were treatable, given time—and so should not be written off prematurely as incorrigible).[9] As Peter Wildeblood's fictionalized character Sir Geoffrey Weston, QC, put it sardonically, 'The magistrates, bless their dotty hearts, took the line that they should punish people for being queer under that age, and let them off if they were over thirty, because if they were still doing it then they were past redemption'.[10]

Most of the individual metropolitan magistrates who submitted memoranda [iii(b–f)], however, were staunchly opposed to reform.[11] Sir Laurence Dunne, Chief Metropolitan Magistrate, for example, posited a version of what the committee came to call 'A Rake's Progress': that appetites were progressive; that if homosexuals were allowed to do what they wanted with adults in private, they would become sated and turn to youths or boys to spice up their jaded sex lives [iii(b)]. In his interview with the committee, J. P. Eddy, QC, former Stipendiary Magistrate of East and West Ham, emphatically supported this general line: '[Homosexuality] lives, I believe, on corrupting youth, because I believe that, in general, an adult homosexual is not particularly attracted by an adult homosexual. He wants youth: he wants boys.' Both he and Dunne, in a defiant display of English superiority, cared not that they did things differently elsewhere and that in many European countries homosexual acts were legal: 'We do not take our morals from the Continent.'[12] Chief Constable C. C. Martin of the Liverpool Police made a similar point about foreigners with the surprising observation, 'I do not think the moral standards in other countries are anything like they are in this country, except probably Canada and Fiji'.[13]

The psycho-medical model—that homosexuals needed psychological treatment and not prison—was fervently supported only by Claud Mullins, who had pioneered such approaches during his time on the bench in the 1930s and 1940s [iii(d)]. In his interview, Paul Bennett, Metropolitan Magistrate at Marlborough Street, was scathingly dismissive of this approach: 'The poor fellow needs treatment…It is the same excuse with shoplifters. They are all ill, every one of them. They are ill: their doctor says so. Here is the certificate.'[14]

The opinions of the Sheriffs-Substitute, Recorders and High Court Judges were similarly sharply divided [iii(g–k)]. Lord Goddard, the Lord Chief Justice, was not alone in being able to countenance the thought of decriminalization but only if this generosity did not extend to the revolting and depraved crime of buggery [iii(j)]. He would, in effect, retain the provisions of the 1861 Act but strike the words 'in private' from the Labouchère Amendment. (This was a matter that was going to concern the committee to a considerable extent: since the exist-ing legislation prescribed harsher penalties for buggery, ought this distinction to be reflected in the committee's recommendations?) Richard Elwes, the Recorder of Northampton, was interestingly different in that, in sharing two case studies with the committee, he made clear his repugnance for the Labouchère Amend-ment and regarded mutual masturbation, in public or private, as either trivial or not an offence at all [iii(h)]. But his touching faith that sodomy was not involved in either case, that the instances were isolated and that the men involved would go on to lead exemplary heterosexual lives appears to be remarkably naïve.[15]

In the memoranda from those involved in the correctional services [iv], espe-cially the prison doctors, we begin to see a more sophisticated discussion of aetiologies—a plethora of different ways in which to render the homosexually inclined population manageable by breaking it down into ever-smaller categories. Dr W. F. Roper, Senior Medical Officer at HM Prison Wakefield, provided a par-ticularly elaborate scheme of classification [iv(b)]. In their interviews Roper and Dr Matheson, Senior Medical Officer at HM Prison Brixton, subdivided the queer population still further according to sexual practices. 'The more fastidious man will not commit buggery', Roper asserted. For Matheson, 'I think the bugger is a worse type of person than the homosexual who goes in for mutual masturbation or something like that, and it does strike me too that the bugger is the worst. Just as in schizophrenia the schizophrenic who starts playing with his excreta you feel that he has gone down very, very far and is very, very ill, so with the bugger.'[16]

Other points raised in the memoranda of the correctional services included fre-quently expressed reservations about the wisdom of locking up homosexuals in all-male institutions if reformation were the objective; the doubtful efficacy of treatment (satisfactory 'adjustment' to one's condition, rather than a conversion to heterosexuality, was the best that therapy could hope to achieve); and the potential benefits of doses of oestrogen to dampen troublesome libidos. In his interview Frank Foster, Director of the Borstals and Young Persons' Division of the Central After-Care Association, added some rather different thoughts. Let peo-ple have homosexual intercourse, he said, 'whenever they can get their doctor to certify that they are sexually impotent'. This would serve the greater cause of sublimation:

> I do know homosexuals who are living a fully constructive life in the community—and we all do, I am sure—who are in no way giving way to any physical side of their homosexuality at all, and I do know too heterosexuals—head mistresses, head masters, people in all walks of life who live a per-fectly chaste life and give a contribution to the community, and surely it is perfectly—yes, natural—that the homosexual should be asked—we realize your position, it carries an additional responsibility—to make a sacrifice to the

community. Heterosexuals make this sacrifice with perfectly good adjustment and without any obvious frustration. We should extend it and say: 'If you would make the same sacrifice, not voluntarily as they do, but for the sake of the community... [*sic*]'.[17]

Since celibacy was probably beyond most men, the rationale for having 21 as the age of consent, as so many advocated, was often based on concerns about the susceptibility to corruption of 18- to 21-year-olds undergoing National Service. The memoranda from the armed forces [v] echoed these fears—arguing that a change in the law would sap the will of young men to resist homosexual advances and that discipline in the ranks (not to mention the moral fibre of the nation) would be undermined. These submissions listed the rules and regulations outlawing homosexual practice in the army, air force and navy, and suggested that—in spite of Guardsmen in London providing sexual services for money or gifts—the problem (including lesbianism) was under control and that no changes were advised. And, although many of Wolfenden's witnesses were quite insistent that most men who had sex with men were not of the 'flaming pansy' type, the Admiralty Fleet Orders [v(b)], which prescribed in excruciating detail how to do a physical examination of suspected homosexuals, instructed doctors to look for identifiable types as corroborative evidence: their general appearance, their feminine gestures, their use of cosmetics and the like.[18] This basic confusion about who the homosexual was—essential and marked in his features or 'as normal as you and me' except for his distasteful desires—was to be a persistent theme in much of the evidence presented in committee room 101 of the Home Office.

*

i.   Police and Prosecution

(a) HO 345/7: Memorandum Submitted by the Home Office

HOMOSEXUAL OFFENCES

...

I. The Present Law

2   The following offences known to English law may be regarded as "homosexual offences":–

(a) Sodomy;	Offences against the Person
(b) Attempted sodomy;	Act, 1861. Sections 61 and
(c) Assault with intent to commit sodomy;	62[19]
(d) Indecent assault on a male person by a male person;	
(e) Acts of gross indecency between male persons;	Criminal Law Amendment
(f) Procuring acts of gross indecency between male persons;	Act, 1885. Section 11.[20]

(g) Attempting to procure acts of gross
    indecency between male persons;
(h) Persistent soliciting or importuning   Vagrancy Act, 1898.
    of males by males for immoral   Section 1.[21]
    purposes.

There are no provisions in English law under which homosexual acts between women constitute criminal offences.[22]

...

II   The extent of the problem

10.   There are no reliable means of assessing the prevalence of homosexual activities. The only reliable statistical evidence available relates to the number of indictable offences which come to the notice of the police and the number of persons convicted of such offences...

...

12.   The figures indicate that since the end of the war there has been a considerable increase in the number of indictable offences. The increase, which is between four- and five-fold over pre-war figures, may not correspond exactly to the actual increase in the prevalence of such offences, but there can be little doubt that the prevalence of such offences has increased substantially and the increase is of such an order as to give cause for concern. The reasons for the increase are not known.

III.   Proposal that the existing law should be amended

13.   There is a considerable body of opinion which regards the existing law as antiquated and out of harmony with modern knowledge and ideas. There seems to be general agreement that the criminal law, in dealing with homosexual, as with heterosexual, acts, ought to provide effectively for the protection of the young and for the preservation of public order and decency, but it has been suggested that the law should confine itself to those objects and that unnatural relations between consenting adults in private, which are not criminal in many other countries, ought not to constitute criminal offences. The proponents of this view maintain that what consenting adults do in private is a moral issue, and that the criminal law ought not to concern itself with such acts unless they can be shown to affect society adversely; it is also represented that even if homosexual acts between men can be said to have anti-social consequences, these consequences can be no more anti-social than those resulting from homosexual acts between women, which the criminal law ignores; and that they are no more anti-social than fornication and adultery, which undermine the fundamental unit of society—i.e. the home and family.

14.   While it is true that the criminal law extends to acts committed by consenting adults in private, there is reason to suppose that the number of prosecutions in respect of acts occurring in these circumstances is small, and that such prosecutions will usually be found to relate to cases in which there are unusual

or aggravating circumstances. This is borne out by an enquiry into sexual offences recently conducted by the Cambridge University Department of Criminal Science, which covered all sexual offences reported to the police in 1947 in 14 police areas.[23] The cases investigated included those of 982 persons convicted of homosexual offences. 253 of these were convicted of indictable offences and 729 of non-indictable offences. The 253 indictable offences involved 402 male victims or accomplices, the great majority of whom were young people under the age of 16; only 11% of them were 21 years or over. There was only one case in which an adult was involved in an offence committed with another adult in private, and this was a case which came to the knowledge of the police when one of the two persons involved attempted to commit suicide. The non-indictable homosexual offences... are, by their nature, offences which occur in public places (mainly in parks and urinals) so that a public nuisance is involved.

15. Until the passing of the Criminal Law Amendment Act, 1885, acts of indecency between males committed in private were not criminal save in so far as they constituted offences against section 61 or section 62 of the Offences against the Person Act, 1861, or the common law...

...

The "Labouchère amendment" has been criticised on the score that its application to acts committed in private provides opportunities for blackmail. While there is no doubt that money has, from time to time, been extorted from men by means of a threat to expose acts of gross indecency, it is probable that the blackmailer relies on his victim's fear of social exposure rather than his fear of criminal proceedings. It is known, for example, that blackmail connected with homosexual acts takes place in countries where the acts are not criminal offences; so, also, in this country blackmailers sometimes extort money by means of a threat to expose sexual conduct which is not criminal (e.g. adultery).

...

31. It has been suggested from time to time that there are frequently medical or psychological causes contributing to homosexual offences, and that treatment should be remedial rather than punitive. The Home Office and the Prison Commission are aware of the importance in many of these cases of psychological factors, and recognise the importance of treatment in appropriate cases. Visiting psychotherapists have been appointed at certain prisons, and prison medical officers submit to the Prison Commissioners, for transfer to those prisons, the names of any prisoners serving sufficiently long sentences whom they think likely to benefit by psychotherapy. There is also a scheme under which prisoners serving sentences too short for treatment to be undertaken at one of the prison psychiatric centres, but nevertheless thought to be suitable for such treatment, can be seen by visiting psychiatrists from the regional hospital board, and as a result of such visits treatment is often started with a view to continuation by the same psychiatrist after release. The Prison Commissioners also have in mind the building of a special establishment for mentally abnormal, but not certifiable, prisoners, and some of

those who would go to such an institution would no doubt be prisoners convicted of homosexual offences.

32.   The problem cannot, however, be solved merely by substituting psychiatric or other treatment for punitive methods. Psychotherapy is not something which can be imposed and brought to a successful conclusion against a person's will. Many homosexual offenders are consciously or unconsciously unwilling to submit to treatment which may succeed in modifying their desires. There is also the consideration that the deterrent effect of punishment is important, whether the question is considered merely from the point of view of what is best for the offender or is considered from the wider point of view of what is best for the protection of society. It is suggested that the problem must be looked upon as one in which neither the considerations of therapeutic treatment nor the considerations of punishment can be disregarded. There must be effective methods of punishment and custody for the protection of the public, but the application of these methods should not exclude the use of therapeutic treatment in all suitable cases.

*

(b) HO 345/7: Memorandum Submitted by the Scottish Home Department, October 1954

HOMOSEXUAL OFFENCES

...

I.   The Present Law

2.   The following are "homosexual offences" at Scots law:–

(a) Sodomy	Common Law[24]
(b) Attempt to commit sodomy	
(c) Indecent assault on a male person by a male person	
(d) Lewd and libidinous practices and behaviour (if the practices or behaviour are homosexual)	
(e) Acts of gross indecency between male persons	Criminal Law Amendment Act, 1885, section 11
(f) Procuring acts of gross indecency between male persons	
(g) Attempting to procure acts of gross indecency between male persons	
(h) Persistent soliciting or importuning of males by males for immoral purposes	Immoral Traffic (Scotland) Act, 1902, section 1 and Criminal Law Amendment Act, 1912, section 7.[25]

(The Offences against the Person Act, 1861 and the Vagrancy Act, 1898 do not apply to Scotland.)

...

II. The Extent of the Problem

8. Statistics for the past 25 years giving the number of unnatural offences known to the police... suggest that there has been some slight increase since the war in homosexual offences; the annual average in post-war years of unnatural offences made known to the police was over 29, compared with an average of 13 in the 1930's; but the average of persons proceeded against was 13 as against 10. The figures are so small that caution is clearly necessary in making deductions from them.

III. Proposals for amendment of the Law

9. There has been less public comment on this subject recently in Scotland than in England and Wales and few proposals for legislation have been ventilated. In Scotland prosecutions for homosexual acts committed by consenting adults are taken only where the offence is aggravated by the circumstances, for example if public decency is affronted or if there is a disparity in age which implies seduction.[26]

IV. The Treatment of Offenders

...

12. The Department, which is responsible for the administration of the Scottish prisons and Borstals, recognises the importance of psychiatric treatment being given in suitable cases, and has made arrangements accordingly. Almost all male prisoners convicted of sexual offences are interviewed by a psychiatrist; most of them are seen by a consulting psychiatrist employed part-time in the prison service, others at a Regional Hospital Board psychiatric clinic. If the offender is suitable for treatment and is willing to undergo it he is during his sentence admitted to a psychiatric hospital as an in-patient or given treatment at a psychiatric clinic as an out-patient. An obstacle to treatment is undoubtedly the short sentence, particularly when imposed on a first offender...

*

(c) HO 345/7: Memorandum by Sir John Nott-Bower, K.C.V.O, Commissioner of Police of the Metropolis, 22 November 1954[27]

...

SECTION I

SODOMY INCLUDING ATTEMPTS

1. (a) Sodomy, including (b) attempts and (c) assaults with intent to commit sodomy—Sections 61 and 62 Offences Against the Person Act, 1861...

The convictions for this class of offence were 27 in 1951, 36 in 1952 and 33 in 1953. It would not be surprising if the number of convictions were but a small percentage of the number of crimes of this type committed, since the majority occur on private premises and are not known to the Police.

2.   Of the 33 persons convicted in 1953, 17 were convicted for offences committed on private premises and 16 for offences committed in vehicles, parks, lavatories and side streets. Of the offences committed on private premises 2 were between consenting adults.

3.   Details of the age groups of offenders and the persons with whom the offence was committed are as follows:–

Offenders	Persons with whom offence committed
11 Adults	Under 17
4 Adults	Youths 17 to 21
15 Adults	With other adults
1 Youth aged 19	A boy aged 15
2 Youths aged 18 and 19	1 Adult
Total—$\overline{33}$	

4.   These defendants were dealt with as follows:

Conditional discharge	1
Bound over	8
Probation	2
Over 6 months up to 1 year	3
Over 1 year up to 2 years	10
3 years	3
4 years	1
6 years	1
10 years	4

. . .

SECTION II

INDECENT ASSAULTS

8.   (d) Indecent assault on a male person by a male person. Section 62 Offences Against the Person Act 1861 . . .

The following arrests were made in the last three years:–

	Arrested	Convicted
1951	137	Not
1952	144	Available
1953	136	129

The 129 convicted offenders in 1953 were dealt with by the Courts as follows:–

Absolute or Conditional Discharge	16
Probation	22
Fine	23
Up to 3 months	14
Over 3 up to 6 months	19
Over 6 months up to 1 year	4
Over 1 year up to 3 years	6
Over 3 years up to 5 years	5
8 years	1
10 years	2
Otherwise dealt with	17

9.   In 1953 321 offences were <u>reported</u> to the Police and the victims came from the following age groups:–

Under 14	253 (79%)
14 and under 17	49 (15%)
17 and under 21	3 (1%)
Adults	16 (5%)
Total	321

. . .

The age groups of persons against whom complaints have been made (including those prosecuted) are:–

Under 14	6 (2%)
14 and under 17	15 (6%)
17 and under 21	8 (3%)
21 and under 30	49 (18%)
30 and over	188 (71%)
Total	266

As will be seen from the tables above, 94% of the alleged offences reported to the Police in 1953 were committed against boys under 17.

10.   The most common form of indecent assault is by a man who persuades or forces a boy to masturbate him at the same time as he is interfering with the boy's private parts. This is sometimes a prelude to an attempt at sodomy. The number of indecent assaults upon adult males is probably much lower because adults behaving in this matter are usually both willing parties so that there is no assault and the offence is then dealt with as gross indecency. (Section 11 Criminal Law Amendment Act 1885.) Indecent assaults upon adult males usually occur when a pervert makes a mistake and thinks he has met a fellow pervert and starts to interfere with him. The person assaulted often takes the law into his own hands.

11.   It is quite frequently found that a man becomes notorious in a neighbour-hood for interfering with small boys and eventually the complaint comes to the knowledge of the Police with the result that he is charged with a number of offences concerning a large number of boys. Usually the defendant has been brib-ing the children by small gifts of money or sweets or offers to take him out on excursions.

Where the victim is a boy under the age of 16 the defence of consent is not open to the defendant, also the case may be tried summarily with the consent of the accused. To avoid calling the boy or boys twice to give evidence, the normal police practice is to ask the Court to deal summarily with all such offences unless they are exceptionally grave.

. . .

SECTION III

GROSS INDECENCY

13.   (e) Acts of gross indecency between male persons, (f) Procuring acts of gross indecency and (g) attempting to procure acts of gross indecency between male persons. Section 11 Criminal Law Amendment Act, 1885 . . .

This offence can only be tried on indictment.

Gross indecency between male persons is usually committed between adult males. The same Section includes procuring and attempting to procure an act of gross indecency, and this Section is used where a man is trying to persuade a small boy to behave indecently . . .

14.   The following arrests have been recorded since 1950:–

1950	1951	1952	1953
262	478	305	335

. . .

In 1953, 240 of those arrested were dealt with under various Bye-laws (Parks, Local Authorities, etc.) and 95 were sent to trial under Section 11 of the Criminal Law Amendment Act, 1885.

The results of the trials of those 95 defendants were as follows:–

Acquitted	13
Approved School or Borstal	2
Discharged Absolutely or Conditionally	13
Probation	11
Fine	39
Under 3 months	2
3 months under 6 months	5
6 months under 12 months	7
1 year	2
7 years P.D.[28]	1

The sentence of 7 years Preventive Detention was passed on a man aged 51 with a very bad record. He was found committing the offence with a man aged 21 whom he had corrupted 6 years previously.

Out of 82 convicted defendants, 17 were sent to prison.

The divisions having the greatest number of arrests were 'A' Division (Hyde Park area) 76; 'E' Division (certain urinals) 63; and 'F' Division (certain urinals) 36. No other Division had over 20 arrests.

15. The offence of gross indecency is most commonly committed in public lava-tories or public house lavatories without an attendant, and normally between two men masturbating each other. Convictions for gross indecency also result where the defendant was about to commit sodomy and the Police stepped in before the act was completed or reached the stage where an attempt could be proved. On occasions three or four men are involved in masturbation with each other at the same time.

The offence seems to be committed by people of all walks of life. There are undoubtedly cases where the two men meet each other for the first time in a lavatory, and having exchanged certain signs, commit an offence, perhaps without any word being spoken. The two men are sometimes of different types, the waiter for example and a professional man.

. . .

17. About half of the offenders are arrested by officers in uniform and the remainder by officers in plain clothes but not specially employed for this purpose. Sometimes, however, a particular lavatory becomes notorious as a rendezvous where perverts indulge in gross indecency and special attention has to be given to it. To give an example in 1953 in a certain lavatory in Greenford a hole about 2" square was cut in one of the partitions between two W.C. cubicles presumably by a pervert, after this a number of men were arrested in these cubicles. The Local Authority placed sheets of zinc on both sides of the partition but holes were then cut by someone in the zinc and further cases occurred including 1 of sodomy. Sim-ilar occurrences have come to light in other parts of London including cubicles at Victoria Station.

The hole so cut is used either to pass notes suggesting meetings or for actual physical contact.

18. When offenders are arrested for gross indecency, the normal procedure is to put on two charges, the first under Section 11 of the Criminal Law Amendment Act, 1885 (not triable summarily), the second under a bye-law where applicable.

The London County Council bye-law of 20th March, 1900 (Every person who in any street or in any open space to which the public have access for the time being shall commit . . . . . . [sic] any act of indecency with any other person . . . . . . [sic] fine £5) is the one most frequently used. If neither defendant has a record for indecency and the case has no specially aggravating circumstances, the Court is informed that the prosecution have no objection to the case being dealt with under the bye-law if the Court sees fit to do so. In the majority of cases this suggestion is accepted by the Court. This procedure has been followed for many years past. It is expeditious, saves time and expense and it seems likely that in cases of mutual masturbation Quarter Sessions would award the same penalty as if the accused had been dealt with summarily under the bye-law.

19.   It will be obvious that this is an unsatisfactory method of dealing with these offences. It means that the question whether an offender is dealt with summarily under the bye-law or sent for trial depends in practice upon a number of factors which have nothing to do with the gravity of the offence which he has committed, e.g. whether the lavatory is in a street or open space (if in a Park or the London County Council area), whether the other party has a bad record, and the views of the particular Magistrate before whom he appears. Another point is that a person convicted of the preliminary act of importuning is liable to 6 months' imprisonment, whereas a person guilty of the actual offence of gross indecency is only liable under the bye-law to a fine of £5. On the other hand if the present practice were changed and all offenders sent for trial under Section 11 of the Criminal Law Amendment Act, 1885, a great deal of Police time and public money would be spent on offences which the courts of trial have shown on many occasions that they regard as comparatively trivial. I would urge that consideration be given to an alteration of the law enabling cases under Section 11 of the Criminal law Amendment Act, 1885 to be dealt with summarily as well as on indictment ... [29]

SECTION IV

IMPORTUNING

20.   (h) Persistent soliciting or importuning of males by males for immoral purposes. Section 1 of the Vagrancy Act, 1898 as amended by Section 7 of the Criminal Law Amendment Act, 1912...

...

21.   Importuning takes place with few exceptions in public lavatories and public house lavatories. There are a number of public lavatories which are notorious for perverts, viz:–

'B' Division	'C' Division	'E' Division
Victoria Station	Piccadilly Under-	+ Brydges Place
South Kensington	ground Station	(closed
Underground Stn.	Leicester Square	February, 1954)
Dudmaston Mews	+ Babmaes Street	+ Rose Street
Clareville Street	(closed after	(now closed
Dove Mews	11 p.m.)	until 1 a.m.)
	Providence Court	York Place
	Dansey Place	Dryden Street
	Falconberg Mews	Galen Place
	Three Kings Yard	
	Grosvenor Hill	

+ The attention of the Local Authorities is sometimes drawn to the unsavoury reputation of certain lavatories and in these particular cases, one was closed and the hours of the other two restricted.

22. Occasionally a man importunes other men in a lavatory by speaking to them, but the normal method is as follows:– A man enters the lavatory, occupies a stall next to another man, exposes himself towards the other man leaning back slightly from his stall, then looks down into the stall occupied by the other man, then smiles up into the other man's face. If the other man ignores his attentions, the importuner turns to the man on the other side of him, or, if needs be, changes his stall. Usually the importuner makes prolonged visits to the lavatory, say a quarter of an hour to twenty minutes, often changing his stall two or three times, and in many cases, several visits are made over a period of about an hour, either to the one lavatory or to several lavatories of the same reputation.

23. There appears to be two types of importuner:–

(1) the male prostitute who is trying to find a man who will pay him for indulging in indecent practices,
(2) the pervert who wants to meet other perverts to arrange a rendezvous where they can indulge in homo-sexuality with no question of payment. A number of this class are found to be visitors to Central London.

Taking these in turn, class (1) is believed to be small. The majority of male prostitutes probably would not take the risk of trying to find a man in a public lavatory where they would know there would be a danger of Police Officers keeping observation.

In the Piccadilly area there have been a few cases of men importuning in the street by speaking to passers-by. These men are male prostitutes. In a recent case the defendant, aged 27, had offered himself for £3 a night to men passing by in Piccadilly. He had been previously convicted of similarly importuning in Piccadilly in 1946, and importuning in a lavatory in Chelsea in July, 1954 and was sentenced to 6 months imprisonment. In 1952 two men were arrested dressed in women's clothing, and were sentenced to three months' imprisonment as male importuners. One had previously been convicted as a female prostitute for soliciting. The two men lived with one another in a flat where they took their clients. One of the men said he made £30 a week.

With regard to class (2), this class is undoubtedly much larger than class (1). There have been cases where perverts have been seen to solicit successfully in a lavatory and then the men have gone to some alley or other place where they can indulge in mutual masturbation, or even in sodomy, and in many cases of sodomy or gross indecency it has been found that the first meeting occurred in a urinal.

24. Importuners appear to come from all walks of society. It does, however, seem that there are more importuners and homosexuals committing homosexual

offences, among men employed in quasi domestic occupations, e.g. waiters, kitchen porters, barmen, and chefs.

25.   To obtain a conviction for importuning, it is necessary to prove that the defendant persistently importuned male persons for an immoral purpose. This necessitates a long observation most of which must usually take place in a urinal. Patrols of two men in plain clothes are specially authorised by a senior officer at New Scotland Yard to observe and detect these offences. These patrols are not authorised until complaints from the public or direct Police observation has shown that importuning in a particular area is becoming notorious. Because of the unpleasant nature of the work, this duty is most unpopular and the officers posted to these patrols are not normally employed for more than 4 weeks at a time and substantial intervals elapse before they are again so employed.

26.   The number of men arrested for importuning since 1946 is as follows:–

1946	1947	1948	1949	1950	1951	1952	1953
188	267	364	441	419	376	344	374

Of the cases in 1953 13 were dealt with on indictment and the others summarily.

   ... [B]y far the largest number of arrests took place on 'B' Division, (149) and 'C' Division, (92).

27.   These 241 arrests on 'B' and 'C' Division in 1953 which amount to 64% of the total in the Metropolitan Police District were dealt with in the following way:–

Dismissed	15
Absolute or Conditional Discharge	36
Probation	16
Fine	136
Imprisonment Under 3 mths	2
3 mths & under 6 mths	13
6 mths & under 1 year	12
1 year & under 2 years	7
2 years	1
Otherwise dealt with	3

   By Section 1 of the Vagrancy Act, 1898 (as Amended) an offender may be proceeded against summarily (6 months) or on indictment (2 years) ...

   The practice in the Metropolitan Police Force is to ask for summary trial unless the offender has a bad record ...

28.   Enquiries have been made to ascertain whether the rise in the number of arrests between 1946 and 1949 reflects a corresponding increase in the number of importuners but it would be unsafe to draw this conclusion from the figures. On 'C' Division the peak year was 1948 and the figures are now (1953) below those of 1946. In this Division a special two man patrol has been employed for every week throughout all the years from 1946. In 1949 there was a considerable

increase in the number of importuners arrested in the neighbouring 'B' Division (245 arrests—Chelsea, Pimlico area). This may have been caused by the shutting of certain urinals in 'C' Division, or Police action may have made the offence too dangerous in the 'C' Division urinals. In the main, however, the increase in the 'B' Division arrests were [*sic*] due to increased Police activity in that year, increased patrols having been authorised because it was suspected that the situation in that Division was becoming worse.

...

SECTION V

COGNATE OFFENCES

29.   ...[H]omosexuals occasionally cause trouble in other ways. For example a certain public house may become known as a haunt of homosexuals. If the men behave properly the licensee has no right to turn them out of his public house and no action is taken. When a large number of homosexuals congregate together there is frequently indecent or disorderly conduct which is extremely offensive to ordinary chance customers. Where there is evidence that this conduct is to the knowledge of the licensee and he is permitting it, action is taken against the licensee under Section 44 of the Metropolitan Police Act, 1839...

30.   To give some examples, in 1953 it was necessary to prosecute the licensees of a public house in Rupert Street, W.1.[30] When the premises were entered by uniform officers after 4 days observation, there were 137 customers in the bar of which 91 were known homosexuals. Only 1 of the 137 customers was a woman and she was drunk and had entered a few minutes before the raiding party with a drunken man. Throughout the observation conduct of a most offensive nature went on between homosexuals and several normal customers entered and left hurriedly in disgust. The resident licensee admitted in evidence that he had been "flooded by perverts and rather stunned by it all". This is probably a correct description of what does happen. By some means the news spreads among homosexuals that a certain public house is a rendezvous and within a short time a crowd of these men descend on the house. In another case where a prosecution was necessary, there had been an article in a popular magazine referring to a certain public house in Blackfriars Road as being "a house of character", the saloon bar of this house was then inundated every weekend with homosexuals many of them coming from some distance away.

...

SECTION VII

GENERAL OBSERVATIONS

...

37.   At present the meeting grounds are confined in the main to 4 comparatively small areas which have become known to perverts...namely Piccadilly

Circus, Victoria Station, South Kensington and Hyde Park. The amount of male importuning in the streets is negligible and except for the lavatories at Piccadilly Circus, Leicester Square, Victoria Station and South Kensington Underground Station, importuning usually takes place in comparative private and is not generally known to the public at large. The very fact that the law can impose severe penalties necessarily imposes a large degree of caution on offenders and, consequently, the activities of these people are not nearly so apparent or offensive to ordinary members of the Public as are the much more blatant activities of the prostitutes who are to be seen daily in many of our main streets.

...

*

(d) HO 345/12: Interview of P.C. Darlington, 'B' Division, and P.C. Butcher, 'C' Division, Metropolitan Police, 7 December 1954

...

[Q635] ... (P.C. BUTCHER): I speak for "C" Division ... The area bounded by Oxford Street, Charing Cross Road, Pall Mall, Piccadilly and Park Lane. That area is divided up for our purposes into the Mayfair area and the Soho area, and in those areas we get two distinctly different sorts of homosexuals. In "C" Division there are always two men. They do a month's duty together in plain clothes on this importuning ... I am not exaggerating when I say that 90 per cent. of the people I have arrested in the Mayfair area are actually in their lunch break. We have followed people from one large block of offices in Berkeley Square. On four days we followed four separate men, from the time they left their office until the time they went back in an hour's time, and they did nothing else but frequent urinals in the Mayfair area ...

In the Mayfair area there are three urinals that are quite famous throughout the world. They are Providence Court, the urinal attached to the public house in George Yard, and the Three Kings Yard urinal which is by the Standard Motor Company's showrooms at the back of Grosvenor Square. Invariably these people we follow do run between these three urinals, one, two, three and back again until they meet someone who is willing to fall in with their wishes, and away they go. During the lunch hour their object is more to make arrangements for a meeting later in the evening. We do not get any gross indecency up there at all. They just meet someone who is of the same way of thinking as themselves, who is willing to try that sort of thing, and they make an arrangement to go for a drink or to go to their flat later on in the evening. It has happened to me dozens of times. They come up to me and say "Are you interested in this sort of thing?" and I can honestly say "Yes" and an arrangement is made, but I do not keep it that is the only thing.

As I was saying these urinals are famous throughout the world and I can quote an example of that. There was a famous Russian dancer who came over to this country a couple of years ago. He went to stay in Tunbridge Wells and he brought his masseur with him. This was the first time this Russian had ever been to this country, and the very next day he was sent out to Charing Cross Road to get a pair of dancing pumps for the great dancer. We saw him in Mayfair, we followed him for 45 minutes and arrested him. We said "How long have you been in the country?" and he said "I came in yesterday". He was asked "How did you know these places existed?" because in this country to a Londoner except a worker in the Mayfair area, these places are placed in such a manner that you would never see them, and he said "I have a map", and he produced a map from his pocket that was given to him in Russia.[31] I arrested another man who heard about it in Hong Kong and another man who knew about the lavatories and was told about them in South Africa. They come to the West End and know exactly where to go... You very rarely, if ever, get the professional male prostitute in Mayfair who does it for gain. The men that use the Mayfair area, if I may put it this way, do it for the love of the thing, to satisfy their own emotions, and that generally consists of the run around in the lunchtime or in the forenoon. I think they get a great deal of satisfaction from the chase, shall I call it, the chase. These urinals have got a certain odour inside them, staleness, and it does excite them, there is no doubt about it. I have noticed it dozens and dozens of times that when a urinal has been cleaned out with dettol and scrubbed clean and smells clean they will not go anywhere near it, but once the smell of cleanliness has worn off you can see these people definitely working themselves up into a frenzy inside, and once they are on heat, that is the way to describe it, it is like the bitch, once they have the scent there is no holding them, they are oblivious to everything else. The 6 ft. 3 ins. policemen whether in plain clothes or in size 11 shoes and Harris tweed sports clothes reeks of being a policeman, and criminals can tell them, but not these perverts. They are oblivious to it, but to my mind the stronger and the bigger the man the more interested they are in getting to know the other side of him. These men in Mayfair are of that type.

On the other side in the Soho area of Piccadilly Circus you are more inclined to get the criminal type of homosexual. He is in it purely for what money he can make out of it. He definitely makes himself up with cosmetics, adopts feminine behaviour and his whole dress absolutely reeks of it. Even the man in the street can tell what he is by walking alongside. They adopt the mincing gait, the plucked eyebrows, they wave their hair and dye the hair, paint their finger nails and use cosmetics. They leave the people they accost in no doubt as to what their intentions are, and I think I am safe in saying that at the present time the minimum charge is £2. They treat it very much from the prostitute angle, they are nothing more than male prostitutes, they frequent certain public houses. There is one, the Fitzroy public house in the Tottenham Court Road area I think is the most famous one, and that is known throughout the world. I think that came about because it was used a great deal during the war by the fighting services, and these people cottoned on to that and moved in. These people solicit men in the same manner as female prostitutes, that is in the urinal when there is no one else there. They will leave you in no doubt at all when they are in the urinal, but since lots of them have been arrested by the police and they have been sentenced to quite severe sentences they have got a little wary and they tend to frequent certain parts of Piccadilly now. If things are quiet they will solicit in that manner, but otherwise a mere lift of the eyebrows. I watched one of them for an awful long time, many days, and never saw him speak to a soul. When I did eventually get him and asked him how he accosted these men he said "I can tell if a man is interested in me by just looking at him." There is a certain look among these people, and if they look and the look is returned that is quite enough and they will enter into a conversation on the way.

The trouble with these professional prostitutes is that they will pick up these clients, and they go home to their clients' flats or houses, and invariably when they leave the next morning the client's suit, his clocks or anything that is hanging around handy goes with him. These thefts are very very rarely reported, because the loser just could not bear to go to the police station and give the facts, so a lot more of this stealing from the premises goes on than is actually reported, a great deal more, and if you do catch these people with the stolen stuff on them the loser is very unwilling to charge, and it is very unusual to get a case arising from that sort of thing.

The trouble with the West End is that we get a great deal of homosexuals from other countries. Apparently the view in European countries is not the same as in Great Britain, and these people think they are still in their own countries. They know where to go and they are arrested more easily, much more easily than the English pervert, because they do it more openly. I have had people from Scandinavia, Switzerland, France, Germany, Belgium all the European countries and from America.

The professional is the lowest form of animal life existent in the West End. Other thieves stick together and there is a sort of esprit de corps, they have certain codes, and they will not squeal on each other, but the biggest insult you could do to a thief is to call him a homosexual, or arrest him for importuning. That is the biggest insult you could give a criminal. These layabouts I call them, that is the term they use in the West End, these low intelligence people, they have not got the intelligence to do a straightforward burglary or steal from a car, it is beyond them. They are too lazy to work and they find it is easy to do this sort of thing. It means a night's sleep for them and money as well, and they sink to that, and that is where you get the beginning of your professional prostitute.

As regards the sort of people we get, we get them from every walk of life. I have had serving soldiers, members of the clergy, particularly from any occupation or profession that has an air of artificiality about it, like the acting profession, the creative professions like hairdressers, dress designers. These people that sort of live in a world of their own, they adopt that manner in their business and they finish up like that. The majority of our people do come from those walks of life. It is very rarely that one arrests a coalman or a dustman or anything like that. The manual labourer never seems to come into that sort of thing. What more often happens is that a manual labourer will go into one of these urinals, get accosted by a person of the homosexual type and he will just hit him. He leaves him in no doubt he does not want to know him, he takes the law into his own hands and hits the chap. Then, of course, these people never report the assault, they take it and bear it as a part of their occupational risk, I suppose you could call it...

...The average person in London does not know of the urinals. They are out of the way, and I think that is the trouble. They get washed out once very early in the morning and they are left unattended throughout the day...The lighting in the evenings leaves a lot to be desired. Most of them now are lit by electricity, but it is still not enough. It is still dim in there, and that is conducive to these people carrying on their practice. Some in the Soho area are lit by gas, and at the first opportunity someone puts the gas out, and then you have to get the Westminster City Council, or we generally do it ourselves to light the gas again, but with the gas-lit urinal the first opportunity there is for the chap to be in there on his own he will put the light out and these places are a menace. When you get a Westminster Council urinal like, for example, the Leicester Square one, the one by the Irving Statue[32] and the one in Piccadilly by Green Park Station they are large and can easily be seen from any part of the urinal. There is an attendant in attendance all the time and you very very rarely see homosexuals using those places. Piccadilly Circus they do use, but they more or less use the circle area of the underground station outside. They stand there and watch the people going in there, and if they see someone they think would be interested they may go in there, but on the whole with the large toilets where most of them have white tiles, the modern sort of convenience, they do not like them, and they do not use them. It is the old-fashioned type that these people use...

...

[Q641] DR. SCOTT: The four men you spoke of coming from the office to go out to lunch together, what first attracted you to them? You spoke as if you had followed them from the office, but did you first spot them in the urinal?—A. To prove a charge of persistently importuning it must mean persistent importuning. What had happened in those cases—before we do any real following of people we spend a couple of days sitting in an old house or in a garage and then watch the people who come and go. There is only one way to get to know these people. If you spend too much time in the urinals you give the game away so we watched four, and we thought this man was going to another urinal, and instead of that he went back into his office. We found by discreet enquiry that he worked there, we found out his lunchtime and when he came out three others came out as well with him that we had already seen using these places a great deal more than the average person would, and it was just a matter of following them from their office round and waiting for them to go back. There was persistency in visiting lavatories, they must have made ten visits to these lavatories in that time.

[Q642] SIR HUGH LINSTEAD: Did you get convictions in those cases?—A. I have only ever lost one case at court. I think the answer to that question is that to accuse a man of importuning male persons is very nearly as serious as accusing him of murder, and it is the most awful thing that could happen to a man. I know for myself I make doubly sure that I have a case before I arrest. I would let him go with half an hour, 40 minutes observation because there has been a little element of doubt, I was not quite satisfied and I say he will come another day. I say that when I have arrested a man for importuning he is done to rights, as it were, there is no doubt whatever in my mind that he is up to what I think he is, because it is a very serious thing, and the greatest of care must be taken...

[Q642]...(P.C. DARLINGTON): In 'B' Division, Chelsea, which I am more conversant with, we get it in the evenings. We get very very little, if any, during the lunch hour. We very seldom keep observation outside the urinal. We are two a month detailed off to do this and we decide to actually go to certain urinals, and we go there, and we take it in turn to go in. One goes in and the other stays outside. We never speak to each other, we have a certain look, a code where we can give certain signs. You go in and you sort of look as if you are going to urinate. We make it our business that we just stand there as if we are urinating and look about us to see what we can see. These people come in and one out of every 50 urinates, you get occasionally one that does, but it is one in 50, the man on 'B' never does. They stand there and there are three or four or more and you are all standing there, and it is deathly quiet. There is no sound of anybody urinating at all. Somebody is in there for some other purpose. You see these men, they start to look about them and give each other the glad eye. They nod their head, they sometimes speak and reach out and touch one another, and practically everyone you see will be masturbating himself. In 'B' Division they do it quite openly. They are in there for two or three minutes and some of them are cunning. Some go in for three or four minutes and will not do a thing, they just stand there, and you cannot tell what they are doing, but the majority of them do it quite openly. They masturbate themselves, that is the main thing. You see somebody comes up and starts to do the same thing and through that we get gross indecency. Then we change round and the other one comes in. We do two or three minutes and then we go out because you cannot stay in too long. The only time we follow the individual is if he stays in this place for three or four or five minutes and then goes out and he becomes a suspect and we follow him. They go to these different urinals, but it is pretty well all the same on 'B', all in the evening, they all do practically the same thing, and it is with a view I think either to getting together, some of them, in these urinals and committing gross indecency, to masturbate with each other, or picking up a friend to take back to their flat or their home to indulge in what we call the finer arts—what they are I do not know. We have not got any male prostitutes on 'B'. I have never had one to deal with myself. There is a little point, that my colleague mentioned about the dark and dirty urinals. The three main urinals on 'B' at Chelsea that we look after are brilliantly lit. In South Kensington Underground is the main one, and that is very very well lit. You can see everything that is going on there, and it does not seem to deter those individuals, not one bit. South Kensington Underground is split into stalls and sometimes these individuals stand close to the stalls, and if they take a fancy to someone, or by some nod of the head or by the wink of the eye they get together, they stand back and sort of show each other what the idea is. Some members of the public do show their disgust by saying "You dirty old so-and-so" and walk out, but they are pretty well all the same, they all masturbate.

[Q643]...Some of them go in for only a couple of minutes. That is nothing. We cannot do anything about that. They go in for a couple of minutes and they get together and there are quite a few places on the Chelsea Section where this obviously takes place behind locked doors. It is the same faces you see. They see you and we see them, and they get to know us, and we get to know them. There is nothing we can do about that sort of person, but the majority are importuning in urinals for either gross indecency or going to their homes. I do not think there is much I can add. As my friend says it is the thrill of the chase, and if they take a fancy to you, or take a fancy to someone they will follow you around for hours, I should say. Several times I have been in these urinals and I have been importuned. I come out, and my colleague goes in, and the man you have under observation, he has taken a fancy for you, and when you come out he comes out, and when you go back in he goes back in, and that sort of thing. If they take a fancy they are oblivious to what you are, because on one particular month that I had, I was working on shift for three months—6 ft. 4 inch policemen, they say you can smell a policeman, I do not know whether you can—but there was one instance where we were standing either side of this importuner in South Kensington, he was only a little chap looking up at me and the other one, but it did not seem to make any difference. When I arrested him I said "Couldn't you tell we were policemen?" and he said "I didn't even think". They do not, they are so engrossed in what they are doing they do not pay any attention...(P.C. BUTCHER): I might just tell you a little story about this following. I had one one day and he was interested in me, and I said to the chap at work with me that if he will follow me to such lengths he will follow me to the police station, and he did. I gave him a smile, I turned and I walked from Piccadilly Circus up Regent Street, Vigo Street, Old Burlington Street, and even walked in the back door of the West End Central Police Station. I there arrested him, and he said "I thought we were going to your place", and that was true![33]

...We were in no doubt, we could have arrested him there and then, but if you can arrest a person discreetly it saves an awful lot of trouble, and it saves giving the game away to other people who use the urinals.

...

[Q648]...There was a man arrested in Curzon Street the other day. His clothing was one hundred per cent. feminine, everything to high heeled shoes in fact. Two uniformed police officers watched him and they thought he was a prostitute they did not know. They watched him and they went up to him and arrested him for using insulting behaviour, which is the power we use—you know about that—and they said "What is your name?" He said Helen somebody or other, and they took him to the police station and put him in the female charge room. Then somehow or other, you know what it is, you look at a person, they thought "This is odd" and they tackled him about his sex. He persisted his name was Helen and they decided someone had to search him. Anyway it came about that he was a man. I saw this chap in there and I had a little chat with him. I said "Look, these people that you accost and you agree to go in their cars or go to their place"—he had no place of his own—"they go with you under the impression that you are a woman"— if I may say so he was rather a dainty fellow, his shape—"What happens to the people when you disclose that you are a man, because it has got to come out?" He was quite frank about it and he said that four out of every five men that went with him up there stayed and had what they wanted with a man. That rather shook me, because I did not think it was quite as bad as that. He said four out of every five men stay.

. . .

Q

[Q652] CHAIRMAN: [With reference to an earlier point made by Butcher about discovering books of phone numbers on some of those arrested] You said at one stage that you saw this man's diary . . . With telephone numbers and the like, but at what stage are you in a position to see his diary?—A. [Butcher]: Prisoner's property. Every prisoner is searched. They have to be searched so that we would see if there is anything relevant to the charge, we just have a look through to see if there is anything obscene or showing the sort of mind this man has . . .

[Q653] Q. And do you come across cross connections in diaries?—A. Yes, the same old 'phone numbers keep cropping up.

Q.   And what happens at that point, when you see the same old 'phone numbers cropping up, what do you do about it?—A. Sometimes it is behind locked doors— there is not very much that you can do, is there?

Q.   It is a question I should rather ask you I think.—A. I see your point, but with these 'phone numbers there it is, the chap may have homosexual tendencies, but you do not waste an awful lot of time by following him. They may go to his friend's place, they may go to a hotel. It is a very slender thread to follow up when there are so many others doing outright importuning, and those are people importuning on the streets.

[Q654] Q   Would it be someone else's job to follow up clues that you might get?— A. Some of them have been followed up. There is the classic example of the place in Curzon Street where the S.I.B.[34] collaborated, we went in together, where they had the guardsmen riding around in a harness and they were chasing them with whips. These guardsmen were being paid quite large sums, they were in the nude and they had a harness on, and these perverts were chasing them around with whips to get the satisfaction.[35]

. . .

[Q656] DR. CURRAN: When I was in the States some years ago the mouth was very popular, I believe still is much more popular in the States than here. Is that very popular here?—A. Yes, because the West End of London is a regular haunt of American soldiers—you probably know that as well as I do—Coventry Street, we call it the "Standard front", that is what we call it, that is Coventry Street from the Prince of Wales to Piccadilly Circus. From 11 o'clock onwards 75 per cent. of the people there are Americans and we have had several gross indecencies in doorways.

Q. But it is an American thing more?—A. It seems to go more with them, yes.

[Q657] DR. SCOTT: You said you have only failed once to get a conviction. Do you find much protest, or do they usually admit, when you charge them, do they usually admit right away? Are the cases often defended?—A. Yes. We can divide them into two different sorts of people. The ordinary chap in the street who gets arrested for importuning ... he knows himself that he has been done fairly and squarely, and if he is going to get any good defence—certain well-known counsel seem to specialise in this sort of defence—it is going to cost big money. When he weighs it up in his own mind he realises that he is guilty and he has got very little to lose anyway. He will just go into the court the next morning and plead guilty. They know that some courts deal more severely than others. We have got the example of our division now. They know more or less what to expect. They are quite prepared for their £10 fine, whereas if they employed counsel, who would invariably ask for lots of remands and juniors and all the rest of it and make a big show and really put the police through their paces, it costs quite a lot of money. Those sort of chaps do not seem to dispute it. They know they have been done to rights, as it were, and they say "There it is, I will go to court and plead guilty". Then you meet the professional man who has got a career. He has got an awful lot to lose, he loses an awful lot of face as well amongst his class, he will fight tooth and nail, and they have counsel. The favourite thing is doctors, and it is most amazing the diseases that these people develop.—(P.C. DARLINGTON): I think what the majority fear is going to prison. If you arrest a man who has been convicted twice and been fined £10 first time and £10 the second irrespective of what it is he thinks "If I come up the third time I am going to prison" and he will fight tooth and nail.—(P.C. BUTCHER): He has got something to lose then, but if he knows that the chances are ninety nine to one that he is going to get a fine that time up he is quite resigned to it.

MR. ADAIR: He gets back to enjoy his same companion that same day without having lost any time?—A. That is quite right, but the defence is purely medical I have found, and the number of diseases that these people can be found to be suffering from is shattering, they are not fit to be on the streets, let alone in urinals, but they put up a good show. These doctors do cost money, and it all costs money for these people.

...

[Q664] DR. SCOTT: If you see one fellow importuning another fellow who seems to be responding to him do you arrest both?—A. Not on just that evidence, no.

Q. You would have had to see the other fellow importuning too?—A. The charge is persistently, Sir.

MR. ADAIR: It has to be persistent, it cannot be persistent with one individual.

DR. SCOTT: You would only arrest the importuner? There is no possibility of an innocent man standing in the stall beside an importuner finding himself arrested along with the other?—A. No chance whatever.

...

*

(e) HO 345/7: Memorandum by the Director of Public Prosecutions, Sir Theobald Mathew, K.B.E., M.C.[36]

1. Homosexual Offences

1. In this particular field my Department conducts only relatively few prosecutions...

2. The cases prosecuted by my Department are normally those of special gravity, difficulty or importance, and the majority are of persons in authority, who have used their position to make victims or accomplices of those in their charge or under their influence.

Therefore any views that I may express are no doubt influenced, to some extent at least, by the fact that I see the worst of these cases.
...

4. I agree with the Home Office view that, although the criminal statistics may be unreliable as a means of assessing the prevalence of homosexual practices, there has been a considerable increase in these practices since the end of the war.

5. Although the reasons for such increase must be a matter of speculation and there are, no doubt, a large number of contributory causes, I would suggest that the complete change in the life of the male adolescent, due to the extended period of education and to National Service, creates an atmosphere in which these habits can be easily acquired and become ingrained.

6. Assuming that homosexuality is still to be regarded as potentially harmful both to the community and to the individual—a view with which from my experience I would certainly agree—I do not consider that any relaxation in the existing law would be justified, particularly at a time when the prevalence of this conduct appears to be increasing, unless there is good reason to believe that this would tend to cause a decrease in these practices.

7. I am aware of the criticisms of the provisions of section 11 of the Criminal Law Amendment Act 1885. I believe these to be largely theoretical, in so far as they refer to the commission of offences in private between two consenting adults, for the reason that, in these circumstances, there is generally speaking no complaint, and

little possibility, in any event, of obtaining the necessary corroborative evidence to prosecute.

...

8. I have no evidence to support the criticism that this provision has provided special opportunities for blackmail...

...

10. In my view there is little that the criminal law can do to help in this matter. Its powers, for practical purposes, are confined to punishing those who have in fact corrupted or attempted to corrupt the young, or have "procured" their accomplices in these crimes, or who have offended against public order and decency. By the time conduct of this kind has been brought to the notice of the police the accused is, in most cases, past reformation, and serious harm has been done to the victims.

11. In my opinion this problem must be tackled at a much earlier stage. A decent and healthy community, in this respect, can only be created and maintained by inculcating into the young a clean sex outlook, both physical and mental, through parental influence and by education.

12. In my view the educational authorities have a special responsibility in this matter, particularly in residential schools, because they have charge of the young at or about the age of puberty and immediately afterwards.

The history of many accused has started with being themselves taught minor homosexual practices, when at school, by masters or by older boys. They in their turn have taught younger boys, and have acquired a habit that requires considerable strength of character to eradicate.

13. I have an impression that, at the present time, minor homosexual practices amongst boys at some schools are not being treated, as a matter of discipline, with the same severity as they were a generation ago, and that the supervision of the boys at night and out of school hours is not so strict. Moreover I have had a disturbing number of cases in which masters suspected, on good prima facie evidence, of interfering with the boys in one school, have been allowed to resign and have been given excellent references to another school, with disastrous results.

14. I realise that there are in every community a number of, what I may term, "genuine" homosexuals, who because of physical or mental abnormalities present an intractable problem. But, in my experience, these are only a very small percentage of those who come before the courts.

I do not believe that most accused, whatever they may say, have this excuse. They have, in many cases against their better instincts, through curiosity, hero worship or cupidity, allowed themselves to be initiated into these practices, and have persisted in them, until their capacity to control or limit their homosexual desires has ceased to exist.

In my opinion, therefore, it is of paramount importance (a) that boys and young men should be taught that these habits are dirty, degrading and harmful, and the negation of decent manhood, and (b) that every practicable precaution should be taken to ensure that those in charge of boys and young men are themselves free from this taint.

. . .

*

ii. Lawyers

(a) HO 345/8: Memorandum of the Council of the Law Society,[37] June 1955

. . .

II. HOMOSEXUAL OFFENCES

3.   It is believed that the main question which the Departmental Committee are considering under this head is whether homosexual practices (buggery and gross indecency) between adult consenting males should, if committed in private, cease to be treated as criminal offences. Being well aware that this question raises difficult medical, ethical and psychological problems, upon which they are not qualified to express an opinion, the Council have preferred to test these offences against Bentham's rules for criminal legislation set out in Kenny's "Outlines of Criminal Law," which they regard as still supplying a useful and practical guide despite the great advances in penology which have taken place since Bentham's time.[38] The effect is that the legislature should be satisfied upon six points before passing criminal legislation to suppress an objectionable form of conduct—and the Council suggest there can be no doubt that the community does regard homosexual practices as objectionable—namely:

(1) The objectionable practice should be productive not merely of evils, but of evils so great as to counter-balance the suffering, direct and indirect, which the infliction of criminal punishment necessarily involves;
(2) It should admit of being defined with legal precision;
(3) It should admit of being proved by cogent evidence;
(4) The evidence should be such as can usually be obtained without impairing the privacy and confidence of domestic life;
(5) It must be reprobated by the current feelings of the community; and
(6) It must be a practice against which adequate protection cannot be secured to the community by the milder sanctions which Civil Courts can wield.

4.   Judged by these tests, the Council think that both buggery and gross inde-cency should remain criminal offences even when committed in private between consenting male persons of full age. The offences are productive of great evils (point 1) inasmuch as they (i) tend to reduce the inclination to marry; (ii) militate against the procreation of children; (iii) are calculated to result in damage to the

State if they get too strong a hold; (iv) are likely (if legalised in private between genuine homosexuals) to contaminate others (and particularly the young); (v) may well, if allowed to go unchecked, result in male brothels; and (vi) probably tend to spread venereal disease.

The offences can be defined with reasonable legal precision (point (2)); they are capable of proof by cogent evidence (point (3)); male persons living together do not constitute "domestic life"—it does not therefore impair the privacy and confidence of domestic life for such proceedings to be taken (point (4)); the offences are undoubtedly reprobated by a large majority of the community whose views, the Council suggest, should not be discounted merely because of the views to the contrary expressed by a vociferous minority of opponents (point (5)); and adequate protection against these offences cannot be obtained through the Civil Courts (point (6)).[39]

5.   While struck by the very severe maximum punishment (imprisonment for life) which can be imposed for buggery even for a first offence—a punishment which they feel to be generally out of keeping with modern thought, as evidenced by the sentences which the Criminal Courts in fact impose nowadays for the offence— the Council feel that a severe maximum penalty has to be retained in order to deal with aggravated cases. This maximum punishment also strangely contrasts with the much milder maximum penalties which can be imposed for this offence in those continental countries such as Spain, Germany and Italy where buggery between consenting adults in private is a criminal offence. This contrast is even more marked when comparisons are drawn with the laws of France, Denmark, Sweden and Switzerland where such practices are not treated as criminal offences at all. The Council have accordingly carefully considered whether buggery and attempted buggery should continue to be treated as crimes separate from, rather than as forms of, gross indecency. On balance they think it desirable that these offences should continue to be treated as separate crimes. The present line of demarcation is both convenient and practical, because, whereas gross indecency may occur through a temporary lapse (e.g., on having a little too much to drink), buggery and attempted buggery are more deliberate crimes and merit different treatment...

6.   The Council have considered whether homosexual offences between females should in any circumstances be made punishable offences; their view, however, is that there is no public demand for any such legislation and no problem of any significant proportions.

...

*

(b) HO 345/9: Memorandum by the General Council of the Bar,[40] December 1955

Part I

THE LAW AND PRACTICE RELATING TO HOMOSEXUAL OFFENCES

SECTION I. INTRODUCTION.

...

4.   Religious Attitude.

We recognise that there are many who, upon religious grounds alone, would condemn any of the practices dealt with by the law in its present form. We have the deepest respect for the views of such people and we accept that such practices are morally wrong and deserving of censure. But our acceptance does not, of course, conclude the question as to whether or not a change in the law and/or practice is desirable.

5.   The Present Position.

Statistics show that prosecutions and convictions have increased in the past three decades so that in 1952 there were six times as many cases for trial and nine times as many convictions as in 1926.

Some of the reasons for this increase may be:–

(a) a general decline in accepted standards of sexual morality and increasing indifference to the Christian Ethic.
(b) The fact that during the war thousands of men were compelled to live for long periods in an all-male environment either on outlying stations, in ships or in prisoner of war camps. It is probable that many who hitherto had never practised homosexuality took to it during those years.
(c) The fact that there were also many both in the Services and otherwise who became practising homosexuals while serving or working in the Near and Far East where this vice is rife.

6.   It is unanimously agreed by all members of the Council that the practice of homosexuality is, under any circumstances, objectionable and an evil to society as a whole, both on ethical grounds and for other reasons.

Nonetheless, the first question is whether or not any change ought to be made so as to exclude from liability to criminal process adult consenting males who indulge together in private in

(a) Buggery or
(b) Attempted Buggery or
(c) Gross indecency.

Upon this question there is a clear divergence of opinion within the Council and accordingly we find ourselves unable to give any useful answer.[41] We therefore confine ourselves to suggesting certain alterations in procedure and penalties and to making some recommendations regarding the disposal of convicted persons.

SECTION II. ALTERATIONS IN PROCEDURE AND PENALTIES

. . .

1. Importuning for immoral purposes.

   . . . In respect of this offence we recommend changes in the law to provide for:–

(a) The right of an accused person to elect trial by jury when proceeded against as a rogue and vagabond.
(b) The maximum sentence of two years to apply to cases in which an accused person does so elect.

Our reasons for these recommendations are as follows:–

i) First, a conviction of this offence can and often does have disastrous consequences to the accused apart from any penalty which may be imposed. We therefore see no reason why this offence should be made an exception to the general rule that a person accused of a summary offence which is punishable with more than three months imprisonment should have the right of trial by jury.

   Secondly, in cases of this kind the evidence for the prosecution, which nearly always consists of that of two police officers, is often particularly open to criticism. The officers are on special duty for the purpose of finding persons committing this offence and their minds are apt to be filled with suspicion as a result of which a guilty construction is placed upon every movement of a person under observation. However honestly officers may be performing their duties mistaken inferences can easily be drawn. Moreover in some, albeit few, cases the evidence of police officers is not—as has been demonstrated in recent cases—wholly reliable.

   Thirdly, Magistrates and particularly Metropolitan Stipendiary Magistrates do tend to accept the evidence of police officers too readily in this type of case.

   In our opinion, for one or more of these reasons, miscarriages of justice do occur and we are gravely concerned that this should be so.

ii) As the maximum punishment on indictment for this offence is already two years, when the prosecution proceeds in that way, we think that the same punishment ought to apply when the accused elects trial on indictment. We wish to state that in our opinion importuning in public lavatories and other public places, which is now so rife, constitutes a real offensive public nuisance. The gravaman of this offence is not so much the homosexual motive as the nuisance and outrage to public decency by the actual solicitation. While we accept the fact that the fear of punishment may not to any great extent prevent persons who are naturally homosexual committing homosexual acts we are of opinion that the fear of serious punishment does deter them from publicly soliciting and importuning.

. . .

SECTION III. RECOMMENDATIONS FOR THE DISPOSAL OF
CONVICTED PERSONS.

...

3.   The primary objective of the criminal law must be the protection of the interests of society at large. Under the existing system reliance is placed primarily on punishment as a means of discouraging such acts and to a much lesser extent upon attempts to cure the individual offender. Underlying the former is the hope that punishment will deter not only the individual from repeating his offence but others from yielding to similar inclinations. Underlying the latter is the belief that certain individuals, by suitable methods of treatment, can be weaned from their homosexual tendencies. Underlying both is the theory that homosexuality spreads like a contagion from established homosexuals to more or less normally sexually constituted men or boys. The validity of these theories requires critical examination. There is, however, a deplorable lack of reliable evidence available for such examination.

4.   We desire to make the following comments on the existing methods of dealing with homosexual offenders:–

(a)   Imprisonment. Nothing would seem less likely to correct a homosexual deviation than confinement in an exclusively masculine community. How far an individual can be deterred from seeking, what has become to him, his normal sexual outlet must be a matter of considerable doubt. It is, however, not unreasonable to suppose that a homosexual can be deterred at least to some extent from obtaining sexual gratification with boys. Moreover, the knowledge that these tendencies may lead to punishment may induce a minority of homosexuals to seek treatment. The available evidence, however, tends clearly to show that many homosexuals have no wish to be relieved of their homosexual tendencies. It is understood that generally speaking psychological treatment is not available in prisons, and, moreover, that prison is an unsuitable environment for such treatment which cannot be undertaken without the willing co-operation of the person concerned.

There is little or no evidence to show whether or not the threat of punishment deters persons from embarking on homosexual activities for the first time.

(b) Probation Orders: In cases in which homosexual behaviour only becomes overt under stress or as a result of drinking, probation appears to be a satisfactory solution...

(c) & (d) Conditional discharge and Fines are appropriate only to isolated cases or minor offences and require no further consideration.

5.   The following alternative methods of dealing with homosexual offenders require consideration:

(a)   Segregation in a suitable institution: It will be generally agreed that there is a group of persistent homosexual offenders who in spite of sentence of imprisonment and attempts at treatment continue to commit homosexual offences against boys and, more often than not, with large numbers of different boys.

A recent example of this kind of offender we have in mind in this context was the accused man in R. v. Hall who was tried at the C.C.C. for a number of acts of gross indecency with young men. In no case was it alleged that he committed or attempted to commit buggery but the conduct which was alleged, which included oral masturbation, was as filthy and disgusting as the full offence. As a result of the influence which this man exerted over very much younger adults whom he obtained as participants and corrupted by continuous conduct of this kind he had utterly degraded them and ruined them. From his record, personality and the circumstances of the offences of which he was convicted, it was essential in the public interest that he should be sentenced to a long term of imprisonment. Any hope of reform or medical care was clearly out of the question and, so long as he was at large, he must always present a serious danger to society. In the event he was sentenced to 5 years' imprisonment, there being more than one offence charged.[42] But had he been only charged with one offence under Section 11 the maximum sentence of 2 years would have been wholly inadequate. The only effective way of protecting society against such persons is to segregate them from the community. For such persons suitable institutions on the lines of a mental hospital should be created. No person should, however, be committed to such an institution unless he has been convicted of homosexual offences involving men or boys under 21 years of age at least twice. Skilled psychological treatment should be available in such institutions for the minority of inmates who may still be considered suitable for it.

(b) <u>Suppressive treatment</u>: A method of drug treatment is available, the effect of which is to suppress all sexual desire so long as the drug is taken. Sexual desire and capacity return unchanged when the drug is withdrawn. Such treatment appears to offer at least a partial solution of the problem. At present, however, it is less effective in cases coming within the criminal law than it could be because there exists no effective method of ensuring that the offender continues to take the drug although clinical examination will reveal the fact that the drug is not being taken. A suitable form of condition to a probation order could be devised after consultation between the medical and legal professions. Imprisonment could, of course, always be imposed for failing to continue the treatment or if it were refused.

6. <u>Castration</u>. In certain countries (including Denmark) castration of persons convicted of homosexual offences is authorised. The available evidence tends to show that so far from it being effective many castrates retain considerable sexual desire and capacity. In any event we would be opposed to its adoption.

7. The choice of the most suitable method of dealing with any individual offender is a matter of difficulty requiring special knowledge. We accordingly put forward the following suggestions: The ultimate decision in every case must remain with the Judge who tries the case…Judges, however, have neither the special knowledge and experience required nor the means of investigating the personality and background of the individual offender. Medical assistance is consequently of considerable value, provided that it is skilled, reliable and independent. At present there is no assurance that the medical evidence put forward in such

cases is not incompetent, tendentious or worse. It is therefore submitted that a panel of doctors having special knowledge and experience of these problems should be set up on the lines of the "approved" medical practitioners under Section 5 (3) of the Mental Treatment Act, 1930[43] ... All sexual offenders should be examined by a member of the panel who should report in the first instance in writing to the Court...

...

*

(c) HO 345/16: Interview of N. R. Fox-Andrews,[44] P. A. O. McGrath, R. Ormrod[45] and R. E. Seaton,[46] on behalf of the General Council of the Bar, 20 February 1956

...

[Q4780] ... (MR. SEATON): I think the best thing will be for me to read the arguments against a change in the law which those on our sub-committee recommended—those on the Bar Council who were against a change of law occurring...

"1 ... [N]o action ought to be taken which might possibly result in any increase in the incidence of homosexuality.

2   Although the fact that acts of gross indecency constitute a criminal offence and are punishable may not have any great deterrent effect, in our opinion it does have some. It is not by any means only the "invert"—to use that word in the sense in which it is used in the Anglican Clergy's Report[47]—who practices homosexuality. There are many others who are not naturally homosexual who may to some extent be deterred by the sanction which the law now provides."

...

[Q4782] 3   In our opinion the effect of changing the law must suggest to the average citizen a degree of toleration by the Legislature of homosexuality... The people of this country have been educated at home, at school and through their religious teaching to regard any form of homosexuality as not only thoroughly objectionable but also as a criminal offence. As a result the average man has a healthy and instinctive abhorrence of it in all its forms... What can be the only possible effect upon that man's mind—particularly in the case of the younger generation—if the Legislature now suddenly says in effect that, although homosexual practices are to be discouraged, nevertheless they are not so objectionable when committed by consenting adults in private as to be worthy of criminal prosecution? ... In effect Parliament will appear to be saying that homosexuality in these circumstances, although reprehensible, is so much less so than riding a bicycle on the foot-path... It has been suggested that the fact that homosexuality is now more generally discussed than it was indicates that public opinion is changing and that homosexuality is now regarded as less objectionable than it used to be. In our view this argument is fallacious. Although these matters are now discussed more freely, the average decent man of today has no less horror of his son becoming homosexual than had his father. Furthermore, if in fact there is such a change of public opinion, we regard it as a deplorable thing and one which ought, as far as possible, to be stopped...

...

5   We believe that the possible risk of the practice of homosexuality increasing as a result of any change in the law is all the more serious because of the consequence which must follow any such increase:–

(a) Every homosexual is a potential danger to youth and any increase in the number of practising homosexuals or in the ease with which they may commit acts of homosexuality must inevitably lead to the danger of more widespread corruption of youth.
(b) In addition to the actual corruption of youth, there is a considerable amount of procuring of young adults for homosexual purposes, e.g. service personnel in London and elsewhere. Many of these procurees are not naturally homosexual; they are either bribed or made drunk. To make the commission of homosexual acts between consenting adults in private lawful must make procuration of this kind easier to the procurer...
(c) If, once back in a private apartment, a man can practise homosexuality with impunity there will also be a greater incentive for importuning, a habit which is already far too prevalent and which already causes a very real annoyance and outrage to public decency.
(d) If homosexuality between adults in private were to be lawful, we believe that there would be a real danger of male brothels being set up...

6   There are...certain other difficulties which we see in any proposed amendment of Section 11.

(a) ...The difficulty arises as to who is to be regarded as an adult in this connection... [T]here are many persons of an age between, say, seventeen and thirty who are frequently procured for homosexual purposes and the corruption of persons of that age who might otherwise never resort to those practices is little less objectionable than the corruption of younger persons.
(b) If Section 11 is to be changed so as to make homosexual acts by consenting adults in private lawful, is buggery as well as acts of gross indecency to be made lawful? There can be no logical reason whatever for distinguishing between those two acts committed between consenting adults in private... [T]o make buggery—that offence which has been regarded as the abominable crime over the centuries[48]—lawful, would all the more emphasise the suggestion of official tolerance towards homosexuality which we believe must follow from any change in the law.

7   ...As we understand them, the main arguments put forward for a change in the law are (a) that harm is done to nobody by two consenting adults committing

homosexual acts in private and (b) that the existing law causes hardship and injustice to that section of the community who are, through no fault of their own, naturally homosexual. First, as we have pointed out, in many instances the practice of homosexuality involves persons who are not naturally and originally homosexually inclined and in all such cases harm is in fact done. It additionally involves persons up to the age of 27 to 30 of homosexual proclivities who are curable if treated in time and who should, so far as possible, be kept away from the risk of contamination by confirmed homosexuals. Secondly, the injustice which the existing law causes is negligible. The number of prosecutions which take place in respect of acts committed by consenting adults in private are fairly few and far between ... The number of persons therefore who can be said to suffer any injustice by being prosecuted for doing something which they cannot help doing and which is natural to them is insignificant ...

That is the conclusion of the arguments against a change in the law. I might just perhaps add this if I may, by way of enlargement, on the question of all homosexuals in our view being a potential danger to the young. I speak from a certain amount of personal experience thinking of what one finds so often with the confirmed homosexual. I draw a very sharp line of distinction between the confirmed and the other person who takes it up for a variety of reasons. In the confirmed homosexual you find in the make-up of the individual just that touch of the natural, that is to say, the ordinary heterosexual man's outlook upon sex. There is a craving for something that resembles a woman, without wanting in the least to have successful intercourse with a woman. If you have that make-up, which almost amounts to a contradiction in terms, what one finds is you get very often the homosexual who picks up a boy. In order to start seducing that boy he will show him lewd pictures of men and women and that sort of thing. He picks up a boy because the boy is the nearest thing to womanhood imaginable—high voice, smooth skin, etc., nearer than the ordinary grown man. It is because the homosexual—I speak, of course, subject to correction for I know there are medical men here—in my experience, for what it is worth, generally speaking appears to crave after something that approaches the female. Indeed, you might not only find it with boys. Everybody knows the homosexual divides into two classes, the catamite and the patient. The patient, of course, is the one who shows various female traces in voice, and in the pronounced case I can remember one case in which I prosecuted a Norwegian ballet dancer known as Lulu. His position was such that he is a classical example of the confirmed homosexual. His development of the female in his psychology, and so on, was such that he disliked being a bugger because he had to lie on his stomach. He much preferred to lie on his back. There is a man who approaches to the woman, the adult male. Nevertheless, our view is primarily the average confirmed homosexual has got that curious streak. He looks for something that approaches a woman but in fact is not a woman, and a boy is the answer to his idea ... Of course, I know there are people who say there are many homosexuals who have not touched boys at all. That may be and probably is so, but we feel there is always that risk.

[Q4783] (MR. ORMROD): The arguments for a change in the law...

...

[1]   We take as our guide the fundamental proposition that the State is not justified in inflicting punishment upon individuals unless it can clearly be shown
that the acts for which they are to be punished are seriously injurious to Society.
In other words moral obliquity per se should not be cognisable by the criminal
law, the purpose of which is the protection of the interests of Society and not the
inforcement [*sic*] of a particular moral code.

2   Experience shows that attempts by the State to invade the individual conscience, however highminded, always fail and frequently do serious harm. The
Volstead Act[49] in the U.S.A. is the best recent illustration. It follows that where
criminal sanctions are shown to be failing in their purpose it is at least as important to question the validity of the law as to invent new and more drastic
penalties.

3   We believe that the time has come to re-examine the law relating to homosexual offences between males. In common with the other members of the Council
we consider that homosexual acts committed with boys should continue to be
punishable. We also consider that homosexual acts committed in public, including
importuning for homosexual purposes, conducting male brothels and procuring
men or boys to commit homosexual offences for gain should continue to be
criminal offences.

4   We differ from the other members, however, in thinking that homosexual acts
between consenting adults in private should cease to be criminal offences...

5   It is at least doubtful whether homosexual practices between adult males
are more damaging to Society than adultery with its grave consequences to the
children whose homes are broken, and to the abandoned spouse.

6   The present law with its exceptionally severe sanctions derives from traditions
the origins of which are obscure.

7   There is little or no historical justification for the view, often relied upon by
those who would punish every homosexual, that homosexuality causes the society
in which it is practiced to become decadent. In Sparta it was thought to promote
courage in battle. Its prevalence in ancient Greece is enough to throw grave doubt
upon the idea. The statistics for this country show that between 1926 and 1938
convictions for all homosexual offences more than doubled yet the country with-
stood the severe trials of 1939–1945 no less well than it endured the 1914–18
War. According to the Kinsey Report[50] ...37% of American males have had at
least one homosexually induced orgasm and 25% of the male population have
had "continued and distinct" homosexual experience over periods of three years
and more.

8 ... There is no evidence of which we are aware that the more tolerant attitude of other countries has had or is having an adverse effect. In this country before 1885 gross indecency between adult males in private was not a criminal offence and no adverse consequences are reported.

...

9 There is no evidence of which we are aware to suggest that if the Law as to adults was relaxed it would lead to an increase of homosexual offences against boys. If such had been the experience of the countries of Western Europe, their laws would presumably have been amended. It is not unreasonable to suppose that many homosexuals might in fact be deflected away from boys by the knowledge that they could satisfy their desires in safety with an adult.

10 The criminal statistics for the period 1936–1948 for this country show a marked and progressive rise in both the numbers of persons charged and in those convicted of homosexual offences. Whether these figures reflect a real increase in homosexuality or a change in the attitude of the police forces towards such offences is a matter for careful enquiry. Assuming, for the sake of argument that there has been a real increase in homosexuality, these figures suggest that the present penal system is failing as a deterrent. The figures show a sixfold increase in convictions over the period. If the risk of detection and severe punishment is an effective deterrent, it follows that there must have been an even greater increase in the number of men who experience active homosexual desires which are successfully resisted. It is difficult to believe that so marked a change in the sexual urges of the men of this country has taken place in the period of 20 years. We believe that the increase of convictions for such offences is a reflection of a general fall in standards of sexual morality and probably corresponds to a similar increase in fornication and adultery. We do not find any convincing evidence that sentences of imprisonment effectively deter homosexuals from obtaining the gratification which they desire.

11 We have carefully considered the argument of those who disagree with us that these statistics indicate that the present is not the time for a change in the law but we cannot accept it. If the statistics showed a steady or declining incidence in convictions it would be argued, no less powerfully, that the present was not a suitable moment to amend the law because it was clearly working effectively.

12 We are impressed by what seems to us to be a savage injustice to individuals in many cases. Men's lives are utterly broken by the publicity which attends their trials no less than by the long sentences of imprisonment which often follow. The disproportion between the penalty and the damage done by the offender is sometimes appalling. We do not think that our legal system should lightly accept such a situation even if the individuals concerned are relatively few. Our impression, however, is that this number is substantial, and that considerable numbers of men are prosecuted every year for homosexual offences committed between consenting adults in private. We believe that the misery so inflicted serves no useful purpose

and we do not think that the situation can be met by changing the methods of dealing with such persons. We think that the fact that psychiatric treatment in the present state of medical knowledge seems able to do little for the great majority of such people should be squarely faced.

13   We believe that the existing law is an encouragement to blackmailers. Section 11 of the Criminal Justice Act 1885 has been described as the "Blackmailer's Charter".[51]

14   We therefore recommend that the law should be so amended that homosexual practices between adult consenting males in private should cease to be criminal offences. For this purpose an adult should be defined as a person aged 21 years...

I only wish, I think, to add one thing and that is this. The arguments stressed by the opposition that if the law were to be changed it would appear as if the authorities regarded homosexual activity in private between consenting adults was less serious than riding a bicycle on the pavement, and therefore no such change should be made, we regard as an extraordinary argument from the point of view of principle. If the law is wrong it should be changed. If the law is doing an injustice it should be changed. The fact that it might surprise many members of the public in our submission is irrelevant to the consideration of whether it is right or wrong. It may be perhaps they do less damage than riding bicycles on pavements. I do not know.

...

*

iii. Magistrates and Judges

(a)   HO 345/8: Memorandum of evidence from the Magistrates' Association[52]

...

I. HOMOSEXUAL OFFENCES
Adult Male Homosexuals

...

6.   From the point of view of the courts, homosexuals may be divided into the following three groups:–

(a) Those who indulge in homosexual practices with adults in private;
(b) Those who commit homosexual offences in public places with adults;
(c) Those who corrupt boys and youths by committing homosexual offences with them.

7.   At present the fact that homosexual practices are indulged in with the consent of the other party is no defence to a criminal prosecution. English law differs in this respect from that of most European countries, under which homosexual

conduct with consenting adults is not a criminal offence.[53] We have given careful consideration to the question of whether English law should be brought into line with continental law in this respect.

8.   We recommend that the law should be amended and that homosexual conduct between consenting adults in private should no longer, within certain limits, be a criminal offence. This recommendation was made to the Council of the Association by two of its Committees, and it was passed by 41 votes to 33. In putting it forward the Council wishes to make it clear that it in no way departs from the general view that homosexual practices are undesirable and dangerous, both for individuals and for the community. The Council was influenced by the fact that there are many evils, such as adultery and lying, which have to remain outside the reach of the criminal law unless there are additional circumstances which make the intervention of the criminal law necessary...

9.   We realise, however, that care must be taken in defining what we mean by "adult" in the recommendation set out in paragraph 8. We consider that many persons of 21 and even 25 or over are still emotionally immature and capable of corruption, and that therefore the age of consent should not be fixed too low. The very fact that such conduct is illegal may well have some deterrent effect on younger people. After considering the matter at length we recommend that the age of consent should be 30. We have been influenced in this decision by reason of the fact that the Criminal Justice Act provides for sentences of preventive detention at the age of 30 or over, which leads us to suppose that those responsible for drafting the Bill, on the advice of experts, were of the opinion that not until that age could a man be considered as set in his ways.

...

Disposal

24.   Criminal courts are primarily concerned with the protection of the public and the prevention of crime. It is generally felt, however, that the most satisfactory means of protecting the public, and particularly where homosexual offenders are concerned, is in most cases to provide the best available treatment rather than to inflict punishment. We consider that incarceration in an institution, where perverted tendencies may be aggravated, should always be a last resort.

25.   As regards treatment, homosexual offenders would seem to fall into two categories:– (a) the true invert for whom the possibility of cure is very slight, and (b) those suffering from arrested development, many of whom can be treated successfully.

26.   Especially for the lesser offences we regard as undesirable a merely penal sentence of detention. In these cases there is often some hope of rehabilitation. We consider it is essential that the offender should be given psychiatric treatment if he agrees to co-operate. The sooner treatment is given the more hope is there of cure, or at any rate, of improvement and of greater self-control. We are convinced

that a large number of these offenders can be satisfactorily dealt with by a probation order with a requirement to submit to treatment, if they are willing to comply.

. . .

Women Homosexuals

30.   It is commonly believed that homosexual conduct between women is not an offence. The Criminal Statistics for England and Wales show that women have occasionally been convicted of offences under section 52 of the Offences Against the Person Act, 1861. In the case of Rex v. Hare (Criminal Appeal Reports, 1932–34, Volume XXIV, p.108) the Court of Criminal Appeal stated that in their opinion the word "whosoever" in section 52 includes a woman, and that "there can be no reason for saying that a woman cannot be guilty of indecent assault on another female".[54] Convictions of women under this section are, however, rare, and it is probable that those recorded relate either to indecent assaults on girls under the age of 16 years, or to cases in which a woman has aided or abetted a male in an assault on a female.

31.   We understand that the Home Office has no information on whether proceedings have been taken under this section in respect of homosexual acts between adult women, but as the section relates to indecent assault, on a question of law it would seem unlikely that the section could be regarded as applying to homosexual conduct between consenting adult women.

32.   We think that the problem of women homosexuals may be more extensive than is generally realised, and that more young girls may be corrupted than is appreciated. As, however, we have no experience of the subject as magistrates, we are unable to recommend an alteration in the law in this connection.

. . .

*

(b) HO 345/7: Memorandum Submitted by Sir Laurence Dunne, M.C., Chief Metropolitan Magistrate[55]

I. HOMOSEXUALITY

. . .

4   The two classes of case concerning homosexuality most commonly brought to police courts are those charging either (a) persistently importuning (male persons) by male persons for an immoral purpose, or (b) charges of gross indecency between male persons...

5   There is another class of case which frequently comes before the court with a charge of persistently importuning for an immoral purpose, but in which I believe it is safe to say that a substantial percentage of the accused are not homosexuals...This is the type of case consisting of importuning by exposure, coupled usually with self-masturbation in public lavatories. Curiously enough, and I make

no attempt to explain it, the great majority of these cases from [*sic*] the Brighton and Chatham sections lavatories at Victoria Station. There is usually no verbal communication from the accused to those he accosts. In the vast majority of cases he is a respectable professional man with no criminal record of any sort. There is no attempt in most cases to make any assignation for grosser practices outside. The defendant appears to satisfy his appetite by exposure there and then. In these cases, I draw a sharp distinction in penalty from cases of undoubted homosexuality, and I am content to treat them as simple cases of indecency, at any rate for the first offence, and to impose only such a fine as the defendant can pay, coupled with a warning that any future offence will be dealt with severely.

...

9   The chief haunts of the male prostitutes who cater for the desires of perverts are the public lavatories and urinals and the streets adjacent thereto. Piccadilly Circus Underground and Leicester Square together with the urinals in Brydges Place, Rose Street, Babmaes Street and the Adelphi are particularly notorious. It is not too much to say that the West End street urinals are plague spots after dark, and any respectable person using them goes into real danger of molestation if he is forced to do so.

10   The male prostitutes are by no means all homo-sexuals. Some are degraded creatures who pander to perverts for purely mercenary reasons. I am happy to say that the old unholy traffic between soldiers of the Guards and Household Cavalry and perverts in the Royal Parks is now a thing of the past.[56] It may be that education and a higher moral sense has played its part, but I fear that the abolition of the old tight overalls worn by other ranks walking out is a strong contributory factor; battle dress or khaki serge lacks the aphrodisiac appeal of the old walking out dress. These men discovered that it was easy to earn money in this way, and the matter was nearly a public scandal. I believe it no longer exists.

11   There was a curious by-product of homosexuality during the early days of the war. A large number of practising homosexuals lived in the George Street, Seymour Street, Bryanston Street and Paddington area. They frequented public houses, and were busy offering their services to service men, and in particular Royal Marines. They would take the men back to their flats. This became well-known, and a number of cases were ultimately heard in court where the service men, with no intention of submitting to any homosexual practice had gone home with these male harpies, committed very severe assaults on them, and walked out with any portable property or cash they could find. A few fairly severe sentences solved that problem.

12   The male prostitutes come from various walks in life. A very large number of those arrested are employed in domestic work, waiters, kitchen hands and domestic servants. There are a number of young vagrants who arrive in London with no work or pied-à-terre who drift into the traffic. The number of persons charged is a most unreliable index as to the number who should be charged. Police strength

is strained to the utmost, and I fear that charges are numerically a truer pointer to the number of police available for this duty than to the number of practising perverts.

13   It is impossible to produce any reliable figures to prove whether the traffic is on the increase, static or declining. Few things remain static in this world, and I see and hear nothing to indicate that the numbers of perverts are decreasing.

14   Another aspect of the homosexual community, and I fear it is a community, is afforded by summonses to publicans for allowing unruly and indecent conduct on their premises. These are not numerous, and they all exhibit a regular pattern. These pests descend like locusts on some licensed premises, drive out the respectable clientele and literally take over the custom of the house. The licensee is immediately in a dilemma. Before he realizes it, the damage is done; his respectable customers have deserted him, and he has either to accept the custom of the perverts or put up his shutters. The perverts in mass are even more noisome than singly. They often wear articles of feminine clothing, answer to feminine names, and use the filthiest of language and innuendo. If appealed to the police can drive out the perverts, and do so, but they cannot re-introduce the proper custom of the house. The result is almost certainly ruin for the licensee.

. . .

16   It is often said that homosexuals should be treated by doctors, and not by the criminal courts with prison sentences. If homosexuality, or rather its physical practice, is to remain an offence, I profoundly disagree. I have in the course of my time on the bench heard a great deal of medical evidence in these cases. I have never yet heard clear and explicit evidence as to what form the suggested psychological treatment would take. Analysis and a label cure nothing. I am sure of one thing. The cases where psychological treatment have any chance of success are rare, and that success can only be obtained where the defendant is ashamed of what he has done, and not merely ashamed at finding himself in the dock. The vast majority of homosexuals are not only not ashamed of their conduct, but actually look down on those not similarly addicted as intellectual inferiors. The vast majority of professional or semi-professional male prostitutes are the lowest of the low, and there is nothing to appeal to in them save to exploit the fear that they too may, in their turn, suffer the punishment inflicted on men convicted. Though it is painful and difficult for magistrates to be uniform when men of position are accused, I am firmly of opinion that sentences for this form of offence should be uniform and severe. As to whether any form of physical treatment, by hormones, for instance, is available or likely to be effective, I am not in a position to say. The Committee may feel inclined to explore this with help from the medical profession.

17   A series of prosecutions involving homosexuality of men of position has, unhappily, focused public attention on this topic, and one hears a good deal of opinion, mostly quite uninformed, to the effect that save for the corruption of youth and possibly public scandal, physical expressions of homosexuality between males should no longer be offences against the criminal law. My personal view is

that to make any such alteration, certainly at the present time, would be a mistake. I decline to be impressed by the view taken in other countries. They have their problems: we have ours. At present, I believe that this aberration is on the increase, and I believe it would weaken the hand of the law in dealing with what is admitted by all, save addicts, to be an evil. To countenance homosexual practices in private is playing with fire. Appetites are progressive, and a homosexual sated with practices with adults, without hindrance, will be far more likely to tempt a jaded appetite with youth. A great deal of encouragement is already given to these perverts by unthinking people who affect to find something funny and not reprehensible in their conduct. I think it would be disastrous to give further tacit encouragement by altering the law in their favour.

. . .

19 Homosexual practices between females are unknown to the law. That this is so is, of course, due to the accident that it is not the subject of Levitical injunction in the book of Deuteronomy,[57] and was unknown to the ecclesiastical courts whose jurisdiction was handed on to the civil courts of justice. In any case, it was probably almost unknown until the emancipation of women in the last few decades has allowed opportunity for its development. Beyond that it exists, and I believe is growing, I know nothing of it, save that there are clubs and licensed premises where those who practice it foregather.[58]

*

(c) HO 345/7: Memorandum by Mr. Paul Bennett, V.C., Metropolitan Magistrate, Marlborough Street, 18 December 1954[59]

. . .

Homosexuals

This offence has also increased enormously. It is due to some extent, in my opinion, to the much publicised medical and psychiatric approach to this subject. An aura of semi-respectability now surrounds it. So much of this publicity in newspapers draws no distinction between the "condition", which cannot be helped, and the "practices", which show a lack of moral control. I exclude, of course, the medical profession from this criticism.

I thought it deplorable for certain well-known figures, on Television, with an audience of 3,000,000, (all ages), to suggest that such practices might well cease to be a criminal offence. If both parties were adult. They added the rider that "seduction of minors" should still be an offence.[60]

To attain both objectives, by legislation and by due process in a criminal court, I should say would pass the wit of man. Such proposed legislation would automatically make homosexual practices respectable. In my view, the wisdom of centuries on this subject is just as wise to-day.

*

(d) HO 345/7: Memorandum by Mr. Claud Mullins, Metropolitan Magistrate 1931 to 1947[61]

I submit this memorandum because during my years as a London magistrate I did not send sexual offenders to prison, but in all possible cases invited them to submit to psychological treatment. If they refused such treatment, I imposed fines, not because I regarded fines as satisfactory, but because my visits to prisons had convinced me that prison life might well result in offenders becoming greater menaces to society than they were before. When I was appointed a magistrate there was no psychological treatment in prison and during my term of office only the beginnings of such treatment came into being.

Homosexual offenders

1   The pioneer of the use of psycho-therapy by courts was Sir William Clarke Hall,[62] but he only sent children for treatment. To quote Dr. Edward Glover's[63] Introduction to my book 'Crime and Psychology',[64] 'Mr. Mullins was not indeed the first lawyer to accept with enthusiasm the teachings of modern psychology, but he was the first magistrate to apply them systematically in ordinary police-court work'.

2   Magistrates cannot try those accused of the major homosexual offences, but often the police would charge men whom they suspected to have committed such offences with minor offences triable by magistrates. The reason was probably their inability to prove the greater offence. Thus many men, suspected of serious homosexual conduct, were brought before me on charges of soliciting, assault, or even of indecent conduct contrary to an L.C.C. [London County Council] bye-law, the maximum penalty for which was a fine of £5.

3   Whatever the charge, my practice in such cases was to direct full social and medical investigation and, where possible, to offer psycho-therapy. My experience was that the nature of the charge became unimportant. If with the help of a Probation Officer I could win the consent of the offender to the wisdom of his having treatment, it mattered little that the legal sanction that lay behind a Probation Order was only a short term of imprisonment or a fine.

. . .

5   This was only possible because I worked in a big city. There was the old Tavistock Clinic (under Dr. J. R. Rees)[65] and later the Institute for the Scientific Treatment of Delinquency, since re-named. There was also the Maudsley Hospital. I was also fortunate in having in S.W. London a general practitioner (Dr. A. C. Court) who had had experience in psycho-therapy. Without such facilities as these, my pioneering work would not have been possible.

6   I became so convinced of the value of psycho-therapy in sexual cases that I was able to arrange such treatment for many of those whom I suspected of being guilty of homosexual conduct, but had to acquit because of inadequacy of proof...

. . .

8   Conditions have changed greatly since 1947 when I retired. There are now methods of treatment besides psycho-therapy. There is much more scope for treatment in prison. The 'special institution' for mental offenders is being erected.[66] Magistrates generally, both professional and lay, are becoming more familiar with the uses of treatment. The National Health Service opens up possibilities of free treatments. Generally there is a healthier attitude among the public towards mental treatment, though the old prejudices can still be found. None the less, I hope that the experience of one who pioneered in this invaluable movement may be of some assistance to the Committee.

...

*

(e) HO 345/7: Statement by Mr. Geoffrey Rose, M.C., Metropolitan Magistrate, Lambeth[67]

1   There is one point on which I might be able to interest if not help the Committee, namely the connection between importuning and homosexual indecency on the one hand, and the existence or provision in large towns on the other of public urinals or conveniences—miscalled lavatories.

2   During the war, especially near the commencement, this court had very many indecency cases of a homosexual nature which all derived from a particular convenience in Upper or Lower Marsh. It was the rendezvous of many homosexuals on their way home from the West End. I suggested to local police that this convenience (of a very old fashioned and ill-lit kind) might well be destroyed... The result was an immediate diminution in this class of case.

3   The same thing occurred some years later in relation to Archbishops Park. This at one time was producing some dozen or more prosecutions for importuning and indecency per week. This old-fashioned convenience was likewise destroyed... and the cases ceased altogether.

4   The result of removing these two insanitary and unguarded conveniences was to reduce cases of importuning and actual homosexual indecencies, the subject of prosecution at Lambeth, by probably 95 per cent.

5   It is true that some of this importuning and indecency may have been driven elsewhere, but the statistics remain. If you destroy its rendezvous, you destroy much of the opportunity and the occasion for this class of crime.

...

*

(f) HO 345/7: Memorandum by Mr. Harold F. R. Sturge, Metropolitan Magistrate, Old Street, November 1954[68]

...

6   The Report[69] proceeds to point out a much more serious "legal anomaly" and continues "In no other department of life does the State hold itself competent to interfere with the private actions of consenting adults", and the word "private" is defined as "not anti-social". This seems to be the main argument advanced in the Report for regarding the law relating to sodomy and gross indecency as unjust and it is to my mind an odd one.

7   Every one would concede that there is no need for the State to interfere with actions which are not anti-social whether they are between consenting adults or not. I should, however, have thought that to engage with another person, even in secret, in actions which are morally wrong, physically dirty, and progressively degrading is grossly anti-social even between adults who are more or less of equal status in life. Is not this conduct even more anti-social where the participants are not of equal status (e.g. The artistic genius and the younger man needing his professional help; the man of wealth or title and the friend from a simpler walk of life)?[70] How many consenting adults may be present before the conduct ceases to be private or not anti-social?[71]

...

11   ... [T]he general level of moral standards has, I think, for long been above regarding sodomy and gross indecency as deserving of toleration. If we have learnt more about inversion we have learnt nothing to justify a lessening of the penalties upon those who engage themselves in spreading mostly among younger people the degradation of homosexual actions...

...

14   In my view, if the law were to be amended so as to abolish sodomy or gross indecency, even between consenting male adults and in secret, as criminal offences (a proposal which the Report appears to be designed to recommend), it would inevitably lead to an increase in these degrading practices.

*

(g) HO 345/8: Memorandum prepared by the Association of Sheriffs-Substitute

I   Prostitution and Solicitation in relation to Males

The Association has no comment to make on the law and practice relating to these topics. Prostitution and solicitation for immoral purposes are rarely, if ever, the subject of prosecution in the Sheriff Court, and it is thought that they do not constitute in Scotland the difficult problem which they appear to do in some parts of England.

II   Sodomy and Indecency between Consenting Adults

Apart from indecent conduct in a public place, prosecutions for sodomy or indecency between consenting adults rarely, if ever, occur in the Sheriff Court. The suggestion which is to be made regarding a change in the law on these matters is therefore put forward with considerable diffidence, and is the result of impressions formed from experience of other allied matters, viz:– Assault for the purpose of committing a homosexual crime, blackmail arising from an earlier homosexual association, and crimes against boys and young men. Crimes against boys and young men constitute a substantial part of the homosexual crimes dealt with in Sheriff Court, and it is particularly in relation to these crimes that a change in the existing law, it is thought, would be beneficial.

It is suggested (1) that homosexual acts between consenting adults, taking place in private, should be put in the same position with regard to the criminal law as the analogous hetero-sexual acts of adultery and fornication, which are not now punishable as crimes: and (2) that the statutory offence created by Section 11 of the Criminal Law Amendment Act, 1885 should be correspondingly altered or repealed. If the law were to be changed in this way, consideration would have to be given to the definition of "adult". It is thought that during their habit-forming years all young men should be discouraged from active participation in homosexual practices. It is suggested, therefore, that, for the purpose of such a change in the law, adult status might be regarded as commencing at the age of twenty-five years, and that the law might remain unchanged in respect of all males below that age.

It is thought that such a change in the existing law would have the following beneficial results:– (1) the adult homo-sexual would be free to work out the difficult problems which, in a predominantly hetero-sexual Society, must necessarily be his, in accordance with his own conscience and the religious or moral code to which he subscribes, without the additional distraction which may arise from the fear of being charged, justly or unjustly, with a contravention of the criminal law: (2) The opportunity for blackmail, which now exists, would be reduced: (3) The full weight of the criminal law would be brought to bear upon the subject which, in Scotland at least, most needs its attention, viz:– the protection of boys and young men from adults who seek to corrupt them.

. . .

This Memorandum was prepared in draft by Sheriff A. G. Walker, who has been dealing with a large part of the criminal prosecutions in the Sheriff Court in Glasgow for a number of years.[72] It was then circulated to selected members of our Association, who were all against the alteration in the law proposed in Section II of the Memorandum.[73] Their views and reasons are sufficiently covered by the comments of Sheriff Kermack of Glasgow and Sheriff Hamilton of Aberdeen which appear below and which, it is hoped, the Committee will find helpful.[74]

. . .

Comments of Sheriff Kermack[75]

I am afraid I don't agree entirely with the proposed memorandum. In particular I think it would be a retrograde step to abolish indecent behaviour between males as an offence, and that it would tend to shock the public conscience...I have had not infrequent charges in the Sheriff Court of this offence...

I would like to think that abolishing it as an offence would make more "offenders" and victims take medical treatment but I know of no such evidence; in fact the result might be just the opposite and make them think there was no need to bother about treatment...

...

Comments of Sheriff Hamilton[76]

I   I do not agree with the proposals in this section.

II   While charges of sodomy are infrequent, charges of gross indecency between adult males are quite often before the Sheriff Court here; certainly more of them have occurred in public places but charges arising from such behaviour in private also come before the Court from time to time, e.g. males lodging together and an older man forcing his attentions on younger men or on men of immature mental years. Experience has shown, I think, that of two males charged with gross indecency one is often more determinedly disposed to such behaviour than the other, and it is a fair inference that debauchery takes place of one who might, but for such association, never have yielded to these abominable practices. The process appears to be a cumulative one, for those who have been debauched seem to seek opportunities to debauch others; one only has to consider the number of adult persons connected with boys' and youth organisations who are convicted of such offences and who undoubtedly attached themselves to such organisations because of their homosexual tendencies. I do not pretend to explain the physical or mental causes of such tendencies but I am convinced that they are capable of subjection by the individuals themselves, and are not manifestations outwith their control as might be said of the conduct of an insane person. It is my opinion that if individuals are not prepared to inhibit such tendencies on moral grounds the law must provide sanctions as a deterrent. Homosexual misconduct is unnatural and I do not consider that it can be described as analogous to adultery or fornication both of which are unconventional modes of otherwise natural behaviour. It is fashionable to say that homosexual offenders have a "mental kink" and require treatment but apart from the odd case of physical ailment such as prostate gland enlargement, the only 'treatment' which may be beneficial is such as will strengthen the willpower to resist offending. Is the same not applicable to most offenders against the criminal law? I cannot subscribe to any proposals which take homosexual misconduct outside the scope of the criminal law.

...

*

(h) HO 345/9: Cases forwarded by Richard Elwes, Q.C., Recorder of Northampton and Chairman of Derbyshire Quarter Sessions[77]

Annex I: Letter of Elwes to Wolfenden, 28 July 1955.

I hope that you will forgive me for adding to your burdens as Chairman of the Departmental Committee by sending you the enclosed transcript of two cases which came before the Derbyshire Quarter Sessions last Easter. It seems to me that your Committee might be interested in them as contemporary illustrations of police prosecutions under Section 11 of the Criminal Law Amendment Act, 1885.

In the first, Dennis, Cluskey and Hill, the first two defendants were, three years after the event, prosecuted at the age of twenty in respect of a comparatively trivial incident which, as all my fellow justices and I thought, should never have been made the subject of a charge, least of all after so long a delay.

The second O'Connell and Pearce, exemplified another modern tendency. Both Defendants were of full age, 28 and 50 respectively. The Police were set in motion as a result of thefts by the younger man on leaving the house of the older, who had complained. In the course of their enquiries they visited the house, saw that there was only one bed and on their own initiative interrogated each man separately—normal police practice. Admissions of mutual masturbation were duly obtained and prosecution followed.

...

Annex II: Regina v. Ralph Dennis, Lewis Arthur Cluskey and Stanley Schofield Hill, 6 April 1955.

...

MR. [T. R.] HEALD [for the prosecution]: May it please you sir. These three men have pleaded guilty to these counts of gross indecency. Each pleaded guilty to a count of gross indecency with the other. In fact, all these counts arise out of the same incident which happened a long time ago.

THE CHAIRMAN [Elwes]: Nearly three years ago.

MR. HEALD: Yes. The matter came to light because the witness, Donald Dodds, was interviewed by the Police on another matter, and then he made a statement in which he said he had seen these three men and another man, who is not in custody, committing the acts of gross indecency, as set out in the charge, apparently in some hut or changing room besides a children's paddling pool. As soon as the accused were interviewed this year, they all made statements fully admitting the offence, although it had taken place so long ago.

...

THE CHAIRMAN: A comparatively trivial affair. Before 1885 no-one would have dreamed of describing it as an act of gross indecency, and the prosecution comes about by a statement made by a man three years after the event. These two men, Dennis and Cluskey, now perfectly respectable men, earning good wages—miners—have had to come here and plead guilty to a charge of gross indecency on facts which happened when they were 17 years old. It seems to me a most extraordinary procedure to have taken, and from my point of view, I am thankful that the newspaper strike has prevented the facts being reported against these men, who have had to stand in the dock on charges like this...

...

THE CHAIRMAN: Ralph Dennis and Lewis Arthur Cluskey, nearly three years ago when you were 17 years of age or perhaps less, it happened that in a secluded place, with two other people, you were parties to what your Counsel has very properly described as a trivial sexual experiment in your adolescence... During the three years that followed, I am quite satisfied that you forgot all about that, that you grew up normally and became what you are now, entirely creditable young men, and then suddenly for reasons which I certainly cannot conceive, it has been thought right to bring you before the committing justices, who committed you to Quarter Sessions charged with acts of gross indecency. If you had been lucky enough to live in 1884, the wisdom of our forefathers in those days would have left you completely untouched,[78] and you would never have been within 100 miles of a criminal court, but because in 1885, without any proper consideration at all, the House of Commons stuck into an Act intended for different purposes, this Section 11 under which you are prosecuted, you have had to be brought here and have had to endure all the scandal and disgrace of standing in a criminal court pleading guilty to what is called an act of gross indecency.[79] For my part, I express disapproval of the proceedings against you. It seems to me to be absolutely unnecessary, and has exposed you and your family to pain and embarrassment which none of you have deserved.

...

Let them be released at once, and let them be given an absolute discharge.

Stanley Schofield Hill, your case is not quite the same. What has brought you here, you being 34 years old, was this same incident, nearly three years ago. You have told the Police, with complete candour, the whole of your private life since three years ago, and that redounds greatly to your credit and convinces us that you are entirely sincere in your wish for the medical treatment which you have already voluntarily started to undergo.

It seems probable that there is a perfectly good hope for you to become a perfectly normal person in the sense that you ought not to be troubled any more with temptations to abnormal sexual conduct. So far as that goes, the fact that you have been proceeded against did give you the necessary push in the direction of the doctors, and now the doctors have got you, we all hope that they will be successful in treating you...

...

In your case, we give you a conditional discharge...

Annex III: Regina v. Wilfred Pearce and Daniel O'Connell, 6 April 1955.

...

MR. [M.] NEWELL [for the prosecution]:... Pearce is 50 and O'Connell 28 years of age, and O'Connell went to live in Pearce's home as a lodger. They slept in the same bed, and these unpleasant offences were committed. Pearce was seen by a police Officer on the 28th February of this year, and the Police Officer said to him, "I have reason to believe you have been sleeping with O'Connell, and some acts of indecency have taken place", and Pearce then made a statement and O'Connell also made a statement. Both of them quite frankly admitted the facts...

. . .

Q. THE CHAIRMAN: Officer, how did this matter come to the notice of the Police?—A. [Det. Sergeant Angus MacDonald] In the first instance O'Connell was charged with stealing property, and he was arrested in Manchester and brought back, and it was then found they had been sleeping together in the same bed, and they were questioned, sir, and they admitted these offences.

. . .

THE CHAIRMAN: Daniel O'Connell and Wilfred Pearce, each of you is of good character. You, Pearce, are aged 50, and there has been nothing against you in your life before. On the contrary, you seem to have behaved well, certainly as a son, because you devoted yourself to the care of your ailing parents for as long as they lived.

You, O'Connell, are 28 years of age, and apart from an offence of dishonesty, which was properly recorded as a minor offence and is completely irrelevant for the present purpose, have also enjoyed a good character.

Now, it happened that you were thrown together because the elder of you offered shelter to the younger, and there was only one bed in the house. In these circumstances, being in bed together, you committed a sin against personal morality of mutual masturbation. You did that in private, and this offence is very far removed from the kind of homosexual crime which has given many people anxiety in our time. This is another illustration of the kind of case which would not have been a punishable offence until 1885...

... There is no element of corruption here. There is no element of public outrage, and so far as I can see, the limit of your offence has been a private sin against personal morality. In these circumstances, we are satisfied that justice does not demand anything except that we should give you a conditional discharge.

... We are quite sure we shall not see either of you again. It is quite obvious that these lapses are not characteristic of either of you, and the probability is that you will both be able to forget about it.

. . .

You, Pearce, we hear with great satisfaction, are hoping to exchange your solitary life, which is probably not a very satisfactory one, for a normal married life with a woman who has agreed to marry you, and whom you hope, anyhow, will overlook it. Standing in the dock and pleading guilty to an offence of this kind is, of course, a very shameful thing and scandalous position to be in, but in the particular circumstances of this case, the woman whom you hope to marry would be perfectly right to forget about this entirely and to regard you as a normal man, as we hope she will.

. . .

Annex IV: Memorandum of Det. Sergeant R. Woolley, Alfreton Division, Derbyshire Constabulary, 13 August 1955.[80]
[Re Dennis, Cluskey and Hill case]

...

The offences with which these men were indicted came to light when Donald Dodds, 20 years, of Hockley Cottage, Colliery Road, Alfreton, was being interviewed regarding other offences, in respect of which he subsequently appeared at the same Quarter Sessions. One of the offences with which Dodds was indicted was an act of Gross Indecency with Hill. Dodds had witnessed the offences committed by these three men ...

The facts so far as the offences charged are concerned are that ... Dennis, Cluskey, Hill and another man whom it has not been possible to interview were in a shelter on the Watchorn recreation ground at Alfreton ... After some conversation about sex these men exposed their persons, and after feeling at each other some masturbation took place, and in Hill's case this resulted in an ejaculation. Hill says that it was either Dennis or Cluskey who masturbated him, but both the latter say that it was the other man ...

In addition to the above offence all three offenders admitted a further occurrence when they felt at each others [sic] persons over the top of their clothing ...

In addition to the offences already mentioned, Hill asked for eight other offences to be considered. These were all in respect of offences of Gross Indecency, in some cases masturbation, and in others by placing his penis in some persons [sic] mouth or vice versa. These offences ranged from the beginning of 1948 until the end of 1954, and cover a large number of indecent incidents, in several cases one offence taken into consideration covers several different acts committed with the same person ...

Hill was without doubt the ringleader in these episodes and was on the lookout for youths and boys with whom he could commit offences. Cluskey is inclined to be effeminate and is the brother of John Bernard Cluskey, who in January, 1955, at the Alfreton Juvenile Court was brought before the Magistrates as being in need of care and protection. The circumstances being that since the age of 13 years he had been indulging in homosexual practices with a number of men in this area.

Neither Hill nor Dennis had been previously convicted, but, in addition to a conviction for a summary offence, Lewis Arthur Cluskey has the following convictions:–

| Alfreton Juv. Ct. 9/6/44 | Indecent Assault on Male (2 cases) | Pay £1/-/5d. costs in each case ... |
| Alfreton Juv. Ct. 14/9/45 | Stealing fruit value £1/17s/7 1/2d. | Probationary period extended for 12 months... |

Annex V: Memorandum from Supt. E. Mallis, County Police Office, Renishaw, nr. Sheffield, 11 August 1955.

[Re Pearce and O'Connell case]

...

In October, 1954, Daniel O'Connell, 28 years, of 78, Sudell Street, Collyhurst, Manchester, a labourer by occupation, obtained lodging accommodation with Wilfred Pearce, 50 years of age, at 38, Station Road, Halfway. Both men were single and they slept together at that address.

At 7.0 a.m. on the 22nd February last, Pearce went to work leaving O'Connell alone in the house. Later that day Pearce returned from work and found that O'Connell had left a note saying that he was leaving the area. Subsequently Pearce found that a quantity of clothing and footwear, together valued £9, were missing from the house.

O'Connell was traced to Manchester where he admitted the theft of Pearce's property.

At Renishaw Magistrates' Court on the 28th February, O'Connell was fined £5 or one months [sic] imprisonment for this offence, and, being unable to pay the fine, the Magistrates ordered that he be committed to prison forthwith.

Local enquiries were made before O'Connell was traced and it was learnt that both men, who were single, were sleeping together, there being no other sleeping accommodation in the house. This was regarded with suspicion and as a result of further enquiries Detective Sergeant Macdonald and Constable 284 Keeble interviewed the prisoner and the complainant at Renishaw Police Station at 1.30 p.m. on the 28th February, 1955. Both men admitted mutual masturbation whilst they were sleeping with each other at Pearce's home. Each made a statement, under caution, saying that the other made the initial approach and that afterwards similar conduct took place on about six occasions.

...

*

(i) HO 345/7: Statement by Mr. John Scott Henderson, Q.C., Recorder of Portsmouth, 10 January 1955[81]

1   I have been practising for 27 years in criminal courts, mainly on the Western Circuit, and during the last 10 years I have been Recorder of Portsmouth.

2   There have always been numerous cases of gross indecency from Portsmouth, but my recollection is that there was a marked decline in numbers during the 1939–45 war. The following table shows the numbers of men indicted for homosexual offences at Portsmouth Quarter Sessions during the last 10 years.

Year	
1945	8
1946	3
1947	Nil
1948	8
1949	19

1950	23
1951	22
1952	21
1953	24
1954	<u>30</u>
Total	<u>158</u>

3   The explanation of the increased committals since 1949 may be that two policemen were detailed to act as a vice squad in 1949, and the explanation of the increase in the figures in 1954 is that in that year a fresh team was appointed to the vice squad.

4   Cases of homosexual offences committed by two naval ratings together are rare, the great majority of these offences being due to visitors to Portsmouth acting together with young sailors. These visitors with very few exceptions were men of hitherto good character, and my conclusion is that they were confirmed homosexuals who came to Portsmouth with a view to finding young sailors who might be prepared to indulge in homosexual practices. Invariably the sailor was picked up in, or after he left, a public house and after he had consumed a large quantity of drink. I never had any evidence which indicated that the sailor was a prostitute who was trying to earn money by permitting the other man to indulge in homosexual practices.

5   I have said that the visitors were men of good character. They had many occupations and some were in responsible positions, e.g. Lieutenant Colonel, Staff Captain, Warrant Officer, R.A.F., clergyman, lay pastor, business men.

. . .

*

(j) HO 345/7: Views Submitted by the Lord Chief Justice of England (Rt. Hon. Lord Goddard)[82]

. . .

ANNEX B

HOMOSEXUAL OFFENCES

ANSWERS TO QUESTIONS SUBMITTED ON 17th NOVEMBER 1954[83]

I   Subject to what I have to say regarding buggery, in my opinion private acts of indecency between adult males ought not to be the concern of the criminal law. They have none of the attributes generally considered to be the constituents of a crime[84] except that they excite disgust and repulsion. Many moral offences are far more anti-social and harmful but are not crimes, e.g. the seduction and desertion of a young girl, or adultery which breaks up a home when there are children.

II   I do not think the draftsman of an amending statute would find any difficulty in making clear what was meant by "in private". He might do it by using such words as "except in any place to which the public have access". A favourite place for this form of beastliness is a W.C. in an underground lavatory, and though the act may be screened from observation by the door it is obvious that decent people using such places would be affronted if two men appeared coming out of a W.C. I do not think any Court would find difficulty in holding that a W.C. in a public lavatory was a place to which the public has access even though the door was closed.

...

The age of consent presents a problem on which there is likely to be wide divergence of opinion. After much thought I incline to 21. My reason is that I believe National Service is to some extent responsible for the undoubted increase. Then from conversations with officers attached to London Command I believe that elderly male perverts do tempt young soldiers to lend themselves to these practices for money. The police know more about this than I can pretend to, but I feel it would be advisable not to put the age of consent too low. One difficult question to be faced is, are the Courts to deal with cases between two males each under the age of consent who are detected operating in private? On the whole I would say that they should be left to parents, schoolmasters or Commanding Officers at least if Parliament could be induced to allow the latter to order a moderate chastisement as a schoolmaster would.

...

VI   It may not appear logical but I feel strongly that buggery ought always to be treated as a crime. It has been a felony for 400 years, and before it was made a temporal offence it had always been treated as a grave offence in the ecclesiastical courts which had in those days extensive powers. It is such a horrible and revolting thing and a practitioner is such a depraved creature that he ought in my opinion to be put out of circulation. I believe, though I recognise that I may be quite wrong, that there is often, and I would venture on generally, a wide difference between the decadent young man who finds or thinks he finds satisfaction in good looking youths to the extent of masturbation and the bugger who is nearly always an habitual. I recognise that as a recent notorious case showed men who have good war records and can properly be classed as brave may be addicted to this vice[85] but, if they are, they are in my opinion such public dangers that they ought to be segregated which can only be effected by imprisonment...

...

One matter germane to this enquiry to which I would direct the Committee's attention is whether the penalties on male persons for importuning should not be drastically increased. The existing law is enough for some cases of casual importuning which one hears of from time to time. The male prostitute, and there are many of them, ought to be liable to a really long sentence. For the young practitioner a whipping would probably be the best thing but it is no good advocating corporal punishment for anything—it can no longer be given for that most detestable of offences—living on a woman's immoral earnings.[86]

*

(k) HO 345/6: Views of certain of Her Majesty's Judges on the laws relating to homosexual offences[87]

Mr. Justice BARRY[88]

While I consider that there are some doubts as to the wisdom of legislature in creating this offence—when committed in private—I think that it would be a most unfortunate moment to introduce any change in the law. Even a short experience of the work on Circuit or at the Old Bailey teaches one the prevalence and the widespread corrupting influence of this vice. Any amendment of section 11 of the Act of 1885 must, I think, be interpreted by those addicted to it as a charter of freedom and an indication that the State no longer regards this conduct as reprehensible.

. . .

Mr. Justice BYRNE[89]

. . .

Gross indecency between adults in private should be treated as a crime although of course the sentence must depend upon the circumstances of the particular case.
In my view there are few more serious offences than corruption of the young.
Let the law remain as it is!

. . .

Mr. Justice DEVLIN[90]

(1) I think that gross indecency by adults in private should not be treated as a crime; and by "in private" I mean that which does not amount to a public nuisance.
(2) I think that where the element of corruption of the young comes in, it should remain a serious crime.

. . .

Mr. Justice DONOVAN[91]

I think that men and women should be treated alike, i.e. that homosexual acts between adults in private should be left to the disapprobation of decent-minded members of the community and not treated as criminal acts.

. . .

Mr. Justice FINNEMORE[92]

On the whole I think gross indecency should be a matter of private morals rather than public crime. I cannot see why we should have to spend time dealing with these cases of men fooling with one another in private.

...

Mr. Justice HALLETT[93]

(1)...

In these days ... the fact that an act or omission is morally wrong or disgusting is not regarded as a sufficient reason for making it punishable by the State. The avowed ground for making anything punishable is that it is detrimental to the interests of the community.

It may be that in former times anything tending to lessen procreation was regarded as detrimental to the interests of the community, but even if this is still right—as to which I express no opinion—it is clear that there are many ways other than gross indecency of gratifying the sexual instincts without furthering procreation, e.g. self-abuse, unnatural acts between women, and even the use of contraceptives.

I cannot myself see any logical reason why abuse of the sexual instincts, if it takes the form of gross indecency committed in private and between persons of full age, should be selected for punishment.

...

Mr. Justice McNAIR[94]

(1) For my part I would not advocate any change. Clearly the gravity of the offence varies enormously but this is a matter which can be dealt with by appropriate sentence or binding over after sentence. It seems to me that it can be left to the good sense of those responsible for instituting prosecutions not to prosecute in the case of gross indecency between adults in private. Personally I have never had such a case before me in my short experience.

...

Mr. Justice SELLERS[95]

I would regard buggery as a serious offence wherever it is practised, for I believe it indicates, if widely indulged in, real depravity in a state or race. It is so widespread now that I think, in any case, it would only give encouragement and licence if the law were in any way relaxed.

Gross indecency might be limited to such acts in relation to young men and boys and in public places.

Some years ago when I did the Crime at Manchester some ten or more cases revealed a men's brothel in Blackpool and the evidence of what took place, actions and language, was revolting.

It may be that in some cases there is an element—perhaps a strong element—of physical make up which accounts for the conduct, but I think it is hard to dissociate this from vice.

...

*

iv. Correction

(a) HO 345/15: Interview of Miss D. G. Anderson and Mr. F. R. Groom, on behalf of the Association of Managers of Approved Schools, and Mr. Headley Chamberlain, Mr. J. H. Clarke, Mr. J. H. Bennett, Miss M. M. Brown and Mrs. M. M. Jackson, on behalf of the Association of Headmasters, Headmistresses and Matrons of Approved Schools[96]

...

[Q4119] (MR. CHAMBERLAIN):...I think that, with boys of 11, 12, 13, 14, one is tolerably certain to find in any community of boys, no matter where they are selected from, experimentation—if no more, and probably very little more indeed...It is because we feel that it is a question of experimentation, lack of knowledge and, quite often, a sordid background that we tend to play down in our schools, so far as the junior sides are concerned, any suggestion that homosexuality provides a major problem nationally within our schools...There is a suggestion in our memorandum[97] ...that any offences as between boy and boy where both are below the age of 15 might not be considered a matter for police intervention or charge...We feel that most often the headmaster in the school, or a member of his staff, is best able to deal with the kind of problem that we feel is likely to arise...[T]hose of my colleagues who are in senior schools are with us in viewing the problem, difficult as it is in senior schools, in an educational light too, and they share our misgivings about the treatment of homosexuality in a school as a criminal matter...

...

[Q4126] ... [MR. CLARKE:] ...It is found in senior schools that you have boys of the age where their biological urges are at their greatest, and they experiment; and also we sometimes find that there is a case sent to us where there has been this problem of perversion beforehand. That comes from the association with older people, and if they have had this perverted attitude we find that they are difficult of supervision because they wish to go on with their nefarious practices, and thus it is there that our greatest remedial work has to be done. I would say it is done best by the chaplain of the school and by the housemaster particularly, and by the whole of the staff in general. We need to have a very careful system of supervision of the dormitories, particularly after the boys have gone to bed; and should we find that the boy still wishes to experiment we can then call in the Medical Officer, who very often refers the boy to the psychiatrist...

[Q4127] Q. [CHAIRMAN:] And then, of course, you have the difficulty to which you made reference—which those of us who have been to boarding schools of the ordinary kind know to be a difficulty anyway, but you have it very much more— of what happens in a dormitory after everybody is in bed. How in practice, if it is not being too inquisitive—how, in practice, do you organize that?—A. From the time the boys go to bed they are visited at irregular and infrequent intervals. They go to bed at 8.45 in winter and 9.15 in summer; from then onwards, until about 12.30, which we find in practice is the most awkward time, they are supervised by someone going through the dormitories...

...

[Q4128] CANON DEMANT: And when you find a boy in bed with another boy, what do you do?—A. I take him downstairs first of all and talk to him, and then I bring into being all the suggestions that I have made before—the School Medical Officer, the Chaplain, myself and, if necessary, the psychiatrist.

...

[Q4142] CHAIRMAN:... We have had it put to us more than once now on medical and psychological grounds that in the great majority of cases—I think it is not an exaggeration to say that—in the great majority of cases the seduction of a boy by an adult does not necessarily have any lasting effect on that boy in the way of turning him towards fixed homosexual behaviour later on. It would seem that your own experience and your own strongly-held opinion would be quite contrary to that view. Is there any distinction to be drawn between (a) the seduction of a boy by an adult who was friendly and who was coming at it from the idea of helping the boy and bringing him along and then fell into the trap of homosexual relations with him—that on the one hand, and (b) on the other hand a seduction which has in it as one of its basic features lack of consent on the part of the boy and a degree of force that might almost—if it were a heterosexual act—come near rape. Have you seen any difference between those two things, or has the distinction not presented itself in practice?—A. [MR. CHAMBERLAIN] My experience is that it is awfully difficult to detect that difference, because if the seduction is primarily beastly it is always preceded by the kind of behaviour which might add a little respectability to the whole thing. In other words, the boys with whom I have had difficulty, personally, in my school have been boys whose reclamation has been made almost impossible by things which have gone on before—they have been seduced by much older males who, under a cloak of respectability, had only one end in view...

[Q4143] Q....There is also, unfortunately, a number of youngish boys who, in fact, enjoy this kind of activity and operate as young prostitutes. One gets cases from time to time in the courts, and you must be very much more familiar with them than I am—of men who have indeed committed this offence and have been sent to prison for having committed the offence, but of whom it is said by the Judge or the Recorder, or whoever it is, that these men were in fact seduced by

11-year-old or 12-year-old boys. It has been said in public a good many times in the last few years. Have you not come across such boys?—A. It is outside my experience.—(MR. CLARKE): I have come across one in my experience, Sir. The boy was seduced by a man on several occasions and he continued to endeavour to have other boys commit homosexual offences with him. Eventually with the help of a psychiatrist, we were able, as we thought, to persuade him to lead a normal life, even to the extent of getting a girl friend after he had left the school, which I think is always a good thing. Unfortunately however, he had left a record of all the men who had given him money for committing these offences, and it became a matter for the courts. I was asked if I would be present when the trial came along, and I was also asked to speak. I said—and I have never departed from this—that I did not think it would do him any good to be sent to prison. I have kept in touch with that lad since. I visited him in prison and my wife did also, and I have kept in touch with him since he left the prison; and still he is one who invites homosexual practices. I think that, if he had been allowed, when he left the school and was living a very normal life, to continue to lead that life he would have finished then. Unfortunately he had to be returned to the court...

...

*

(b) HO 345/9: Note by the Prison Commissioners for England and Wales[98]

...

4   Segregation—It is sometimes suggested that homosexual offenders, and other offenders known to be prone to homosexual conduct, should be segregated from other prisoners. The Commissioners have never felt that such a course was either necessary or desirable in the general interests of the prison, while on the other hand they feel sure that it would be contrary to the interests of the homosexuals themselves: and these prisoners must not be put in a position of "less eligibility" in respect of training and treatment in prison because of the nature of their offence. If they are to derive any benefit from imprisonment we must, in the words of a prison governor of great experience, "try to treat homosexuals as normal people and trust that they will react, and in the main they do, certainly outwardly".

5   The Commissioners would go so far as to say that they would view with revulsion the idea of a prison, or part of a prison, in which every inmate was a homosexual...

...

7   Special treatment...

8   It is sometimes suggested that homosexual prisoners are specially favoured by the prison authorities. It is no part of approved policy so to treat them. The fact

that a prisoner has been convicted of a homosexual offence is not the whole truth about his personality, and the type of work given to each prisoner is related, so far as possible, both to his suitability for it and to his personal needs. A great many homosexuals, particularly those who are first offenders in training prisons, are persons of superior education and intelligence, and if their personalities are otherwise sound they are therefore likely to find themselves in sought-after employments. This would apply equally to a non-homosexual prisoner of superior character and cultural level. Homosexual prisoners in local prisons tend to be recidivists and less suitable on personality grounds for good jobs.

9   On the other hand, it is sometimes suggested that homosexual prisoners are constantly jeered or sneered at, or "picked on" by both staff and other prisoners. If this does happen, it ought not to happen: but Prison Officers, like other people, vary in their reactions towards homosexual persons, and there may be occasions when a Prison Officer who feels strongly repelled by them makes little or no effort to conceal the repulsion. Another type of officer may feel sympathy with the offender, though not with the offence. The male prostitute and the corrupter of youth are likely to be objects of contempt in most societies, and in this respect prisoner societies are no exception. The obviously effeminate type inevitably comes in for at least as much derision in prison as he would excite anywhere else; while the type of homosexual who tries to parade a fancied intellectual superiority to the common herd is certain of exquisite unpopularity. Consultation with experienced governors provides no evidence that homosexuals are otherwise treated by other prisoners in any exceptional manner.

. . .

11   <u>Behaviour and influence</u>—It might be urged that while the segregation of homosexual prisoners may not be desirable in the interests of the homosexuals themselves, it ought to be carried out in the interests of the remainder. Such a suggestion is an over-simplification of a complex problem.

It is by no means the case that every homosexual prisoner is a danger to others. For instance, the first offender in a training prison rarely is so. Among active homosexuals there are many who are interested only in boys, and these, apart from some unstable and mentally abnormal prisoners, are not in general a danger to other prisoners. Among the active homosexuals who are attracted by adult males, it is only an aggressive type that is a danger. In general, in the better type of prison the homosexuals are a stabilising and beneficent influence, and greatly assist the smooth running of the prison.

12   The passive homosexual tends to be a great nuisance in prison, though he is not invariably so. The prisoner who has been a male prostitute outside may not find any customers inside whom he regards as worthwhile, and may therefore behave fairly well; the more purely mercenary his outlook, the less trouble he may be. It is the temperamentally female type (not necessarily of effeminate experience though often so) who is the canker; supervision may be sufficient to prevent any actual misconduct, but such types tend to create an unfortunate emotional

atmosphere. The suggestion is sometimes made that they receive exceptional indulgence, and are allowed to parade their 'charms', enhanced by cosmetics, in an unseemly manner. To this we can only say that such types are extraordinarily persistent and ingenious in their efforts in this direction, but that so far as our knowledge and experience goes they are never 'allowed' to do it. Any officer who failed to suppress such manifestations would be failing in his duty.

13   It is inevitable that despite all the efforts that even the best staff can make, a prisoner, at least in a closed prison, will be more conscious of the existence of homosexuality than is a person outside. In the first place there is a much higher concentration of homosexuals in a prison than there is outside; the Governor of one of the large London prisons estimates that of the total "Star" adult population of his prison nearly one-fifth are homosexual.[99] Secondly, as in other mono-sexual types of institution, some persons who are apparently normally heterosexual will turn to homosexuality through frustration arising from lack of the normal sexual outlet. Thirdly, a low type of man may turn to homosexuality out of boredom, or sheer incapacity to occupy his mind with anything worth while. Despite all this, the Commissioners doubt whether prisoners who were not homosexual before reception become permanently homosexual as a result of the contacts made in prison.

14   Such homosexual conduct as takes place in a closed prison can but seldom take the grosser forms; supervision is too continuous for that. For the most part it is a question of unhealthy friendships or associations. The Commissioners would not maintain that no physical misconduct ever takes place; indeed they have some evidence to the contrary. If this should happen, it would be due to some failure of supervision against which is it virtually impossible to guard.

15   Homosexuality in Borstals[100]—This is a separate and different problem. Borstal inmates are of an age at which sexual consciousness is great and may in some cases be dominating; and supervision cannot hope and indeed does not seek to be as continuous and effective as in prisons. There is undoubtedly plenty of opportunity for homosexual indulgence ... Remedies are difficult in Borstals where the inmates are lusty adolescents, too often sexually, and sometimes homosexually, experienced before they reach Borstal. Individual guidance by doctor, housemaster and chaplain, each in his respective sphere, is always going on; and the Commissioners put great store by the countering value of good house and group tone, hard work, a full and busy routine, healthy recreations, and watchfulness on the part of the staff.

...

APPENDIX I

Note by the Director of Medical Services (Dr. H. K. Snell[101])
on the Medical Approach to Homosexuals in Prisons and Borstals

...

Etiology

7   A real difficulty which arises in discussing homosexuals lies in the fact that there is no satisfactory or generally accepted classification of homosexuals. Classification by type of offence may be acceptable from the legal point of view but does not meet the medical requirements. A differentiation is often made into congenital and acquired homosexuality. Congenital homosexuality is regarded as being due to a genetic aberration or to an endocrine disorder. So far as I am aware no proof has been demonstrated by endocrinologists or others that would show convincingly that homosexuals are born and not made. (Hermaphrodites are not under consideration.) A study of identical twins might help to throw light on this matter.[102] Acquired homosexuality is presumed to arise in consequence of environmental influences.

8   For practical purposes it seems reasonable to assume that there may be to a great or lesser extent a constitutional bias towards homosexuality, but that in the majority of cases the part played by early environmental influences has probably been of greater importance. In considering these environmental influences, the family situation is usually significant; the history of the possessive mother, and mother identification, for example, recurs so frequently that the association cannot be ignored. Another factor which appears repeatedly in the history of men charged with homosexual offences is that of themselves having been homosexually assaulted in childhood or early adolescence. While it is not easy to assess how far such an assault has been decisive, or even contributory in determining homosexuality, I have little doubt that once such an introduction has been made and has been followed by a further indulgence in a predisposed individual with a weak heterosexual interest a conditioning or habituation to homosexual practices may well be set up. Other contributory factors are general emotional immaturity and personality inferiority leading to shyness in the presence of the opposite sex and fear of venereal disease. Inadequacy in relation to their fellow men may also lead to association with those younger than themselves and go on to indecent behaviour.

9   All these factors must be considered in relation to the normal development of the sexual instinct. It is recognised that in the adult male the strength of the heterosexual urge varies within wide limits between individuals, and at different stages in life in the same individual, and once again this may be related to constitutional and environmental influences. The sexual instinct is widely regarded as passing through a homosexual phase in its development and this phase is gradually replaced by adult heterosexuality; the influences to which the growing boy is subjected during the formative years help to determine the extent of his maturation to heterosexuality. Thus we can see the bi-sexual as the victim, as it were, of incomplete development of the normal sexual urge and it is useful to assess his sexuality along the lines of a feminine/masculine ratio.

10 A third group, the pseudo-homosexual, is probably composed of those who have outwardly and in the absence of stress or contingency, developed normal heterosexuality; their earlier homosexual urges are but feebly repressed, their veneer of heterosexuality is thin. It is in this group that the denial of hetero-sexual opportunity, for example, among seamen, prisoners of war and occasionally in our own establishments, results in homosexual behaviour. Further, it is to be noted that alcohol, by not only removing inhibitions, but by increasing sexual urges, plays an important part in this group as indeed it may do in other forms of homosexuality.

11 The prostitute may be recruited from any one of these groups but is probably most frequently found in the bi-sexual. The true invert is more usually of a character which would not lend itself to prostitution and, further, it is not infrequently observed that the over feminised passive male prostitute can, if he chooses, shed his feministic mannerisms to a very marked degree.

...

Treatment

...

20 The idea that treatment means the conversion of the homosexual into a heterosexual person by an analytical or other procedure is attractive but, with perhaps a few exceptions, the possibility of doing so is doubtful. If the true invert is indeed born and not made, then one would not expect analysis or any other form of psychotherapy to change the direction of his sexual urge. The most that can be hoped for in a majority of the cases accepted for treatment is a better understanding of their condition and a better adjustment to society. The forms of treatment offered include individual analysis, group discussions and therapy, together with ancillary techniques such as psycho-drama and physical methods when applicable.

21 The questionnaire sent out by the Commissioners to all establishments revealed that among the convicted prisoners and inmates the number expressing a desire for treatment was, in fact, remarkably low—out of the total 1065 cases on whom returns were made 81.0% had no desire for treatment. Further, the replies to the questions showed that only 13.7% were regarded by medical officers as possibly suitable for treatment, and only 6.1% of the whole group of 1065 were accepted by Wormwood Scrubs and Wakefield for it.

...

25 One approach to treatment would be an attempt to abolish the sexual urge. Some chronic homosexual recidivists, especially those who show delinquency in other directions and are of a type who do not exhibit personal responsibility, sometimes ask for a surgical operation, usually meaning castration: how far they would persist, if it were possible to call their bluff, is debatable. Moreover, castration of the body does not mean castration of the mind, and it is at least dubious whether such an operation would have the desired effect if performed after the individual had reached maturity: further, carried out after sexual maturity it would not alter the direction of the sexual impulse.

26   At this point it is natural to consider oestrogen therapy. This is a form of treatment which prison medical officers have been instructed not to use because it is not possible to state categorically that irreversible changes in the testes would not occur. From the point of view of the prevention of crime oestrogen therapy, by temporarily reducing or abolishing the sexual urge, would be of value provided the prisoner continued to take the treatment while at liberty. During custody oestrogens might facilitate treatment by curbing the patient's preoccupation with sexual thoughts or indulgence in masturbation. In one or two recent cases (not necessarily homosexual) this form of treatment has been approved when carried out by and under the direction of an outside clinic...

...

28 It is common knowledge that the Commissioners plan to set up a special establishment on the lines recommended in the East/Hubert Report on The Psychological Treatment of Crime, 1939,[103] and they have obtained a site for this purpose at Grendon Underwood in Buckinghamshire. It is hoped that the psychiatric treatment of offenders, among whom doubtless will be some homosexuals, may be carried out under more favourable conditions than is at present possible, and that there will be improved opportunities for research which will result in a better understanding of the problem of homosexuality and other psychological disorders.

APPENDIX II

Note by Dr. J. C. McI. Matheson, D.S.O., Senior Medical Officer,
H.M. Prison, Brixton

...

(1) Should the Law be altered?

I do not think so. It is sometimes argued that two adult men indulging in homosexual conduct, both being willing partners in the homosexual act, are not harming anyone else, and therefore are not committing a crime, i.e. doing some act which is forbidden by Law and the State has no right to interfere.

This argument is unsound because:

(a)   how is it to be decided whether one or both of the partners in the act is not being harmed?
(b)   the State has an interest in the proper use of the sex instinct, viz. to procreate children. Each child born is of potential value to the State, economically and socially; economically for the productive value to the State when the child becomes an adult; socially for the contribution which can be made to the general welfare of the State by each member of it.

Can, in this present time of striving for world peace and world disarmament, the psalmist David be quoted:

"Lo, children are an heritage of the LORD: and the fruit of the womb is his reward.

As arrows are in the hand of a mighty man; so are children of the youth. Happy is the man that hath his quiver full of them: they shall not be ashamed, but they shall speak with the enemies in the gate."[104]

If it is admitted that the State has an interest in the proper use of the sex instinct, and if, as I believe, individual choice precedes every act then the State has a duty to employ the sanction of the Law to try and ensure that the right choice is made.

. . .

(2) Is the persistent homosexual always interested in partners in the same age group similar to his own or, as he grows older, is he attracted more to younger partners?

I have made a survey of prisoners at Brixton who were diagnosed as homosexuals over the period from 1.1.54 to 31.5.55:

69.0% were attracted to partners in the age group similar to their own or other adult age groups;

27.7% were attracted to boys;

3.3% were attracted to boys and adults.

. . .

I do not consider seduction in early years is a cause of homosexuality in adult life. It is one often given by homosexuals as an excuse for their crime.

If seduction in early years was a valid cause there would be many, many more homosexuals than there are at present.

In those homosexuals who have been seduced or assaulted in early years there are always other factors present of more significance in determining their homosexuality in adult life.

. . .

I do not think a true homosexual can be "cured" but psychotherapy can help him to make a better and healthier adjustment to his disability and so avoid the sanction of the law invoked to control him.

. . .

APPENDIX III

Note by Dr. J. J. Landers, O.B.E., Senior Medical Officer, H.M. Prison, Wormwood Scrubbs [*sic*].[105]

. . .

2   It has often been suggested by psychiatrists that it must be almost impossible to tackle the problem of the confirmed homosexual in a homosexual environment such as a prison, even if there were enough therapists to give each patient a few years analysis. This is not altogether true. A number of homosexuals have been, and are being, successfully treated in prison . . .

3   I think that over-classification of the different types of homosexual adds nothing to our knowledge and, if anything, leads to confusion. We do not really know if homosexuality is constitutionally determined, and there is very little evidence that it is. We are, however, fairly certain that homosexuality is conditioned early in life, probably in the first five years.

4   My concept of the true homosexual, often referred to as the "invert" or "constitutional homosexual", is a person who has rarely, if ever, experienced any heterosexual desires, who is attracted to men like himself, and he sometimes forms a strong emotional attachment to his partner. He is rather selective in choosing a partner—much more so than the bi-sexual, and one does not often find a true homosexual who is attracted by, or seduces, young boys. The form of his sexual deviation is dependent upon his unconscious personality. It is difficult to say how far psychotherapy in prison produces any pronounced trend towards heterosexuality in the type under discussion, but in a minority of cases some movement in that direction has been observed, such as a better tolerance of female company. Psychotherapy produces a lessening of anxiety, and thus a better social adjustment. It gives the patient a better insight into his difficulties, and enables him to accept the fact that he is homosexual. Sending a sensitive true homosexual to prison may well do harm in a number of ways. It may bring him into contact with coarser forms of homosexuality. I do not think it does him any good unless it is associated with psychotherapy.

. . .

12   I am not in favour of an age of consent or of any relaxation of the present laws in relation to homosexual offences. One of the most important methods whereby young people's characters and morals are formed is by example. To see young male prostitutes, or lovers, above an age of consent, living in luxury would have a bad effect on the decent-living industrious youth of the nation. There would also be the demoralising effect on young servicemen from being aware that older comrades and perhaps men of superior rank were indulging in homosexual practices privately with impunity. Homosexuality propagates itself unless checked by improving moral standards and would tend to increase the unmarried group in the population.

Even if an age of consent is put as high as 25 years, my personal opinion is that conduct which is morally wrong and unnatural at 24 years should not be permissible a year later.

. . .

APPENDIX IV

Note by Dr. W. F. Roper, Senior Medical Officer, H.M. Prison, Wakefield

1.   Material and Experience

The following remarks are based upon an experience of homosexual offenders of all types in prison since 1926. I have been more especially interested in them since coming to Wakefield Prison in 1946 because I have been in charge of the Psychiatric Treatment Unit there and because one-sixth of the population of 720 are homosexual offenders.

...

2. The Patterns of Homosexuality

The patterns of homosexuality are very diverse but may be broken down to variation between pairs of opposites along the following dimensions.

a. Obligate and Facultative

Obligate homosexuals are those who have no other possible shared sexual outlet because of a total lack of desire for the opposite sex; possibly the term obligate is wrong because they have always the alternative of chastity. At the other extreme are men who prefer women but will take on men when it is convenient to do so, as may be the case when no women are available. In between lie all gradations of bisexuality.

b. Paederasts and Adult-seekers

At the one extreme are those who have no interest except in pre-pubertal boys; at the other are those who desire their partners to be fully mature. There tends to be an overlap in the middle since many paederasts are prepared to encroach into early adolescence and many adult-seekers like youths, provided that they are past puberty.

c. Active and Passive

Some have no use for homosexuality unless they can take the dominant male role and others will always take the female role. But in the majority of cases there is a good deal of interchange, depending on the age and mutual relationship of the partners.

d. Cultists and Non-Cultists

A few derive their greatest thrill from coupling in company; others require the normal conditions of privacy. Some couple in private but like to belong to a homosexual circle in which some interchange of partners occurs and which serves to recruit new members.

e. Promiscuous and Non-Promiscuous

Some men are entirely promiscuous and never seek the same partner twice; they are often forced to take great risks by frequenting urinals and the like in the course of their search for new partners. Others remain with the same partner as long as they can and are rarely seen in prison unless they favour juveniles or adolescents.

f. Profit seeking and Non-profit seeking

Some are interested in homosexuality purely for gain and may have a lively heterosexual life as well. Others are accustomed to paying their partners or, if this is not necessary, are very generous to them. Many others are somewhere in between. A person disposed to blackmail is easily led on to the exploitation of the position and some paederasts report that even little boys may engage in petty blackmail.

g. Religious and Anti-religious

In some homosexuals interest in religion is strongly developed and this interest appears to be reinforced by their guilty activities; they appear to feel that if they are faithful otherwise they will be pardoned one favourite sin. At the other extreme are those who profess a hearty contempt for religion and conventional morality. Most fall somewhere in between.

h. Feminised and Virile

Some homosexuals are feminised in their mannerisms and attitudes, others take a very hearty and aggressive line. Most of them lie in between. It does not necessarily follow that a feminised man is also girlish physically though it tends to be so; nor does it follow that the virile man is always aggressive; curious cases occur in which very virile looking men prefer the submissive role.

i. Sodomists and Masturbators

Some men are not much interested in anything less than anal intercourse; others think this loathsome and concern themselves solely with mutual masturbation but most try both. Oddities such as oral intercourse, fetichism [sic] and transvestism are not infrequent and there are cases of pansexualism in which homosexuality is merely one aspect of the determination to explore all means of sexual stimulation.

3 Types of Homosexual

. . .

The classification which seems best to me at the moment is as follows:–

A. Paederasts
1 Maternalistic
2 Inadequate
3 Psychopathic

B. Adult Seekers
1 Aggressive
2 Pansies
3 Inadequate

C. Pseudo-Homosexuals
1 Adolescent
2 Deprived
3 Mercenary

Maternalistic Paederasts

They are men of good personality in all other respects; they have a real tenderness for the young and will do anything for them. They have commonly taken up positions of trust amongst boys in order to serve them as teachers, scoutmasters, clergymen and so forth. They think of themselves as big brothers but they have a habit of fondling which readily becomes sexual fondling and occasionally buggery...

They take imprisonment well despite their heavy sentences and social disgrace. They co-operate readily, have no temptations in prison and are an asset in any prison.

The background is that of mother identification; they take over the maternal attitudes and have usually been their mother's pride. Often their fathers have not had much use for them and they have grown up longing for a loving father; because of this longing they feel that their highest role is to play this part to boys later; but fondling obtrudes into the picture and spoils their work. Some of them marry and the marriage may work well enough if their wife is content with a tepid sexuality. Often they cannot marry because every woman seems in some sort their mother so that they are impotent and lack desire. On a deeper level there is sometimes a concealed resentment against women because of a perception that maternal love has left them incomplete as men.

Inadequate Paederasts

They are usually less mature in personality and less devoted to their boys. The essential defect is their inadequacy amongst men, so that being a man amongst boys makes a strong appeal to them, particularly if they can have a uniform and a position of leadership as well. The background is often much the same as above. Seduction by an older man in boyhood or adolescence is common and seems to leave a permanent impress in the sexual pattern...

...

Psychopathic Paederasts

This is a small but dangerous group in which interest in boys is mingled with sadistic impulses. They may torture and even kill their victims. A rankling jealousy of other children in childhood is normally evident in their history.

Aggressive Adult Seekers

These men delight in having other men or youths submit to them. They are scalp-hunters and may include female scalp-hunting within their sexual activities. They have often been unhappy children who have been bullied and have become bullies and their activities usually stem from school days or adolescence. They may like youths but will not touch the immature because they are not scalp-worthy...

Pansies

These men delight in attracting the attention of other men and have gone over to the feminine role. Some exaggerate it. The background is feminisation in childhood together with the discovery that they can redeem their inferiority by boldly going over the sexual line and developing a female role. They are upset if it is put to them that it is a poor thing to win admiration by the loss of virtue and they resent criticism. They like women in a platonic way and are not infrequently under-sexed though they have to grant favours in order to retain admirers. Despite their defensiveness it is well worth while keeping contact with them in prison; on the whole they despise their homosexual partners as well as themselves and the interest of a normal person in them may give them a sufficient sense of being worth while to enable them to dispense with homosexuality. Some of them are talented in their way, especially as entertainers; many turn to domestic service or hotel work. They all live for notice and appreciation.

Inadequate Adult Seekers

This is a large but somewhat mixed group, mostly of somewhat feminised men, who have not made any determined move towards a definite role, and who pair off as occasion offers. Their background contains diverse elements, with inadequacy as the dominant factor.

Adolescent Pseudo-homosexuals

Some degree of homosexual messing about is not uncommon in adolescents who finally normalise without much difficulty. It depends upon the fashion of their environment.

Pseudo-homosexuals from deprivation

This is the common case of lack of women and the turning of men who are sexually crude to their own sex—a relapse into adolescent mucking about. A sub-group is composed of men who avoid women for a while because of disappointment or because of the unpleasant surprise of venereal disease.

Mercenary Homosexuals

These do it only for gain and for the power it gives them over others. They have a normal interest in women.

4   The Prevention of Homosexuality

To take it very shortly, the important measures so far as we can judge from the background of homosexual offenders are these:–

a. The avoidance in childhood of persistent maternal identification and feminisation.
b. The appreciation by fathers and father substitutes that they ought not to treat cissies and inadequates with contempt or disregard.

c. Care that juvenile rivalry and bullying does not take a homosexual turn.

d. Care against early seduction and an appreciation by parents that their sons as well as their daughters are at risk.

e. A general awareness that some devoted workers for youth have a kink.

f. Care in institutions of all kinds.

g. The break up of homosexual recruitment cells where they can be found.

h. Sufficient sexual instruction, formal and informal.

. . .

APPENDIX V

Note by Dr. F. H. Brisby, Senior Medical Officer, H.M. Prison, Liverpool

. . .

5   Dealing first with the constitutional homosexual, probably the passive type, although small in numbers, gains most prominence in the public eye because they are easily discernible, and whilst it is impossible to assess the problem from the point of view of numbers, they are probably fewer in number than is generally accepted. Awareness and recognition of the conditions and publicity given to this question in the press tends to give the impression that the increase is real, whereas it may only be apparent. They are easily recognised, they are fully conscious of their inversion, but even in this class they vary a great deal. Many of them from an early age display quite vicious proclivities. Moral apperception is clear. In the lower walks of life they soon become male prostitutes, only too eager to capitalise their abnormality. There is no suggestion of mental conflict. In an institution where one has to look after them they are a fruitful source of trouble in a most vicious manner, and with them sex appears to be completely devoid of any emotional connotation but to be limited purely to the physical field. The majority of these people, owing to their propensities for dissemination of vicious immorality, may well be considered truly morally defective as well as sexually perverted.

6   The other type of invert—much more troublesome—is aware, either by training or acceptance of Society's code of conduct, of the disability, very often with mental conflict and remorse when they have given way to their deviationary tendencies. In these cases, except where they are almost entirely physiogenic, they can be helped to a considerable extent by psychotherapy, if even only through their conflict being recognised and explained to them and by elucidating any psychogenic element that may have predisposed to the intensifying of the physiogenic factor. In other words, in many cases where they have a will towards normality they do benefit greatly from what has been so aptly termed a psychiatric "prop" and recognition of the fact that control of an instinct is a more tolerable stress for their mind than the conflict and guilt imposed by catering to it.

7   Apart from the true invert, probably the majority of cases fall into a class with all gradations from the homosexual who may be apparently heterosexual to the heterosexual who may have transient deviationary tendencies towards

homosexuality. Into this latter category fall so many of the cases that are hard to understand, where one sometimes gets people of otherwise good character who may have an isolated episode of this nature, and so often one gets as a precipitating factor a period of prolonged stress and tension (probably from quite extraneous causes such as business worries or domestic strife), where, due to this tension, powers of resistance may be weakened and innate deviationary tendencies may manifest themselves given environment and opportunity. Very often there has been an unwise resort to alcohol as a means of escape from their tension, with the inevitable weakening of their powers of resistance and control.

8   It should be appreciated that occult homosexual tendencies may well be fostered by the congregation of men together, such as obtained in the last two wars, and again, only too often the ill-advised campaigns on the question of V.D., where heterosexuality got almost in inevitable competition with the risk of V.D., may well have played no small part in fostering homosexual tendencies in one so predisposed.[106] I have met not a few cases where heterosexual development has been stifled by a dread of V.D. Such cases, of course, respond very well to psychotherapy, which very often has to be directed primarily towards the factors causing the stress. The chief difficulty is that in many cases they are totally unaware that they ever possessed such deviationary propensities, and they quickly respond to explanation of the mechanism and advice as to how to minimise, if not avoid, the danger of a repetition.

In this—and this applies to treating so many cases of homosexuality—the man's inherent sense of moral values and the general integration of his personality must always be one of the main criteria. Man's ability to view his behaviour objectively as opposed to viewing it subjectively (which is a great fault with homosexuals by and large) is of extreme relevance.

10   [*sic*] Another type met with and where one is entitled to hold out a fair prognosis, given that they are of fair intelligence, is where in the earlier decades of life—say, up to the 30's—there has been an impairment of emotional development...

11   It is never enough to write a man down as a homosexual simply because he has indulged in homosexual practices.

It is, I think, fair to advance the opinion that other offences of a sexual nature may be based on occult homosexuality, and in this respect I have in mind that some at least of our cases of sex offences against young, immature girls are due to such a cause, and a similar mechanism—that is, of an innate tendency towards homosexuality—is very often the basic mechanism at work in some cases of persistent indecent exhibitionists.

12   No matter which of the foregoing types of homosexual one is considering, there is one basic trait of personality common to all and that is an innate sense of inadequacy, probably most marked in the true invert, who tends to

over-compensate by posing and posturing, advertising his inadequacy in his mannerisms and his dress...

...

16   As to the influence of childhood experiences in the development of deviation-ary tendencies, I think it is probably generally accepted now that these have been vastly over-estimated, with this important exception, that where a child is intro-duced to such practices by a person of authority (for instance, by profession or social position, such as the schoolmaster or scoutmaster), the effect may be much more grave, as the child immediately fails to develop any sense of impropriety in such behaviour.

...

18   ...I think it would be wrong for the law of this country to say, at any rate by implication, such practices [as homosexual intercourse] should be tolerated, as in the minds of many people the implication might well be that such practices are right.

The administration of justice in this country does not look more closely at the offender than at the offence, and so long as public decency is not outraged the law is slow to take punitive action against mutual participants, and if such practices are legalised I cannot see how this factor of outraging public decency is going to be safe-guarded. Prosecution of perverts does not mean persecution of perverts, as some seem to infer.[107]

19   ...

On this question of an age of consent, I cannot but envisage chaos in a child's mind if he is told that something is wrong but will become right at some statutory age, say 16 or 21. Legality for the most part connotes propriety, or should do.

20   Again, it is not infrequently stated that prison is harmful to homosexuals and that imprisonment is prejudicial to subsequent psychotherapy...Under prison regime and penal conditions to-day I have never known a case come to harm. I have known very many who have been grateful of the opportunity of assessing their conduct and re-adjusting their sense of values...

21   There are, of course, some homosexuals who go to prison whom prison does no good, but almost invariably those are your confirmed constitutional homosex-uals who do not want good done them. They very often manifest such a degree of moral turpitude that they do not wish to be helped...

22   ...

Any connotation of irresponsibility is fatal to treatment, and that is the reason I deplore the impression sometimes given that every homosexual is a doctor's case. This may be very dangerous. It presupposes disease, something the offender cannot help. The issue of personal responsibility must every [*sic*] be brought before the man, with Society rendering what help it possibly can, and certainly little good is going to come of Society adopting the view that as a condition is something that we find it [*sic*] difficult to cure, therefore let us legalise it.

*

(c) HO 345/9: Note by the Scottish Home Department on Scottish Prisons and Borstal Institutions

...

Behaviour and influence

The behaviour of a homosexual prisoner is in a high percentage of cases as good as, and often better than, that of a prisoner convicted of another offence. Initially he is unpopular with his fellow prisoners owing to the nature of his offence, but, unless he tends to draw attention to himself or was in a prominent position in civilian life prior to conviction, he begins before long to be increasingly accepted into, and often to exercise a stabilising influence in, the corporate life of the prison. The passive homosexual tends to be a troublemaker and an exhibitionist, but ... is kept under close supervision by the staff. Occasionally prisoners have made general statements alleging that homosexual practices occur in prison, but for several years past no cases of physical misconduct have been reported by Governors and no petition has been received from prisoners complaining of improper approaches.

...

APPENDIX II

Note by Dr. W[illiam] Boyd, Consultant Psychiatrist to the Scottish Prison Service

This memorandum is based on the examination of 129 prisoners, in Scottish Prisons, convicted of sexual offences, and also on experience as a Consultant Psychiatrist dealing with patients referred to a mental hospital and psychiatric clinics.

...

General Discussion

Sexual activity is essentially a physical response to a mental stimulus. The response varies in each individual according to his personality which is the summation of many factors including his inborn physical and mental make-up, early environment and training, education and health ... The emotion is normally one of affection and it progresses in a fairly constant way. The infant loves himself and the first sexual interest is aroused when in infancy or early childhood pleasure is obtained from self-stimulation. In this early stage it is a mechanical activity.

During childhood the affections are confined within the family, towards the parents, brothers, sisters and animals. Later there is often an affection for someone of the same sex outside the family for whom there is admiration and hero-worship— a homosexual phase. Towards puberty there is a mixed affection for his heroes and for those who now regard him as a hero. At puberty the boy becomes aware of sex as a positive physical appetite with friendships for the opposite sex and love affairs, leading eventually to marriage and the foundation of a family. Later there is a waning of the sexual power. The phases of sexual development proceed gradually but factors may influence the normal development, arresting it, and this may give rise to so-called "fixation". The cause of the arrested development may be a physical or emotional trauma of some kind.

...

Conclusions

Homosexual Offenders

Homosexuals can be separated into two main groups (1) those who have never experienced a heterosexual desire. In this group there is a history of early homosexual attachments, often the result of an early sexual experience.

(2) Those who are attracted to both sexes with a strong bias to individuals of the same sex. In some homosexuals of this group the homosexual practice began when natural intercourse was unobtainable, and continued either as a homosexual practice or together with normal heterosexual relations. In some cases the practice began while serving in the army or in prison. From the statements of prisoners homosexual practices occur in prison in spite of the vigilance of the staff.

Homosexuality exists in individuals who lead a useful cultured life as well as in those who show evidence of a psychopathic state with other undesirable antisocial characteristics.

Homosexuality is not exemplified by a specific appearance. Although some show a femininity in appearance the majority are masculine and indistinguishable from ordinary normal men.

...Active and passive homosexual practices occur in the same individual. Respectability is no bar to homosexual desires and to obtain gratification an intelligent respectable individual will associate with another individual of lower social status. The true homosexual has strong feelings of affection for his homosexual partner.

Proof of an offence committed in private is dependent on the statement of one of the offenders and this may give an opportunity for blackmail. It has been stated by a legal authority that nearly 90% of blackmail cases are cases in which the person blackmailed was guilty of homosexual practices.[108]

The incidence of homosexual practices between females is difficult to determine. They take place in private and it is my experience that there are homosexual tendencies rather than practices.

Homosexuality is a sexual perversion which is regarded by many homosexuals as expressing their sex feelings and they do not regard the practice as one which merits punishment. Many recognise that it is unnatural and seek treatment.

...

Treatment

...

...In regard to homosexual offenders there is some hope of cure for the stable bisexual and for the constitutional homosexual there is the hope that treatment will prevent a repetition of the offence. The unstable psychopath whether homosexual or bisexual seldom responds to treatment and can only be considered with the treatment of psychopaths as a group.

...

*

(d) HO 345/8: Memorandum of Evidence from the National Association of Probation Officers, May 1955

...

Changes in the Law

1.   The members of the Association do not consider that any important changes are called for in the law relating to sodomy, attempted sodomy, assault to commit sodomy and indecent assault by a male person on a male person with the exception referred to in the next paragraph.

2.   We have given considerable thought to the proposal, frequently made, that a change in the law relating to acts of indecency between male persons might be desirable. A majority of members of the Association is of opinion that the law should be changed so that homosexual practices taking place between consenting male adults, in private, should not constitute a criminal offence. The minority opposed to such a change is, however, a substantial one.

The majority is of opinion that in any case such a change of the law should be made only with every possible safeguard against abuse, with the provisions that it should apply only to consenting male persons of the age of 21 years or over; that there should be no offence to public decency or the comfort or convenience of other members of the public, and that neither party should enter into such practices for gain.

Protection of Minors

3.   It is in our opinion essential to protect young people under the age of 21 from indulgence in homosexual practices or from assault, procuration or seduction by those over that age who wish to undertake such practices. It has been suggested

that, in case of any alteration in the law, the age at which consent to such practices might be given should be lower than 21, but we are of opinion that the protection of young men, particularly those undertaking national service training, would make this unwise. A change in the law on the lines discussed above would mean that homosexual practices between young men under the age of 21 would constitute an offence (unless there is a further change in the law) and this would create a most unfortunate situation when such practices indulged in by those over that age were to be allowed without punishment. We are, therefore, of opinion that, if any change in the law is made to provide that homosexual conduct in private between consenting males over the age of 21 should not constitute a criminal offence, children or young persons under the age of 21 indulging in such practices should not be charged with an offence but should be regarded as being in need of care or protection, and brought before the court on that account...

*

(e) HO 345/8: Memorandum submitted by Mr. Frank C. Foster, Director of the Central After-Care Association (Borstal & Young Prisoners' Division)

1 ...

This division is responsible for the After care of all boys discharged from Borstals in England and Wales and for the After care of young prisoners but this analysis is confined to Borstal boys.

2 ...I have taken at random fifty case records...From examination of these records certain dominating factors emerge.

3 21 boys exhibited latent or active homosexuality at a tender age and of these 7 are known to have had homosexual experience before they reached the teens. 6 are experienced in the West End homosexual circles...[B]oth among the homosexual and non-homosexual group the pattern of 'protector–protected' relationship recurs. Repeatedly the picture is not merely one of financial or material gain or sexual satisfaction but a seeking after something more.

...

5 Examination of the 'sample' shows that in only ten cases out of the fifty could home backgrounds be called normal and in three of those family emotional or mental instability is present; in thirty-one cases there are grossly broken homes and in the remainder there is an abnormal family situation.

...

7 ...

With this sort of background it can be assumed that these boys will show a high degree of emotional instability. This is true, but what is more striking is the large number, thirty-three, who can only be called 'emotional isolates' and who seem to have no strong emotional attachment to anybody. It is not difficult to appreciate that to these the homosexual relationship can offer something, however synthetic, that their previous experience has not offered them.

...

9   A strong attraction for the 'emotional isolate' is, of course, not only that homosexuality offers him a quickly established and intense emotional attachment but that it enables him to enjoy material comforts and association with a class of society far beyond that to which he could aspire by orthodox behaviour in his native environment; it is notable that the 'friends' of our boys include not only those in the rather squalid homosexual circles but people whose names are not unknown in the 'fashionable' world, friendships that tend not to outlast the boy's willingness to 'pay' for them.

10   The following extracts from a series of letters from 'friends' to one of our boys, a boy of pleasing manner but of very modest background indicate the dangerous attraction of this mode of life.

'My dear,
    I enclose a cheque for £15 .... [*sic*]

My dear X,
    I will send off a cheque for £10 ...... [*sic*] I will see about getting you a dressing gown. I am so glad you were pleased with the socks, shirts and pyjamas ...... [*sic*]

...

X My Sweet,
    I have booked a double room and bath-room for us at the Y Hotel on Sunday night. I'm just longing to see you ...... [*sic*] Please bring photographs of yourself and of boys on Sunday.
    I am thinking of you all the time and longing for Sunday .... [*sic*]
    Tons of love.

Dear X,
    .... [*sic*] I am taking that one room flat at Marble Arch next week and if you come up you can rest your weary head on my pillow (I said head!)
    God Bless baby (what a size!)'

This boy lived a life of hectic excitement and, to him, glamour without a thought (as is typical) of the sudden descent to neglect and poverty that may be the end of these relationships. So he discovered when one friend wrote terminating their association and another found it advisable to leave the country.

11   To sum up then, the young homosexual tends to be emotionally rootless and isolated, socially and domestically insecure and consequently to be seeking an easily made and intense attachment.

12   Regarding Borstal treatment as an effective method of treating homosexuality there is this to be said. The problem is how to treat the boy so as to enable

him to replace an intense, hot-house, but shallow and sterile relationship with more orthodox and stable relationships. He must be given social skills and social confidence...

...

14   It is clear that existing legislation does little to prevent adolescents turning to a life of homosexuality if they so wish and does little to deter (despite the fact that one boy remarked to me that 'it isn't as easy in the West End as it was. The police are getting too hot.')

15   ...[G]reater protection should be afforded young people from adult homosexuals. It is clear from the cases I have examined that the adult homosexual is often quite ruthless in his desire to procure young people and that little beyond self-gratification lies behind his protestations of affection. The existing law is failing to afford this protection. It should certainly not be relaxed in such a way as to afford greater licence and opportunity of corruption.

The sorry truth is that in too many circles homosexuality is not regarded as reprehensible. The welfare of young people should have priority over the desires of such circles.

...

*

(f) HO 345/8: Memorandum of the Reverend Martin W. Pinker, Director of the Central After-Care Association (Men's Division)

...

3   I have never forgotten some advice given to me by a very experienced Prison Governor soon after I came into my prison work.

Like most beginners in this particular field of service, I was inclined to be somewhat gullible. The Governor had noticed this, and during a friendly conversation he said to me:–

"I would not say a word to destroy the fundamental faith you have in your job, but it may help you to remember that when dealing with sexual offenders a 'homo' is nearly always a homo; a bugger is nearly always a bugger, and a prostitute is nearly always a prostitute."

I have to confess regretfully that after 26 years' experience I do not consider the advice was wrong.

4   I have long held the opinion that changes in our law for dealing with homosexuals and prostitutes are desirable, and indeed much overdue, since (a) I do not recall many instances of which it could be truthfully asserted that a sentence of imprisonment, in itself, solved the problems of these particular types of offenders and rendered it less likely that the offences would be repeated. (b) The law as it stands is not equitable as between the sexes. (c) No matter how abhorrent the practices may be, in my opinion most are, comparatively speaking, not criminal

as we accept the term to-day, since it cannot be said with confidence concerning many practices indulged in at present listed as unlawful, that they are grievously harmful to the community as such.

5   The general public tends to place all homosexuals into one category and regard them as most distasteful and criminally-minded people. I have not found this to be so—indeed, many who have been convicted of these offences, I have found to be quite likeable people in their community life, despite the unnaturalness of their private lives. Many I have known have been kindly and generous as to others' needs, and, in all other ways, normal law-abiding citizens.

6   The <u>educated</u> homosexual, whether passive or active, makes a sad picture: a prison sentence serves no useful purpose for these. One such told me on his release from prison that while the execution of the law had brought ruin to his home, he felt no real sense of shame for what he had done; he was just made that way, and desired the intimacy of his men friends to those of the opposite sex.

This type of prisoner is invariably a good prisoner; but I think that most Prison Governors would agree these men leave prison in a worse state, physically and mentally, than when they were convicted, and no amount of institutional training or treatment has the slightest corrective effect on them.

7   The <u>uneducated</u> type of homosexual is a somewhat different proposition. He is generally a poor, pathetic creature who often has been led into these unnatural practices because he hardly knew better. At his initiation into the mysteries of his sexual life there was no one to guide. What he learned was through the vulgarities of the street corner. Lack of a disciplined, yet friendly, environment in his home surroundings has frequently driven such a one to a companionship of others, leading him into habits which at least seemed to satisfy an inward desire. I believe there is a chance for some of this type to pull themselves together if correction can be given in an early stage. This type is immature, and when the full realization of their habit has been brought home to them, some reveal a sense of shame and inferiority. I believe that in such circumstances some would respond to psychiatric treatment. Here again, I think a prison sentence, given solely for punitive reasons, is likely to do more harm than good.

8   Another type is the <u>delinquent homosexual</u>, and these present an entirely different problem. Often enough they come from the uneducated group, who have found the way to make their practices pay them good dividends. They are unprincipled, and develop into heartless creatures who thrive on blackmail. A change in the law as regards homosexuality would certainly rob them of their most common livelihood.

9   There is also the problem of the <u>pervert who concentrates his attention on boys</u>. Some are mental cases, and, in my opinion, should be dealt with accordingly. Others are, to all outward appearances, quite normal persons who live a sort of Jekyll and Hyde existence. I have no doubt whatsoever that Society needs to be protected against these people, since their behaviour is violently anti-social and as criminal

as carnal knowledge of a girl. Quite clearly, if the law were changed concerning homosexual practices with adults, there should be an age of consent concerning youths, as with girls.

Public opinion demands that Society should be protected against those who commit such offences, and, for my part, I see no alternative to some form of incarceration.

. . .

*

v. Armed Services

(a) HO 345/7: Memorandum from the Admiralty

. . .

2.   Guidance on dealing with homosexuals in the Royal Navy and Royal Marines is given in a Confidential Admiralty Fleet Order... Briefly stated, the Admiralty policy embodied in the Order is:

To foster by positive means a healthy public opinion averse to homosexual practices;
To make it difficult for homosexual offences to take place undetected in H.M. Ships and establishments;
To bring alleged offenders to trial by court-martial whenever there is prima facie evidence of an offence;
To remove from the Service any man whose addiction to homosexual practices has been established beyond reasonable doubt.

3.   It is not possible to estimate the number of homosexuals in the Royal Navy and Royal Marines. Since the end of 1950, 39 men have been convicted by court-martial of homosexual offences: this figure represents an average annual incidence of less than one in 15,000. Just under half of the convictions were for acts of gross indecency, the remainder for more serious offences. The heaviest penalty imposed was two years' imprisonment and dismissal with disgrace...

Besides those tried by court-martial, a certain number of men against whom no offence has been alleged confess voluntarily that they are addicted to homosexual practices... The man is interviewed and sent for medical examination, including examination by a psychiatrist. If it seems from all the information available that the confession is true, the man is discharged... Careful investigation is necessary, because some men are prepared to make a false confession of this kind to obtain their discharge from the Service...

. . .

4.   It is recognised that in some individuals, homosexuality is susceptible of cure. Nevertheless, it is not considered right to retain a known homosexual in the

Service, where facilities for appropriate treatment are not readily available and where the predominantly male environment is unfavourable to a cure and affords exceptional opportunities for corrupting others.

5.   It might be expected that men without women, as in H.M. Ships at sea, would be particularly prone to homosexual practices. However, such figures as are available certainly do not suggest that homosexuality is more prevalent in the Royal Navy and Royal Marines than in the other Services ...

6.   Homosexual offences in which members of the Service are involved with civilians occur from time to time, and there are grounds for believing that such offences often go undetected. It is sometimes suggested that the traditional seaman's uniform ("square rig") is peculiarly attractive to homosexuals. There are no known facts to support this belief (indeed, the incidence of homosexuality appears to be at least as high in branches of the Service which do not wear this type of uniform), or to indicate that homosexuals are more strongly attracted to sailors than to other young men....

7.   The Admiralty are aware that a body of opinion exists which favours amending the law so that homosexual acts carried out in private between consenting adults would no longer be an offence. Any such change in the law would be likely to produce serious difficulties in the Royal Navy. It has long been recognised that the conditions of life in a disciplined Service provide the older and more experienced men with more than usual opportunities for putting pressure on younger men (who may also be inferior in rank) to submit to their desires. As the law stands at present, the younger man is aware that if he consents he may render himself liable to a criminal charge. If the law were changed in this respect, a strong motive for resistance and for reporting the occurrence would be removed.

*

(b) HO 345/7: Confidential Admiralty Fleet Orders, 29 January, 1954.

Section 2.—PERSONNEL
C.A.F.O. 28: Unnatural Offences

...

PART I—GENERAL

...

5.   Apart from the effect of the infliction of severe punishment wherever possible, Their Lordships also wish to impress upon Commanding Officers that much may be done by unremitting attention to the moral well-being of the ships' companies under their command. They in no way under-rate the aid afforded by the various religious agencies in this matter, but they think that, apart from such influence, there is room for the creation of a healthier public opinion amongst the men,

which may be fostered by taking suitable measures to point out the horrible character of unnatural vice and its evil effects in sapping the moral fibre of those who indulge in it. As ships' companies change frequently, the matter should be brought to their notice from time to time as may be necessary.

6.   The help of the steadier and more reliable men on the Lower Deck in resisting the tendency which is too often displayed to treat these matters with levity should also be enlisted. It should be impressed upon the younger members of the ship's company that if they have real cause of complaint as to the conduct of other men towards them they may be assured that their complaint will be fully investigated, that it is their duty to resist any approaches of this nature, and that unless they wish to be considered as participating in the crime they must make complaint at the earliest moment.

7.   Means should also be devised to prevent opportunities for the commission of such offences in the out-of-the-way parts of the ship.

. . .

PART III—MEDICAL

INSTRUCTIONS FOR THE GUIDANCE OF MEDICAL OFFICERS IN DEALING WITH SUSPECTED CASES OF UNNATURAL VICE

. . .

2.   The examination should be conducted as follows:–

(a) *In the case of the passive partner:–*

> 1) Note the general appearances. Look for feminine gestures, nature of clothing and use of cosmetics, etc.
> 2) Visual external examination of the anus for:–

Appearance of bruising or inflammation.
Whether redundancy or thickening of the skin is present.
Evidence of irritation, inflammation or presence of thread worms.
Recent tears, lacerations, fissures and piles, old scars due to previous ulceration, or any other physical sign that might be present and might cause dilation or relaxation of the anal sphincter.

(*Note.*—It cannot be too much stressed that the "classical" appearances described in many books are most uncommon. The "conical" anus occurs only in the confirmed practitioner.)

(3) Examine the anus for size and elasticity (it is useful to measure the size of the opening by some standard measure such as the number of fingers) and note any discomfort or otherwise during the examination. A speculum may be used.

(4) A swab must be taken from inside the anus with the aid of a proctoscope or speculum for demonstration of spermatozoa, and another from surrounding parts for identification of lubricant and spermatozoa.

(5) The man should be examined most carefully for the presence of V.D. The presence of any discharge from either the anus or urethra should be noted and slides and swabs taken for the identification of gonococci. It should, however, be remembered that the G.C. cannot be identified for certain except by culture. The presence of a suspected chancre and its position should be noted and samples of serum exudate from the sore should be collected in capillary glass tubes for the identification of treponema pallidum. These samples should, if possible, be examined immediately after being collected, but further samples should also be sent for examination by a specialist in pathology. When possible all cases in whom a V.D. is suspected, should be sent to a venereologist for examination at the earliest opportunity; this especially applies in those cases where G.C. infection is suspected, since the organism in this case can only be identified with certainty after culture.

(6) If it is alleged that the practice has been carried out recently, the under pants and shirt should be examined for the presence of stains which may still be damp. Any suspected stained articles should be wrapped in cellophane or brown paper and sealed for transmission to a laboratory. If it is possible to collect a specimen of liquid semen from an article of clothing, it is desirable to send this in a suitable container. In some cases the blood group of the donor can be determined.

(7) Other suspicious objects such as tins of lubricants, should be sent to a laboratory for examination for the presence of spermatozoa or pubic hair.

...

(b) *In the case of the active agent:*–

The examination should be planned to establish whether the penis has, in fact, been subjected to friction and is contaminated with faeces. It is obvious that the presence of a mixture of faeces, lubricant and spermatozoa will constitute strong evidence.

The examination should be conducted as follows:–

1. *Examination of the penis* for evidence of friction, for the tearing of phrenum and presence of faeces especially beneath the prepuce if uncircumcised. Also for the presence of lubricant which should be collected on a swab as well as any suspicious material and treated in the same way as other samples. Examination of the base of the penis should be made for contamination with faeces and spermatozoa.
2. *Examination of the clothing*, in this case the front of the pants, trousers and shirt, for fresh stains and again for a mixture of semen and faeces, the clothing being treated as mentioned previously.

3. *Examination of objects* in the possession of the suspected person such as hand-kerchiefs, rags, etc., and also of tins of lubricant, in a similar manner to that already described.

...

*

(c) HO 345/8: Memorandum by the Air Ministry

1   Homosexuality within the Royal Air Force has not presented a serious problem. During the three years 1951–1953, 86 members of the Royal Air Force were convicted by court martial held under the Air Force Act for offences of this kind. This represents an average annual incidence of about one in 9000. In addition, 16 members of the Royal Air Force were convicted by the civil courts during the same period for the same type of offence. The extent of homosexuality within the Royal Air Force which these figures indicate compares favourably with the incidence of the offence among the civil population as a whole, particularly when regard is had to the large number of young men leading a communal life in the R.A.F.

2   ...An analysis of known cases indicates that the offence tends to be more prevalent among airmen in the 18–20 year age group than among older airmen, and that it is almost entirely confined to ground trades. This may be a reflection of the fact that air-crew are all selected volunteers with a paramount interest in flying. Statistics also indicate that the incidence of the offence is greater overseas than in the United Kingdom.

3   It may be that in the course of professional consultation a medical officer learns that an airman has homosexual tendencies. The question then arises whether he should report the matter to the Commanding Officer with a view to the airman's discharge. No rule can be laid down as to whether a medical officer's duty to the service should take precedence over his duty to respect his patient's confidence and each such case would have to be treated on its particular merits. In general, however, it is expected that the medical officer would observe the ethics of his profession and maintain secrecy over a voluntary admission of this nature unless of course he had the free consent of the individual to disclose the matter to the Commanding Officer.

Law and Practice under the Air Force Act

...

8   The normal punishment awarded by an R.A.F. court-martial for buggery is from one to two years' imprisonment with discharge with ignominy, and for gross indecency from 6 to 12 months' imprisonment accompanied by discharge with ignominy or, where detention is awarded, from 3 to 12 months [*sic*] detention. Where the offence is aggravated by the abuse of a position of trust and the corruption of young airmen, the punishment would be increased and might

amount to 3 or 4 years' imprisonment or 18 months to 2 years' imprisonment, respectively.

...

Disposal of homosexuals

11   Quite apart from the moral issues involved, the removal of homosexuals is of vital importance in an armed force. The homosexual cannot exist in isolation; he must have an accomplice, usually several, and in seeking to extend his corrupting influence he is no respecter of rank or person. Homosexual practice brings together men of widely different ranks and position to the prejudice of discipline. There is also a security risk since the man compromised by homosexual conduct may yield to pressure for disclosure of secret information.

12   It is therefore almost the invariable practice for known homosexuals to be removed from the service. If they have been convicted by court-martial and sentenced to a term of imprisonment, discharge with ignominy usually accompanies the sentence in the case of an airman, and cashiering must accompany the sentence in the case of an officer. Where the sentence on an airman is detention, which cannot be accompanied by discharge with ignominy, he is usually discharged administratively "for misconduct" but occasionally "services no longer required"... Where the convicted person was found guilty of an isolated offence in circumstances which suggested he would not repeat it, he might be retained in the Service.

Medical Treatment

13.   Since the practice in the Royal Air Force is to get rid of the confirmed homosexual as soon as possible, the question of medical treatment hardly arises. Sentences of imprisonment passed by court-martial for this type of offence are served in civil prisons and such treatment is a matter for the prison doctor...

...

*

(d) HO 345/8: Memorandum by the War Office

...

THE EXTENT OF HOMOSEXUALITY IN THE UNITED KINGDOM AND ITS
BEARING UPON MEMBERS OF THE FORCES

...

3   In London homosexuality is undoubtedly much more prevalent than elsewhere in the country. Owing to its essentially secretive nature, the problem cannot be well expressed in statistics, but it is relevant to note that during 1954 the Special Investigation Branch of the Royal Military Police investigated twenty-eight cases of sodomy and gross indecency in London District alone, as compared with

one case in the whole of Western Command and five cases in Scotland. The twenty-eight cases in London involved fifty-one soldiers and twenty-three civilians and nineteen of the cases concerned unnatural relations between soldiers and civilians ...

. . .

THE APPLICATION OF THE LAW ON HOMOSEXUALITY TO THE ARMY

. . .

Possible changes in the civil law

11   Under the Army Act, section 18 (5) ... the military authorities have a wide general power to deal with any offence of an indecent or unnatural kind and it might therefore be said that, regardless of any changes that are made in the civil law, the Army could continue to deal under this provision with any form of indecency considered detrimental to discipline. On the other hand, there would almost certainly be a measure of public opinion opposed to the Services continuing to treat as offences acts no longer regarded as criminal under the civil code.

12   Owing to the communal nature of Army life, particularly during service overseas, the War Office feels strongly that acts of an indecent or unnatural kind by officers and soldiers must generally be dealt with as offences, and that they should be dealt with swiftly and with every appearance of severity. They would hope that the Committee will accept this view.

HOMOSEXUALITY IN RELATION TO THE WOMEN'S SERVICES

13   The number of cases of Lesbianism reported in the ATS during World War II was exceedingly small.[109] Though it was not a criminal offence under the civil code, it had a disciplinary aspect when it occurred as it either involved women of different ranks or could lead to the perversion of others. The subject was dealt with in a memorandum prepared in 1941 by a woman medical adviser who distinguished clearly between the adolescent "crush", normal friendships, unhealthy friendships and true promiscuity.[110] This memorandum was issued to commanding officers who requested advice and to other women members who asked for help, but it was not issued widely for fear of creating a problem by drawing attention to it. Cases that did arise were normally dealt with by posting; a very few promiscuous Lesbians were discharged.

14   Known cases of Lesbianism in the women's services since the war have been extremely rare and these have usually been dealt with by discharge as "services no longer required". The War Office does not consider that there is any major homosexual problem in relation to the women's services and they would be opposed to any suggestion that Lesbianism should be an offence in the Army. In the case of women, ignorant but perfectly harmless behaviour may well be misconstrued; to have such cases subject to discipline rather than to guidance and common sense treatment would, it is felt, be far more harmful than the present

administrative arrangements under which only serious cases are dealt with and which, if anything, err on the side of leniency.

Appendix 'A'

Homosexuality in London

Summary of statement by GOC[111] London District
on aspects of the problem affecting the Army

The areas in the Metropolis in which offences involving soldiers tend to originate are CHELSEA, VICTORIA, HYDE PARK, PICCADILLY and the eastern edge of EDGWARE ROAD. In greater London, Windsor is reported to be an extremely bad centre of homosexual activity. Within these centres the actual venue of offences appear to be:–

(i)   Parks
(ii) Stations
(iii) Public Houses and Cafes
(iv) Private Houses and Flats

Importuning and soliciting have, at times, been unconcealed, occurring for instance in Windsor in broad daylight.[112]

. . .

3   It is considered that the contamination of members of the armed forces stationed in greater London is a greater risk than that incurred in the provinces; that is, apart from other large cities. For the average other rank there is, in addition to the separation from his family, which is his normal lot, an environment containing all shades of possible entertainment but all at a very high cost. It is thus possible for him to be perpetually short of money and amidst attractions where complete supervision is impossible. There is also reason to believe that persons afflicted with homosexual tendencies are strongly attracted towards soldiers and particularly towards men of the physical requirements and standard of deportment required by the Guards Brigades, to which the majority of soldiers in this District belong. There have been cases in which soldiers have obviously succumbed to a temptation for easy money in the first place only to have the hold on them consolidated by blackmail. Thus, to summarise, a situation exists in London where the vice and its target exist together in concentrated areas and circumstances which favour the practice of the former and render the latter more vulnerable.

. . .

2
Medical Practitioners and Scientists

Introduction

In the first flourishing of sexology from the late nineteenth century, the traditional period of 'the making of the modern homosexual', pivotal figures such as Havelock Ellis began to implant the notion of homosexuality as a natural, inborn variation exhibited by an anomalous minority.[1] The Freudian challenge to this was relatively slow to take hold in Britain, but by the postwar years scientific and medical discourse was replete with notions of childhood sexuality, universal homosexual phases in adolescence, arrested psycho-sexual development, mother fixations and latent homosexuality.[2] Alfred Kinsey, in his explosive studies into male and female sexuality in the US published in 1948 and 1953,[3] had little patience for any psycho-sexual notion of the 'normal'; but—in advocating the idea of a scale or spectrum of healthy expression and tastes—he too rejected the minoritizing sexological tradition and any discrete categories of 'homosexual' and 'heterosexual'.[4]

The medical practitioners and scientists among Wolfenden's witnesses appealed by name above all to Freud and Kinsey. But what is most noticeable about their testimony is how they attempted to reconcile competing discourses to try and make sense of the many different varieties and apparent aetiologies of homosexuality with which they were familiar. Some homosexuals appeared to be 'true inverts', either born that way or at least whose sexuality was fixed very early in infancy. These people could no sooner be 'cured' than heterosexuals could be turned homosexual; but sympathetic doctors or psychotherapists might be able to help them 'adjust' to their condition and to their social environment. Yet, beyond this 'essential' or 'obligate' or 'genetic' group of inverts, there appeared to be a great number of 'perverts' (the term used if one chose to be disapproving and judgemental) whose homosexuality was 'acquired' or 'facultative' or 'psychological'—one possible explanation being that they failed to make it out of the adolescent homosexual phase because of any number of environmental factors (broken homes, domineering mothers, inadequate fathers and so on). And further beyond (or maybe including) these 'acquireds' there were those who succumbed to 'situational' or 'substitutive' homosexuality (*faute de mieux*: no women being available in boarding schools, borstals, prisons, the armed forces and other institutions) or mercenary homosexuality ('gay for pay', to use an anachronism). These people might be treatable—helped to find their way back to the correct path—or prevented from going astray in the first place.

Testifying to the enduring strength of a moralizing, religious discourse in 1950s' Britain, the memorandum from the British Medical Association [i(a)] included the claim that the prevention of homosexuality was as much a moral and spiritual as a legal and administrative problem. An upstanding citizenry of clean-living, right-thinking Christians needed to be mobilized in the battle against temptation.[5] The BMA had no doubt that male (less so female) homosexuality was a serious social problem; even though homosexuals 'are often charming and friendly people' and only constituted 2% or 3% of the British population (lower than Kinsey's celebrated figures for the US), their practices were repulsive, and cliques of them could establish undue influence in high places. The doctors did support a change in the law—partly because, like some of the other witnesses, they were impressed by F. J. Kallmann's studies of identical twins, which appeared to suggest a strong genetic component[6]—but they felt that clearly everything possible needed to be done to reduce the incidence of homosexual sex and subversion.

Dr E. E. Clayton, appearing as a witness on behalf of the BMA, gave a striking example of the fine line that individuals and the nation had to tread in capitalizing on sublimated same-sex desire:

> [A]s it is agreed that there is a potential homosexual element in everyone, it is what happens to that element that is important. If it goes constructively into things like comradeship, loyalty, friendship and courage, then it is a good thing and something that the country needs. If these intellectual and emotional outlets are diverted, then they go in the opposite direction and they go into the thing that medical people deplore... [W]e believe that it is everyone's responsibility to ensure that there is a sane, healthy, constructive attitude. The absence of that makes it so easy for the essential homosexual... to slide into becoming an enemy of the State really, and becoming attached to an alien ideology instead of to a constructive ideology, which is really the basis and the strength of democracy...[7]

Most of the medical and scientific witnesses were, like the BMA, favourable towards decriminalization for consenting adults in private (usually recommending the age of 21 but sometimes as low as 17), even as they couched their analysis in a language of 'sickness', 'mental deformity', 'immaturity' and 'personality disorder'. Most considered nurture to be more important than nature while tending to believe that sexual orientation was substantially fixed during early childhood. All believed that the seduction/corruption/contagion thesis was much exaggerated; all rejected the notion of a rake's progress; few believed there was much overlap between those men who favoured pre- and post-pubescent males; most did not support the notion that reform would open the floodgates to promiscuous homosexual behaviour—not least because a hostile public opinion would always act as a powerful brake. Many wrestled with the competing minoritizing and universalizing discourses, some even coming close to a logical solution to an obvious conundrum: if a strain of homosexuality is present in everyone, or if everyone goes through a homosexual phase, or if the human animal is basically bisexual, why did homosexual sentiment and expression generate so much anxiety? Perhaps the rational answer was social, cultural and psychological. In other words, the problem with the fear of homosexuality was the fear. The British Psychological Society [ii(i)] and the Institute of Biology [iii(a–b)] invoked such examples as ancient Greece, pre-literate societies and most mammals to demonstrate the naturalness and ubiquity of homosexual sex when not inhibited by cultural (read Judeo-Christian) prohibitions. And the Institute of Psycho-Analysis [ii(h)] and the Institute for the Study and Treatment of Delinquency [ii(l)] repeated a Freudian notion that had become standard in the psychoanalytic community and beyond:[8] that the men who were most virulent in their condemnation of homosexuality were projecting their own anxieties, conscious or unconscious, about their own imperfectly realized psycho-sexual 'normality'—the type of 'methinks he doth protest too much' riposte to homophobes that gay men have always found so satisfying.

Since the question of treatment was part of the Wolfenden Committee's remit, the witnesses proffered a variety of views—whether prison was effective or not, segregated mental institutions desirable or not, deep psychoanalysis and psychotherapy indicated or not.[9] And one theme that persistently arose was hormone treatment—the injection or ingestion of oestrogens such as stilboestrol—not to cure queers but to reduce their libido in a form of chemical castration (real castration was considered but rejected, partly because it was irreversible, partly because it was deemed to be of doubtful efficacy). Excerpts from an article by R. Sessions Hodge of the Burden Neurological Institute, Bristol, and of the Neuro-Psychiatric Department of Musgrove Park Hospital, Taunton, [i(e)] describing his pioneering clinical research in this field, are reprinted here because a copy of the article was included in the Wolfenden archive. Female hormone treatment was in use already in ordinary medical practice and in Scottish prisons but prohibited in English and Welsh prisons because of uncertainty as to whether it would cause infertility or other undesirable side effects.[10] But the Prison Commissioners were reconsidering their position. As Sir Frank Newsam of the Home Office explained to Wolfenden in January 1955, the Home Secretary was favourable to overturning the ban but needed the kind of political cover that an authoritative statement from the Wolfenden Committee might provide. Wolfenden, in response, hedged his bets but reassured Newsam that oestrogen treatment would indeed be considered by the committee and its witnesses.[11] In the event, nearly all of the witnesses were in favour of its use as a temporary expedient in difficult cases, and this was reflected in the Wolfenden Report.[12]

The final document in this section contains notes of an interview with Alfred Kinsey by a subsection of the committee. Kinsey was visiting London, and so the opportunity was taken to meet with him at a special weekend session outside the Home Office arranged by Dr Curran. The Kinsey Scale, as noted, received much

support in the British scientific and medical community, even though some witnesses disputed whether the number of 'Kinsey 6s' would be as high in Britain
as in the US. Drs Curran and Whitby praised its objectivity, amorality and lack
of a theory of aetiology [iii(c)]. Kinsey's star status thus ensured that his views
would carry weight, and he rehearsed many of the conclusions from his studies,
including the idea that sexuality was not an innate matter that predetermined an
individual for life, but that for most people sexual orientation was pretty well fixed
by the age of 16. His observations about the fluctuations in statistics for homosexual offences in New York City gave powerful support to the idea that alterations
in police activity were the primary determinant. But the one area where he agreed
with Freud, the great importance of an individual's first sexual experience to the
point of orgasm in predicting future sexual behaviour, was not easy to reconcile
with the widespread rejection of the seduction thesis by the other witnesses. The
committee would have its work cut out in making sense of all the conflicting
evidence.

*

i. Doctors

(a) HO 345/9: Memorandum of Evidence Prepared by a Special Committee of the
Council of the British Medical Association, November 1955[13]

. . .

I. INTRODUCTION

. . .

Position of Medical Profession

7. Homosexuality and prostitution are essentially social rather than medical
problems, but, since health depends largely on environmental and sociologi-
cal conditions, the Council's Committee welcomes the opportunity of advising
on their nature and causes and of making suggestions for their control and
cure ... Everything which helps to encourage physical and mental health, social
responsibility and stable family life is the concern of the medical profession, for
on these factors are founded the virility and soundness of the national life ...

. . .

Sex and Social Responsibility

9. Homosexuality and prostitution are problems of ancient origin and there is
no short cut or easy road to their solution. They are both sexual activities amounting to social ills and it is therefore necessary, in seeking a remedy, to consider
sex in relation to social responsibility. The proper use of sex, the primary purpose
of which is creative, is related to the individual's responsibility to himself and
the nation, and the Committee believes that the weakening of personal responsibility with regard to social and national welfare in a significant proportion of

the population may be one of the causes of the apparent increase in homosexual practices and in prostitution. While many people would regard gross homosexual acts as more serious than promiscuous heterosexual intercourse in that they are "unnatural" it must be emphasised that it is illogical to condemn the one and condone the other...

...

11.   The attempt to suppress homosexual activity by law can only be one factor in diminishing this problem. A public opinion against homosexual practices is a greater safeguard, and this can be achieved by promoting in the minds, motives and wills of the people a desire for clean and unselfish living.

12.   At the present time doctors observe their patients in an environment favourable to sexual indulgence, and surrounded by irresponsibility, selfishness and a pre-occupation with immediate materialistic satisfaction. There is also no lack of stimulation to sexual appetite. Suggestive advertisements abound on the street hoardings and in the Underground; provocative articles and illustrations appear in the daily and, especially, the Sunday newspapers; magazines and cheap novels with lurid covers frequently provide suggestive reading matter; and the erotic nature of many films and stage shows is but thinly veiled. This background tends to increase heterosexual over-activity, while, for homosexuals, it fans the fire of resentment at the latitude allowed to heterosexual indulgence, when their own sexual activities are condemned and they are regarded as criminals.

13.   Homosexual activity and prostitution are both important problems in human relationships. With prostitution, whether homosexual or heterosexual, the sexual relationship becomes a business transaction. People who are mainly concerned with themselves and their sensations associate together and obtain from each other the physical and emotional experiences they desire. Personal discipline and unselfishness have little place in their thoughts. If this behaviour is multiplied on a national scale the problem to society is apparent, for widespread irresponsibility and selfishness can only demoralise and weaken the nation. What is needed is responsible citizenship where concern for the nation's welfare and the needs of others takes priority over selfish interests and self-indulgence.

HOMOSEXUALITY

II. HOMOSEXUAL PERSONS AND THEIR PRACTICES

...

15.   Authorities on the subject of sex agree that most people, if not all, possess in different degrees both homosexual and heterosexual potentialities. It is believed that the normal development of the sexual drive passes through auto-erotic and homosexual phases in childhood and adolescence before it reaches the normal heterosexual maturity. A preponderantly homosexual drive in an adult and association only with members of the same sex therefore represents some immaturity of development which may be due to a variety of causes. The Committee

distinguishes two main groups of homosexuals in both cases, the Essential (the genetic and the early environmental) and the Acquired.*

ESSENTIAL

(i) Those whose homosexuality is genetically determined

The genetic basis of homosexuality is not entirely accepted by everyone, but doctors do see patients whose homosexual tendencies appear to be so fixed that they seem to be inborn...

(ii) Those whose homosexuality is caused by environmental influences in very early life

As has been stated it is probable that most people, if not all, possess homosexual potentialities. If early environmental influences, outside the scope of conscious memory, are brought to bear on a tendency to homosexuality they may have harmful effects. For example, in some young children the development of their personality is disturbed by exaggerated emotional attachment to one parent, the absence of one parent or some other abnormal factors in their environment.

ACQUIRED

Those in whom the tendency to homosexual practices is predominantly determined by new factors arising in later childhood, adolescence or adult life.

...

16. The two groups, the Essential and the Acquired, present different problems from the point of view of medical treatment. Clinically, the orientation of the essential homosexual is commonly regarded as irreversible and not amenable to treatment, though the individuals may be deterred from overt homosexual activity and helped to make a good social adjustment. Many homosexuals in the acquired group are amenable to treatment and their homosexual tendencies are regarded as possibly reversible.

...

The Essential Group

18. Essential homosexuals are likely candidates for seduction and may easily become habitually practising homosexuals in the company of other homosexuals. The thought of physical intercourse with a member of the opposite sex may be repulsive to them and they therefore compensate by homosexual approaches. At first they may be quite unaware of their condition. Later, when they become conscious of it, they may either attempt to refrain from activity, or they may control their practices by personal discipline, or they may deliberately become practising homosexuals...

19. Many essential homosexuals are discreet and indulge in homosexual practices as an expression of their attachment to a member of the same sex. Some,

---

* In practice it is difficult and often impossible to diagnose to which of those groups an individual belongs.

however, exploit their condition, employing homosexual practices profession-ally for material gain. The essential homosexual person often argues that as he is "made that way" he is not responsible for his condition and should not be punished for his actions. The fact, however, that these homosexuals may be irre-versibly conditioned does not absolve them from responsibility for decent public behaviour. There are many heterosexual people who, because they suffer from some physical defect or other disability, have to adapt their lives to their own limitations and the case of these homosexuals is not essentially different or more difficult.

The Acquired Group

20.   Some individuals may adopt homosexual practices occasionally or habitually for varying motives. Among the numerous ways in which homosexual habits may be acquired are the following:

  (i) The continuance in adult life of schoolboy or schoolgirl conduct. School-boys and girls often indulge in minor homosexual practices (e.g. mutual masturbation) out of curiosity, particularly in a one-sex environment, such as a boarding school. Usually they soon outgrow this phase. A few, however, may carry on such practices occasionally or habitually when opportunities offer and may even ultimately become confirmed practising homosexu-als. A boy or girl who persistently practises homosexual acts is probably genetically conditioned in that respect.
 (ii) Seduction. Boys or girls who are seduced in childhood or adolescence may themselves take up homosexual practices. Whether they do so depends on the circumstances of the particular occurrence and on the child's own reac-tion. It is probable that seduction by an older boy of a younger one causes little or no harm, but the seduction of a youth by a dissolute adult has far more serious consequences. A boy or girl, however, who comes from a good home and has a sound training in character and behaviour will not easily be led astray by a dissolute person.
(iii) Initiation. This is a frequent cause of homosexual practices. Those who have not had sound character training, and weak-willed or mentally sub-normal persons who come into contact with practising homosexuals are tempted to imitate them, especially if they themselves have homosexual tendencies.
(iv) Segregation. Homosexual practices are likely to occur whenever the sexes are segregated. The effects may be only temporary, as with many members of the fighting Services during the war, but it is known that prisons and institutions provide favourable opportunities for homosexual practices.
 (v) Indulgence without responsibility. Many men see in homosexual practices a way of satisfying their sexual desires without running the risks of the seque-lae of heterosexual intercourse. They believe, for example, that there is no danger of contracting venereal disease in homosexual activity.* Other men

---

* This is not so. Homosexual practices can, and do, cause venereal disease.

adopt homosexual practices as a substitute for extra-marital heterosexual intercourse because there is no fear of causing pregnancy or emotional complications in the life of a woman.* Other men shirk the responsibilities of marriage and resort to the sterile practices of homosexuality.

(vi) Defective homes. Conditions in the home are often responsible for the development of homosexual behaviour. Over-crowding may cause laxity in personal discipline. The parents may set a bad example by loose living or may fail to exercise proper parental authority, so that the children imitate their irresponsibility or fecklessness. On the other hand, parental discipline may be too strict, and the prohibition of any sort of sexual contact or friendship with the opposite sex during adolescence may only drive young persons to seek experience with their own sex. He [*sic*] thus becomes an easy prey for the seducer.

(vii) "Cultural aspirations". Some people adopt homosexual practices because they think such activity denotes a superiority of mind and the possession of cultured and artistic instincts.

(viii) Depravity. Debauched persons may seek in homosexual practices new sensations for their jaded appetites.

...

Appearance of Homosexuals

23. Not all homosexual persons can be identified as such by their appearance and manner, for many have no special characteristics. Homosexuals themselves are usually able to recognise each other in various ways, including gestures, smiles and mannerisms, and peculiarities of appearance and habits. In some there is a tendency to self-display in dress and hair styles and in the use of scent and make-up. In the effeminate type of male there is often a certain softness which is difficult to describe but easy to sense. The voice may be high-pitched and facial hair scanty. On the other hand, many homosexuals are virile and masculine. Homosexuals are often charming and friendly people and many of them are well-known to be of artistic temperament.

Homosexual Practices

24. An understanding of what kind of act constitutes a homosexual practice is necessary to any consideration of homosexuality. People have been known indiscriminately to condemn the slightest physical contact between males and to read sexual misconduct into the most normal friendly gestures. Such a restrictive attitude is unjust and may embarrass true friendship. There is nothing wrong in the love of a man for a man or of a woman for a woman. It is only when actual homosexual practices result that it is to be deprecated.

25. Few people are aware of the practices which are indulged in by male homosexuals. Mutual masturbation in private would not be considered contrary to

---

* This also is fallacious, for unrequited homosexual love, jealousy, and the break-up of partnerships may cause just as much emotional stress as heterosexual love affairs.

public order, though most people would feel that it denoted a lack of self-respect or self-control, but when it is carried on in public places it becomes offensive to social decency. Homosexual practices which are anti-social and impinge on public order include soliciting, importuning, and indecent behaviour generally. When such acts as penile friction, mutual masturbation, intercrural intercourse, anal intercourse (sodomy) and oral intercourse (fellatio) take place in public lavatories or at badly lit street corners, they must obviously merit public condemnation.

26.   Not only are their actual practices repulsive, but the behaviour and appearance of homosexuals congregating blatantly in public houses, streets, and restaurants are an outrage to public decency. Effeminate men wearing make-up and using scent are objectionable to everyone.

27.   Homosexual practices tend to spread by contact, and from time to time they insidiously invade certain groups of the community which would otherwise be predominantly heterosexual. Some homosexuals resort to special clubs and certain houses, both in town and country, where their activities are unlikely to be detected, and where they introduce individuals who may adopt similar practices.

28.   Other ways in which male homosexuals arouse the hostility of the public include their alleged tendency to place their loyalty to one another above their loyalty to the institution or government they serve, and, on the part of homosexuals in positions of authority, to give preferential treatment to homosexuals or to require homosexual subjection as expedient for promotion. The existence of practising homosexuals in the Church, Parliament, Civil Service, Forces, Press, radio, stage and other institutions constitutes a special problem.[14]

Homosexuality in Females

29.   ...The following paragraphs (30–38) were written by a committee of the Medical Women's Federation,[15] whose members, by virtue of their practice amongst women, have special knowledge of the subject...

30.   In early adolescence, with the onset of puberty, many girls pass through a homosexual erotic phase as part of their normal development. The attachment is usually to someone in a superior position and is glamorous and romanticised. The object chosen ranges from the captain of the school team or the head girl to an attractive young mistress, often a games mistress, or film star. There is usually little or no personal contact with the beloved object and indeed such contact is hardly desired by the worshipper. In dreams and day-dreams the girl may experience orgasms, but she is seldom conscious of their true nature. This tendency normally plays only a transitory part in the life of the average girl, and sooner or later, encounters a rival, by which it is superseded, in the girl's attraction towards someone of the opposite sex, which again may be highly glamorous and romantic until it is replaced by the more adult type of heterosexual relationship. The age at which interest in the opposite sex develops varies in different social classes, ranging from about 13 in the lower social levels to about 18 in the higher levels.

31.   It occasionally happens that some deviation in this developmental pattern arises. In some instances this may be due to glandular constitution, especially in those women who remain boyish and immature in appearance. This, however, is not consistent, for quite a number of girls of this type develop heterosexual interest early and marry young. A deviation may arise through emotional causes, such as the unwise handling of an adolescent homosexual love affair by the love object or teacher or parents, or some unhappy heterosexual experience which makes the girl afraid of the opposite sex, or the lack of opportunity to meet boys of her own age, or the fears instilled into her by her parents in regard to men and sexuality. It is an interesting fact that there was little homosexuality in the women's forces during the late war, even in cases where the women were segregated for long periods from male company.[16] It does, however, from time to time constitute a problem in women's prisons, approved schools and remand homes, particularly if there is a ringleader who is a psychopathic personality or mentally subnormal.

32.   There is no evidence that homosexuality influences or is influenced by the reproductive complexes. If homosexuals do marry and have children they do not appear to undergo any special disturbances either during pregnancy, childbirth or the puerperium.

33.   Some homosexuals desire to become mothers, and in some cases they marry in order to have children, although they are seldom able to obtain any pleasure from intercourse with their husbands. Conscious or unconscious homosexuality is one of the recognised causes of frigidity in married women. Female homosexuals do not appear to have any greater predisposition to neurotic or psychotic tendencies than the average heterosexual.

34.   On the whole, homosexual women apparently feel no sense of guilt apart from what one might call social guilt. They do not feel that their preference for a member of their own sex is in any way immoral or wicked. Many homosexual relationships, however, do end in disaster, usually because one or both parties become jealous and possessive. In some cases one of the parties falls in love and marries, causing great unhappiness to the one who is left. On the other hand, members of the Committee of the Medical Women's Federation have personal knowledge of cases in which an active homosexual association has been a positive and constructive factor in the lives of the participants. They also know that some women of distinction, who have made a valuable contribution to the life of the community, have been practising homosexuals maintaining a faithful love relationship for many years. The promiscuous Lesbian, sometimes addicted to perverted physical practices, who delights in seducing and corrupting weaker members of her sex, is relatively rare.

35.   The Committee of the Medical Women's Federation is unanimously agreed that the corruption of young girls is uncommon among women, who do not appear to find them sexually attractive.

36.   Female homosexuality as a form of organised vice exists to a limited extent in most large cities. The promiscuous Lesbian who is often incapable of love plies

her trade for money only, and it may be that she arrived at this stage after passing through one or more homosexual experiences which involved some sort of love relationship, and having drifted into this form of life she has degenerated into homosexual prostitution. There is, however, little actual knowledge on this point. Such women are not infrequently psychopaths, and many of them associate their homosexual practices with various sadistic and masochistic activities.

37.   It is a surprising fact that there has been so little public or official condemnation of female homosexuality. Although there are references to it in the literature of the Church,[17] the condemnation seems to have been less strong than in the case of male homosexuality. The secular law has not concerned itself with female homosexuality, perhaps because until recent times woman was regarded in law as only the chattel of man. Moreover, homosexual practices among women seem never to have been regarded as constituting a social danger. The reasons for this may be that they were rare; that, owing to the fact that women do not find little girls attractive as sexual objects, there have been very few cases of seduction or violation of girl children by women; and that such homosexual practices as have occurred have arisen for the most part between consenting adults and have been carried out in private.

38.   The Federation's Committee is strongly of the opinion that there is no case for legal interference with any form of female homosexuality. The negative attitude of the law throughout the ages appears fully justified by the know[n] facts. Female homosexuality has never presented a serious social problem. The very occasional break-up of a marriage, or the degrading behaviour of promiscuous Lesbians, can best be dealt with by social condemnation. The only danger foreseen is that homosexuality may be exalted by the foolish or vicious into a cult, and that could not be prevented by legal intervention.

III. THE INCIDENCE OF HOMOSEXUALITY

...

40.   The frequency of active homosexuality in the male in the United Kingdom has usually been taken to be of the order of 2 or 3 per cent.[18] The Kinsey Report... may indicate the desirability of a re-assessment in this country, but the Committee believes that if a similar study were made here the incidence would be found to be much lower, though, even if the United States figures were reduced by 50 per cent, the number of persons involved would be sufficiently high to represent a serious social problem.

...

44.   The statistics of criminal offences published annually by the Home Office disclose much interesting information... Since 1930 homosexual offences known to the police have increased by 850 per cent[19] ...

45.   These police statistics, if accepted at their face value, may be taken as reflecting a definite and disturbing increase in the incidence of homosexual practices.

The interpretation, however, is not so simple. Part of the explanation may lie in a higher degree of police zeal in tracking offenders as a result of the increased publicity given to this type of offence; or the segregation of men during the war years and the habits then formed; or the post-war communal economic stress and the resultant break-up of marriages. Perhaps also the more tolerant attitude of society towards homosexuality may have led practising homosexuals to be less discreet in their activities with the result that more of them appear in court. The number of detected homosexual acts must be only a fraction of the total number of such acts committed in public and in private, and if undetected acts are increasing in the same proportion as the detected acts, the position is most disquieting.

. . .

IV. AETIOLOGICAL FACTORS

47.   In the view of the Committee it is important that there should be general recognition of the existence of individuals who are wholly homosexual and are incapable of heterosexual relationships...

The Case for a Genetic Basis

48.   The Committee invited Professor L.S. Penrose and Dr. Eliot Slater[20] to express their views on the evidence for a genetic basis for total inversion. It must be admitted, however, that the case for a genetic basis is not acceptable to all observers...

. . .

52.   A considerable number of studies of twins has been made, all of them showing that the uniovular twins of homosexuals are nearly always homosexual also, while the binovular twins of homosexuals are relatively much more normal. The latest, largest and best of these studies is that by Kallmann (1952). His figures are:

Twins of Homosexuals	Kinsey rating of degree of homosexuality			Total
	5–6	1–4	0	
Uniovular twins	31	13	0	44
Binovular twins	2	11	38	51

This shows that there was some degree of concordance in 100% of the uniovular pairs, and a high degree of concordance in 70%, whereas in the binovular pairs there was some degree of concordance in 25%, a high degree in only 4%. If Kinsey's data can be used for comparison, then the uniovular twins show a sharp deviation in the homosexual direction, while the binovular pairs are not much more homosexual than the general American population.

53.   The close similarity of uniovular twin pairs suggests that certain genes lay down a potentiality which, in average circumstances, will lead to homosexuality in the person who possesses them...The marked disparity between the uniovular

and binovular twin pairs ... fairly conclusively excludes the hypothesis of single gene determination. Whatever the genetical influences are, they must be complex.

. . .

56. The evidence summarised in the preceding paragraphs tends, in Professor Penrose's opinion, to support the view that there is a small proportion of the population who are so constituted, perhaps in large part by genetical causes, as to be unable to form normal heterosexual relationships and to be strongly predisposed to form homosexual ones. Variations in sexual polarity might be regarded as a perfectly normal trait, comparable with variation in stature, hair pigmentation, handedness, or visual refractive error. These traits are all probably dependent upon interaction between heredity and environment and the variation within all of them Penrose believes to be probably of degree rather than of kind. He therefore concludes that in the great majority of cases of homosexuality the condition is not abnormal but an example of a natural, and probably inevitable, type of biological variation.

Endocrine Factors

57. The committee is indebted for the following paragraphs (58–62) to Dr. S. Leonard Simpson, its endocrinologist member,[21] who has undertaken some research on the physical characteristics of homosexuals.

58. Doctors have noticed that some homosexual persons have physical characteristics which indicate either a lack of normal virile features or the positive presence of feminine features ...

. . .

61. Dr. Simpson's study is of a series of cases of homosexuals drawn from varied sources with a view to showing whether, and to what extent, a proportion of homosexual patients manifest physical abnormalities or divergences from normal and whether these may be explicable on an endocrinal basis. It is not claimed that a solution of the endocrine aspects is fundamental to the study of homosexuality, for, even though the endocrines may determine some aspects of physical constitution, they are only the mechanism by which this is done, and the endocrine constitution itself is determined by a genetic precursor. Moreover, many normal people, perhaps the majority, differ from the conventional idea of maleness or femaleness in their physical and emotional characteristics. Many heterosexual males have so-called "feminine" characteristics. It is equally true that many homosexuals, some might say the majority, are very virile in their physical aspect. The value of the exploration being undertaken may lie in its possible support of the evidence for a genetic basis in a proportion of homosexuals.

. . .

Early and Later Environmental Influences

. . .

66.   Homosexual patients seen by psychiatrists in prison, or referred by the courts, frequently give histories of strained parental relationships, including marital disharmony, separations, or the absence of one parent in early childhood. The development of their personality in early life is often disturbed by too intense relationships with one parent, usually the female, either as the result of the dominance of the parent figure or of the over-dependence of the child. Psychoanalysts who have investigated many homosexuals intensively also stress this finding of an exaggerated emotional attachment to mothers who are either dominant or over-protective. Such an attachment renders heterosexual adjustment difficult or impossible to the patient because to him intercourse is, perhaps unconsciously, tantamount to incest, carrying with it the incest taboos and phantasy punishments.

67.   The extent to which these psychological factors are fundamental to the production of homosexuals is difficult to assess. Very much may depend on the personality and temperament of the particular child. A Senior Prison Medical Officer who is also a consulting psychiatrist informed the Committee that he found common to all types of homosexual an innate sense of inadequacy. He believes that "if homosexuals can be brought into communion (not necessarily literally) with a fixed body of normal people such as one meets in the Christian community a very great step in overcoming their sense of inadequacy and inferiority will be taken". His remarks in a letter to the Committee seem to be so valuable to an understanding of the homosexual's outlook that they have been reproduced in Appendix C.

V. THE MEDICAL PROFESSION AND THE LAW ON HOMOSEXUALITY:

1. THE DISPOSAL OF OFFENDERS

. . .

69.   . . . The law needs to be both deterrent and reformatory. Existing law and practice do not meet these requirements satisfactorily.

70.   . . . The apparent disproportion of sentences imposed is sometimes greatly disturbing, especially to the medical expert who has full knowledge of a given offender.

71.   . . . If the law is just and proper some punishment with deterrent effect should be inflicted on the offender, but, in the Committee's view, prison is not usually the most suitable place for dealing with him. Many offenders, especially first offenders, could be helped more effectively by medical treatment and moral encouragement outside the prison.

72.   With certain types of homosexual offender a term of imprisonment may even have the reverse effect of what is intended. Hardened offenders find prison life not uncongenial, for it is known that homosexuals often find considerable opportunity in prison to exercise their homosexual activity. Moreover, they may introduce other prisoners, not confined for sexual offences, to homosexual practices. There

is also a danger that a homosexual with a proneness to self-display may have that tendency fixed by prison experience, so that his condition on release will be worse than on admission.

The Question of Segregation

73. The Committee considered whether, when homosexual persons are sentenced to imprisonment, they should be segregated in special prisons or in special accommodation within ordinary prisons, where they could receive appropriate treatment, including constant supervision. Other prisoners would thus be protected from their homosexual activity. It seemed to the Committee that a policy of segregation would be neither practicable nor desirable. It would place a stigma on homosexual offenders which might only confirm them in their resentment against society. Moreover, time and money required for segregation could more profitably be devoted to the treatment of other and more responsive groups of prisoners.

...

VI. THE MEDICAL PROFESSION AND THE LAW ON HOMOSEXUALITY:

2. THE CONTROL OF OFFENCES

...

91. Adverse Effects of the Present Law   These may be summarised as follows:–

(a) The present law creates a background of excessive fear for many homosexual persons, whether practising or not, and this inevitably gives rise to abuses. Opportunities for blackmail are undoubtedly created by existing legislation. An easy weapon is placed in the hands of the blackmailer for all kinds of purposes—financial, personal and ideological. Such experiences of blackmail may produce in homosexual persons an exaggerated degree of nervous tension and strain.

(b) A legal distinction is made which weighs heavily against even the mildest form of homosexual practice as compared with promiscuous heterosexual intercourse, and this discrimination arouses a feeling of resentment in homosexual persons and in their sympathisers.

(c) The law discriminates between the sexes. Male homosexuality is condemned while homosexual practices between women are ignored.

(d) The law treats every homosexual act as an offence *per se* ... It is arguable that a homosexual practice should be regarded as a crime only when it contains a criminal element apart from the mere fact of homosexuality, as in seduction, prostitution, corruption, or public indecency.

(e) The wording of the statutes present [*sic*] certain difficulties in administration. For instance, the requirement that importuning must be "persistent" means that the police have to keep watch and allow repeated offences instead of being able to step in with a warning as they would with a heterosexual nuisance-maker.

(f) The law may often be responsible for encouraging a certain amount of homo-sexual activity, for it provides the glamour of "forbidden fruit" and the excitement of avoiding detection.

92.   Salutary Effects of the Present Law   These may be summarised as follows:

(a) The law as it stands instils in the public mind the idea that homosexual practices are reprehensible and harmful.
(b) It helps to create a public opinion against them.
(c) It protects the young against seduction.
(d) It has value in that a relaxation in any degree might be regarded as an admission that public standards had fallen to a lower level and that the law condoned the lower standard of morality. This might lead youth to think that laxity in sexual behaviour was of little importance, and this in turn could encourage irresponsibility in general and in matters of sex in particular. It might also lead to an immediate increase, though possibly temporary, in the extent of homosexual practices.

93.   Consequences of Relaxation...

94.   The Committee agreed unanimously that if any relaxation of the law should become effective the age of consent should not be less than twenty-one years. The consent of youths under this age is not likely to have the quality of responsibility, and if they form so early the habit of indulgence in homosexual practices it may persist in adult life and influence their general behaviour and social outlook ...

...

96.   The Committee considered the possible effect of a relaxation of the law on the future health of the nation. The incidence of homosexual practices is probably not sufficiently high to have a real and appreciable effect on the marriage rate or the birth rate. Even if there were a relative relaxation of the public attitude towards homosexuality, it is likely that healthy forces in the community would be such as would prevent any substantial permanent increase in the incidence of homosexual practices.

The Law in other Countries

97.   In support of legalising homosexual practices performed in private it is often stated that in certain Continental countries where such practices are not ille-gal no harmful effects have accrued and that there is therefore no reason why they should be prohibited in this country. The Committee is of the opinion, however, that the analogy is not very helpful, as it understands that in some countries there has never been a legal bar to homosexual practices between adults. To legalise homosexual practices in this country would amount to a revolution in an aspect of public policy, and it is possible that it would be followed, perhaps only temporarily, by an increase in homosexual activity.

VII. THE PREVENTION OF HOMOSEXUAL PRACTICES

Home and School Influences

98.   The real safeguard against homosexual activity is public opinion, and measures to increase a healthy attitude towards sex should be promoted and supported by all possible means. The purpose of sex needs clarifying, for in many people its object has become almost entirely selfish. A healthy social environment, springing from secure and happy homes which give a sound background of character training and where sex is kept in its rightful place, would be the greatest prophylactic. Boys and girls from homes where high moral standards are aimed at, where there is parental harmony, sensible, but not over-strict, discipline, a general recognition of personal responsibility and sincere religious faith, will not easily fall a prey to the seducer's invitations to homosexual or heterosexual misconduct. Thus homosexual practices or the tendency for them to spread from one individual to another will be limited.

99.   A healthy public opinion is dependent upon a high level of national morale, which should be nurtured in schools, youth institutions and other organisations. Provision for athletic activities and competition in work and games promotes a healthy atmosphere in which homosexual practices have no place. If the morale of the school or institution as a whole is high any occurrence of homosexual activity will not be difficult to handle. Those associated with the education, training, and activities of youth should be carefully selected. More care on the part of those responsible for such appointments would eliminate the risk of presenting excellent opportunities for indulgence to those homosexuals who, consciously or unconsciously, are attracted to work amongst youth just because of its opportunities. School teachers and others concerned with youth should be able to recognise the type of boy who appears to be liable to develop homosexual tendencies.

100.   Ultimately, the prevention of homosexual practices depends on right human relationships. When parents and the community in general are more concerned about the happiness, welfare and security of each other than they are for themselves and their own security and self-satisfaction, homosexual practices will decrease.

101.   The family doctor may play an important part in prevention. Parents often turn to him for advice over problems of adolescence. By his advice and his personal qualities he may bring his influence to bear when a child is just beginning to get into difficulties. He may, if necessary, obtain for the individual psychiatric treatment at a stage when the patient is responsive.

102.   At school age the importance of child and parent guidance as a preventive factor becomes clear, and it should be developed, together with parent–teacher associations. In public and boarding schools a heavy responsibility lies on the staff. Training in sexual education, properly undertaken, should take its place with training in such matters as table manners, fair play, honesty and all those other elements on which a boy bases his standards of conduct and which enable him

to live on terms of respect for himself and the community. Homosexual practices at school should be treated as would other misbehaviour. They may represent no more than a natural youthful desire for experience. It is desirable, however, that the school doctor should be consulted. In some cases punishment might be of little value and might even do harm.

. . .

VIII. THE TREATMENT OF HOMOSEXUALS

104.   An important object of the medical treatment of a homosexual person is to help him to adjust himself to his condition in as high a degree as possible and to reach a stage where he is able to exercise sustained restraint from overt acts which would bring him into conflict with the law. It will also help him to achieve self-discipline. This will lead to increasing self-respect and enable him to feel that he "belongs" to a society which does not condemn him outright. The more the public understands the nature of homosexuality, the more effective can be the individual treatment by the doctor.

105.   The medical profession has no panacea to offer for the cure of homosexuality, but it is in a position to do valuable work in enabling the individual to overcome his disability, even if it cannot alter his sexual orientation. It must be admitted with regret that some of the advice given to homosexuals in the name of treatment is often useless, simply defeatist, or grossly unethical. Some of it may be even dangerous, as when, with insufficient investigation into aetiology, a confirmed homosexual is advised to marry, no thought being given to the future partner.

106.   Not all homosexuals need, or are amenable to, medical treatment. Many respond to kindly and sympathetic help from friends or those in authority over them. Some adjust as a result of spiritual experience. A large proportion can benefit, to a greater or lesser degree, by medical treatment, though some confirmed homosexuals seem to be beyond psychiatric help in our present state of knowledge. The Committee estimates that a definitely valuable and significant improvement, in the sense of successful personal and social adjustment, takes place in a number of patients, though it must be admitted that completely successful treatment is rare with essential homosexuals.

107.   The limitations of medical treatment should be borne in mind by the courts. It is sometimes wrongly assumed that an offender who is not susceptible to corrective treatment must necessarily be especially vicious and deserving of treatment.

. . .

Forms of Treatment

. . .

119. <u>Analytical Psychotherapy</u>. All the "schools" have techniques for the treatment of homosexual patients. Analytical psychotherapy requires that the patient shall be intelligent, fairly young and able to talk about his difficulties and, especially, shall have enough persistence and desire for cure to carry on with the treatment in spite of the emotional discomfort involved. This form of treatment is very suitable for neurotic patients...

. . .

122. <u>Physical and Drug Treatment</u>. Although physical methods, such as E.C.T.[22] or abreaction,[23] used in connection with psychiatric treatment, are indicated in some cases, they are of value for the accompanying mental illness rather than for the homosexual condition itself. The drugs mostly used in the treatment of homosexuals are the oestrogens (stilboestrol, ethinyl oestradiol,[24] etc.), which produce temporary cessation or diminution of sexual desire in most subjects. When the drug is stopped, normal sexual feeling returns. It is useful (i) as a temporary measure to relieve anxiety in those who are seriously tempted, and to demonstrate that the patient can come and obtain relief if his desires become insupportable; and (ii) as a semi-permanent treatment in older patients in whom there is no contra-indication to severe limitation of sexual life. Its effect varies. It is most effective in a small group of patients who are very highly sexed, and with some of these the dose may be adjusted to reduce, rather than suppress, sexual feelings. Young people are rarely suitable for drug treatment unless the outlook is very poor, but in some who have been repeatedly punished and despair of avoiding further imprisonment, the drug may break a vicious circle and allow them to give up homosexual associations; when the treatment is abandoned later their situation has greatly improved. The method is useful in only a minority of cases, and is really valuable in a very small proportion.

123. <u>Castration</u>. The Committee does not recommend castration as a method of treating homosexuality, although it has been adopted in certain Continental countries. As with oestrogen treatment, it is very effective in a small proportion of cases, but further offences sometimes occur, probably because the homosexual has a need to express his affection, even though the sexual drive is much reduced. Oestrogen treatment can always be abandoned without permanent effect on the patient, but a radical treatment such as castration could only be considered if it were almost invariably effective. There is, moreover, some doubt whether castration always causes a cessation of sex drive.

. . .

125. <u>Treatment of Homosexual Women</u>. The Committee of the Medical Women's Federation writes:– As regards treatment, we know of no evidence that any form of drug treatment is of any use in treating sex deviations of women, except in so far as allied conditions, such as drug addiction or poor general health, complicate a case. The proper approach appears to be indirect, through the avoidance of obvious errors in the bringing up of children, and the encouragement of social

conditions favourable to the development of normal heterosexual love. Recognition of the undesirability (or as some believe, the essential wrongfulness) of this barren way of life is of major importance, for the adolescent may be pushed one way or another by external suggestion, as well as by inward drive.

126.   Psychiatric treatment may be of great benefit in early cases, but is unlikely to cure the congenital or deeply-rooted invert. Even where cure is improbable, however, a psychiatrist may be of great assistance in guiding the Lesbian to adjustment which will not injure others.

...

APPENDIX C

LETTER FROM A PRISON DOCTOR WHO IS ALSO A CONSULTANT PSYCHIATRIST

(see para. 67)

...

Christianity is the only code of behaviour which has unchanging and unvarying standards and values. If one accepts this, then giving up a code of conduct or tendencies to deviationary conduct is easier, even if it means or entails giving up what might be termed natural instincts or deviations thereof, whether perverted or solely deviationary from normal, or if not giving them up, at any rate keeping them in control.

It is surely commonsense that if one accepts that these people who are given to homosexual tendencies or practices are dependent types with a consciousness of their being different, as my observations have led me to accept unreservedly, or put differently, that they are aware of the fact that they are outside "the herd"; that if they can be brought into communion (not necessarily literally) with a vast body of normal people such as one meets in the Christian community, a very great step in overcoming their sense of inadequacy and inferiority is attained. Given this, they are better placed to face up to and assess their difficulties. They are buttressed by the doctrine of ability to do all things by faith. They look for help to the Omnipotent and to understanding by the Divine Compassionate. They realise that as temptation is the normal constituent of human existence, so is resistance thereto...

A belief in a personal helper is essential. A merely intellectual conception of God will have all the shortcomings of the intelligence of the disciples of such a Deity. When I speak of God in this respect I mean the God as propounded in the New Testament and the One Mediator.

Above all, in cases of homosexuality of no matter what grade or type, cure "is a goal only reached after striving and fighting oneself, with victory probably the greatest satisfaction one can experience", as one of my patients put it, but it is a fight which one cannot undertake alone, and if there is any other solution than belief in Christian doctrine and principles and faith then I do not know it, nor do I find myself able to conceive of one.

...

APPENDIX E

CONVERSION AND THE HOMOSEXUAL

Memorandum prepared at the request of the Committee by its Secretary

(The Committee, while not necessarily endorsing in detail the views expressed in this appendix, includes it as a significant contribution.)

I was requested by the Committee to prepare a statement on homosexuals and religious conversion. The evidence on the subject was obtained by interviewing individuals whose names were provided through a number of sources ...

CASE HISTORIES

Sales Representative in Provincial Garage, age 31

Introduced by his Scout Master (a lay preacher) at the age of 12 or 13 years to homosexual practices. He stated that the whole troop used to "do it" at weekend camps. As a youth and young man, although he was a slave to homosexual practices, he had one heterosexual experience. He maintained his special friendship with the Scout Master and practised regularly with him.

When the war came the discipline and excitement of the Army and active Service released him from his habits, but when he became a prisoner of war he reverted to them. On discharge from the Army he took up again with his Scout Master friend, and eventually, to try to reform, he confessed to the police. There was much scandal and publicity and he and the Scout Master were convicted and each given a five year sentence. They were sent to the same prison! He complained to the authorities and was transferred to Dartmoor. He came out after three and a half years on account of good conduct, though he said that "all the prisoners regularly indulged in homosexual practices". After release he gave up trying to reform and was convicted for attempting to seduce a boy in a cinema. He received a twenty-one months' sentence. While in prison he had interviews with the prison psychiatrist, who finally advised him that there was no cure and that he had better find a friend to live with and keep out of the way of the police.

It was also during this sentence that the Salvation Army band came to his prison and played hymns; its members gave personal testimony. Their statements impressed him deeply and he asked permission for a private talk with their leader who explained how he could be "saved". He "gave his life to Christ", and through faith in what he was told found that his homosexual desires were eliminated. Though he has been on occasion subject to temptation, he has been able to overcome this through spiritual help. He has learned to avoid the company with which he previously associated, and also his old haunts. He has made new friends and taken up rugby football, a healthy outdoor game which has benefitted him considerably, in spite of the fact that it throws him in contact with young men. He reports that temptation is getting less and less and he has not succumbed to it since his conversion.

Group: Acquired.

Window Dresser, age 44

He states in a personal report that from the earliest age he can remember he was physically attracted by "my own sex". Brought up by his mother, he played mostly with his older sister and her friends. He commenced to masturbate at 13; this became a frequent habit. He always played girls' parts in school plays and loved dressing up in his sister's clothes. At 18 he went into the drapery trade. At his work was seduced by another, was first repelled and then enjoyed the experience and imagined it with other men. Began to practise frequently, joined "drag parties" and such like. He felt inferior and compensated for this by dress, alcohol and so on. At 26 he tried heterosexual experience, which was only successful by male imagination.

About that time he met people whose quality of life attracted him by its sincerity, reality and effectiveness. He was intrigued by "the diversity of types and their sense of purpose and keenness. They suggested that 'if everyone were a hundred per cent. Christian, the world would be a better place'. I agreed, 'if everyone does it'. 'It must start somewhere', I was told, they had started to try it, why not me?" Later on one of them told him that he had been a homosexual but "when he gave his life to God for Him to direct, God gave him the victory and he began to be different". In the healthy moral climate these new friends created for him he made his decision and he found the habits of a life-time were broken and he was filled with joy, lost inferiority and was able to be real and natural.

During the war he maintained his new discipline and as a prisoner of war he had "a gloriously victorious time and others found new life". After return to England he had lapses, but his friends stuck to him and he later had the conviction he should marry. He then met a childhood friend, courtship went well and they married. Even then there were occasional lapses, but eventually he told his wife everything about the past and since then victory has been complete and intercourse normal. Four years to date ...

. . .

Group: Essential

. . .

Female

As a young woman she came under the influence of a dominating elder woman with whom she went to live as "husband and wife" for 12 years. She had had a religious up-bringing and was so disturbed in her conscience that she became a morphia addict. Through conversion and the faithfulness of a new friend the relationship was broken and she was able to become free and has remained so for 20 years.

Group: Acquired

Salvation Army Sergeant Major for many years in charge of a men's hostel

He had a religious home but practised masturbation secretly as a boy and youth. He was strongly attracted to other boys and youths, but never indulged in homosexual practices with them. Taking a first-class Honours Degree and winning the Gold Medal at his University, he embarked on a career in the Civil Service. He was sent to a Colony where he found himself greatly drawn physically to native boys and men. Being a conscientious young man he suffered greatly from the fear of falling to the temptation of physical practices with them and bringing discredit on his Service. So great was the conflict that he was invalided home with a "nervous breakdown".

He was unable to do anything for some years, during which time he was, in his own words, "truly converted". Since then he has devoted his whole life, over 30 years, to social work among men. There has been no homosexual activity, and his attraction for his own sex has been diverted into selfless channels.

Group: Essential

Engineering Draftswoman, age 36

She was strongly attracted to her own sex from childhood and repelled by the opposite sex. A tomboy as a child, she used to don boys clothes and would enjoy fighting boys and beating them and then revealing to them that she was a girl. She practised with one particular woman and others for some years until she was converted in a Baptist Church, where she was baptised. From this time she refrained from homosexual activity and her temptations became less. She has now lost her dislike of men and remains a disciplined, satisfied personality.

Group: Essential

...

DISCUSSION

Although homosexuality is looked upon as an entity and special laws exist for restraining homosexual offenders, from the practical point of view and for the purpose of dealing with individuals who practise homosexual acts, the problem becomes simplified if it is approached from its moral aspect. There are, of course, other factors but the problem, in my view, is preponderantly moral rather than physical. While essential homosexuals may not be responsible for their condition, they cannot escape responsibility for their behaviour. It would seem, therefore, that to deal fundamentally with the character of the homosexual offender is of primary importance. By so doing he or she can acquire personal discipline and ability to overcome such habits. Homosexuals are capable of reformation and their qualities, wills, intellects and emotions can be directed into selfless and constructive channels.

Love in its widest and noblest sense is the basic solvent of human relationships. It is only when the emotion is exclusive, selfish and sentimental that it becomes dangerous, and when it is expressed physically, indulgently and unnaturally, it becomes reprehensible. Homosexual practices are uncreative, senseless and sensual, and indicate that those who practise them, although they may be the victims of constitutional tendencies or conditioned by others, are selfish and probably undisciplined and self-willed.

I have discussed conversion with many individuals who have experienced it, and who have demonstrated a newly-found freedom, whether it be from sexual (homo- or hetero-) or other types of indulgence or from anti-social behaviour in its many and varied forms. I have no doubt in my own mind that homosexuals can in general be cured of their desires and habits and that the prevention of homosexuality is a moral and spiritual, as well as a legal or administrative problem.

. . .

*

(b) HO 345/7: Statement by Mr. Reynold H. Boyd, F.R.C.S., 52, Harley Street, W1[25]

. . . These views are based entirely on my own observations in 30 years of practice, 20 of which have been spent as a specialist in the sexological field. I am a genito-urinary surgeon specialising in infertility, venereal diseases and sexual abnormalities, and over the years have treated, in private practice and as hospital out-patients, a great number of homosexuals of both sexes; I have, too, a number of friends who are homosexuals.

In brief, however, my views are:–

1 I believe that the practice of homosexuality is not an inherent difference but a bad sex habit and implies, in the vast majority of cases, no physical or mental causal disease.

2 It is, like all bad habits, such as chronic alcoholism, within the power of the individual at the beginning to abstain completely but as time goes on the habit paths become too well used to permit readjustment. The stage of initiation into these bad habits is in the formative years of adolescence. At this time only can the fixation of the habit be prevented.

   (a) By judicious parental control and guidance.
   (b) By co-educational methods.
   (c) By careful screening of people in positions of trust, i.e. teachers, service officers and scout masters and by the condign punishment (segregation or gaol) of those who seduce the young.

3 I do not believe that hormone treatment, even high dosage oestrogen to males, stops the impulse when the habit has been formed, and in my experience the psychiatric treatment of any but the early cases is equally useless, and in the oldest group it verges on the frivolous.

4 I do not believe that any bad sex habit so patently unbiological as homosexuality should be condoned but as it is widespread and in many countries accepted with indifference it does not in itself merit savage punishment or, indeed, punishment at all.

5 It should be possible, therefore, for judges to receive a directive not to apply the rigours of the law where no "public aspect" is involved and no public susceptibilities offended. (Such, I believe, is already the attitude adopted by many judges.) This would not make the occasional or casual homosexual relationship in private an offence but would on the other hand not allow Lesbians or homosexual males to live notoriously to the disapproval of other people in their suburb or block of flats.
. . .

In fine I do not believe in the efficiency of medical treatment—hormonal, medicinal or psychiatric. I believe that the law should be redefined and clarified but not changed to the extent of condonation of a bad and unbiological sexual habit. I consider the continued "restraint" of those who seduce the adolescent completely justified and that prison, however imperfect a solution, is the only one open to us.

*

(c) HO 345/9: Memorandum on Venereal Disease and the Homosexual submitted by Dr. F. J. G. Jefferiss, V.D. Department, St. Mary's Hospital, London, W.2.[26]

. . .

3   Of 1,000 consecutive male patients with gonorrhoea or early syphilis attending this clinic in 1954 ..., 84 (8.4%) admitted that they had had recent sexual contact with another male...

. . .

4   There are several reasons why even this high figure does not represent the true number of homosexuals attending. Firstly, the patient is unwilling to admit the source of his infection, both from natural shyness and fear that his guilty secret might be passed on to the police. Of course, the patient with rectal infection is unable to suggest any plausible alternate source of infection, but the man with a urethral discharge or penile sore can easily say that he has been with a woman and nobody doubts him. I have been told often by homosexuals that there are many men, tough manual workers and others, not true homosexuals, who seek them out and have intercourse with them...

5   As most of our homosexual patients are, by all ordinary standards, physically normal, it is sometimes very difficult to spot them. Some of them, far from appearing effeminate, are lusty he-men, e.g. physical culture experts, athletes, weight lifters and champion cyclists. Also there must be many homosexuals who dare not come to public V.D. clinics, but attend private doctors for a greater sense of secrecy.
. . .

11   My third group of patients consists of 54 whom I had the opportunity to question personally and at length...

They came from every social class, one even from a noble family...

Forty of them (74%) came from broken homes...Only 14 (26%) came from normal happy homes where both parents took part in their upbringing...

12 There was no obvious evidence of any endocrine change, no gynandrism, except in one who had little body hair and only shaved twice a week...

A few were typical "pansies", but most of them behaved more or less normally. It is by some slight characteristic in their bearing and manner, rather than by any physical abnormality, that they betray their homosexuality.

...

13 It seemed to me that these homosexuals could be divided into two groups, the "congenital" and the "acquired". The former included the true inverts in whom the deformity was complete and was so strongly embedded that it could not be rooted out, however powerful the influences were that had been directed on it from within or without. I think that these men are mentally deformed, having men's bodies and women's minds. Also in this group were the incomplete inverts who had been either made more or less homosexual according to whether they had had a good home life and upbringing or not. Then there were the youths who had not given up the common homosexual leanings of childhood. They are immature, but surroundings and opportunity play an important part here too.

Many people do not think that any type of homosexuality is inborn, but consider that it is acquired during the early childhood years from some outside influence, but the study of identical twins by Kallmann supports my view.[27] Whatever is the cause, it is something which occurs very early in life to produce this inversion, complete or partial.

The "acquired" were those who had little or no homosexual make-up of their own, but had been influenced by others or circumstances to practice it. This is the type of homosexuality which may be prevented by strict moral upbringing and the avoidance of the opportunity. In this series 51 of the 54 appeared to be "congenitals" and only three "acquired". But as this group is a very selected one I do not suggest that this is the true ratio. But there is no doubt in my mind that the majority of patients that I have seen are of the congenital type or at least had acquired their homosexuality in early childhood before ever they had any kind of homosexual experience.

...

15 We do not readily spot the "acquired" type, partly because he is nearly always "active" and therefore has penile or urethral disease, and partly because his appearance is quite normal. The few that I <u>have</u> seen go with men because the male consort is less clinging and cheaper, not requiring so much courting and money spent on him. They are also attracted to homosexuality because it calls for less responsibility, there being no fear of making the partner pregnant, or being entrapped into matrimony. Another reason is the widespread, but

mistaken, view that V.D. is not caught from men. Added to this there is the attraction that homosexuality is illegal and cloaked in secrecy and, therefore, exciting to the adventurous, but not too discriminating youth. If homosexuality were treated as a deformity of the mind rather than as a crime I think it would lose much of its attraction for the young men who are, at present, tempted by it.

16   These three of the "acquired" type were similar in many ways to the usual juvenile delinquent. Had they and some of the less strongly developed inverts had the benefit of good family upbringing by both parents they might have been kept on the normal road. I think that the only effective treatment is preventive and it must be started in childhood before the habit develops. A true invert, once he realises his homosexuality is, in my opinion, not open to any influence or treatment...

17   Patients have told me that, having been referred to psychiatrists by the courts, it is to their advantage to pretend that treatment has been a success or they might be sent to gaol. In fact, they say, it has no effect at all and they carry on with their homosexuality as before only with more discretion.

18   Many homosexuals say that they are so placed that, for fear of the law, they dare not be known to have male sex friends. Having no other means of finding partners they have to resort to picking up strangers "incognito" in the "West End" or similar places. They say that, in this way, the law as it now stands leaves no alternative sexual outlet for the homosexual but intercourse with strangers and male prostitutes and so encourages the spread of venereal disease.

*

(d) HO 345/7: Evidence submitted by the Royal College of Physicians

The problem of homosexuality is not only extremely complex, but also one on which our knowledge is in many respects incomplete...

In the first place the distinction between homosexual and heterosexual individuals is not clear-cut: degrees of homosexuality exist, and in any population there are a much larger number of people with homosexual trends of greater or lesser intensity than of those who become social problems. How far such trends become manifest depends to some extent upon the culture of a society: in this respect there are wide differences between ancient-Greek and Islamic culture patterns on the one hand and Hebrew and Christian cultures on the other. Rather than put forward theoretical explanations, which whether they are psychological or physical, are still controversial, we think it more useful to suggest a classification of male homosexuality according to the circumstances in which it manifests itself and the relationship of these circumstances to the need and opportunities for treatment.

(1) Homosexuality may be a symptom of a recognisable mental disorder such as schizophrenia or mental defect, or one resulting from an organic disorder usually of the brain but sometimes of other organs. These individuals constitute a small group and present no great problem. Treatment is required for the underlying primary condition.

(2) Homosexuality may be maturational or transitional. Homosexuality in adolescence is often of this kind. Adolescents are often attracted by members of their own sex and may pursue a course of trial and error in sexual matters. Psychological sexual maturity may, however, be delayed in some cases considerably beyond adolescence; and this type of homosexuality, therefore, may sometimes present itself in older men. Such patients are often amenable to treatment. Available evidence suggests that a prison sentence does not in itself bring about psychosexual reorientation, nor does punishment per se protect society. Probation should therefore be resorted to wherever possible when expert evidence suggests that treatment is likely to be successful.

(3) Homosexuality may be substitutive, that is to say it may appear in individuals who are ordinarily heterosexual, but who manifest homosexual tendencies when placed in an environment in which for some reason heterosexual relationships are impossible, for example in isolated male communities. Homosexuality of this type can usually be successfully treated.

(4) There remains the group of individuals whose homosexuality cannot be explained as the result of any of the foregoing circumstances. They are primarily and essentially homosexual. We do not regard these individuals as ill in the conventionally accepted sense though they may develop psychological symptoms as reactions to the social and emotional difficulties created by their homosexuality. The relationship between psychosexuality and bodily structure is by no means simple; and in many instances no such correlation can be demonstrated. Even if there is such relationship it may be that the physical type does not so much determine the individual's homosexuality as influence his method of dealing with his tendency.

In the present state of our knowledge, homosexuals in this fourth group cannot be sexually reoriented, though they may be helped to reach a better adjustment of their social environment. Many, of course, succeed in doing so without medical assistance.

When homosexuals of the fourth group are repeatedly convicted of offences against juveniles, it might be desirable that they should be segregated in some special institution where they can be detained for a prolonged period for the protection of the public and also to receive treatment.

...

*

(e) R. Sessions Hodge, 'The Treatment of the Sexual Offender with Discussion of a Method of Treatment by Gland Extracts', *Séances de Travail de la Section de Biologie*, 306–17[28]

. . .

[p. 309] Study of a group of homosexual adult males leads me to the conclusion that the great majority are true deviants from the heterosexual pattern in whom there has been from an early age this deviation: [p. 310] that it is as much their form of sexuality as heterosexuality in the other group: and that heterosexuality is as repugnant to them as homosexuality to the heterosexual.

I distinguish between those whose behaviour is, I think, properly described as licentious and vicious and those who are deeply concerned by their deviation and the inevitable restraint it places upon them in their social contacts, and their resultant barren family life.

The first of these groups is without doubt, I submit, a social menace. They have their counterpart amongst heterosexuals and restraint by Society would seem proper and inevitable. But committal to prison means committal to a strictly homosexual environment where, I submit, their deviation would be if anything perpetuated. What guarantee has Society of their reform by this method when applied merely as custodial care for a limited period of time?

The second group presents a difficult medical problem. There may have been no overt homosexual act and certainly no charge against them: but they are always at risk should they yield to their—to them—natural sexual inclination. They are, too, doubly at risk—the risk of arrest and the risk of blackmail. How common this last may be no man can tell, as the subject of blackmail is vulnerable to a degree and his remedy, to him, most difficult of application.

It is my further submission that the true homosexual deviant—whose deviation has been present for most of his life, is detectable in his account of his childhood and is an integral part of his dream content—presents a case of singularly resistive type and not readily amenable, if at all, to psychotherapy.

. . .

[p. 312] Our attention was first called to the possibility of inhibiting sexual feeling by hormone treatment by a series of cases of acromegaly, a disease characterized by overgrowth of the face and limbs (mainly at the cartilage junctions) and due to a tumour of the pituitary gland, treated at the Burden Institute with very large doses of œstrone (female sex hormone).

This method of treatment, which we owe to Kirklin and Wilder [D.L. Kirklin and R.M. Wilder, *Proc. Mayo Clinic*, 11: 121, 1936], aims at the reduction of the pituitary tumour by overwhelming doses of œstrone. Hutton and Reiss [E. L. Hutton and M. Reiss, "The Hormone Treatment of Acromegaly", *J. of Ment. Sci.*, 88: 373, 550–553, 1942] treated a case of acromegaly with injections of œstrone 100,000 I. U. daily. The patient, a well-built vigorous man, exhibited after a couple of weeks' treatment a complete absence of sexual feeling, together with some degree of over-development of both breasts.

In all the male cases of acromegaly treated with œstrone sexual libido had practically vanished after the first fourteen days. Cessation of the treatment after various periods was followed by a return of libido in about two weeks; but in observations lasting as long as eight years it has been found that libido was absent so long as maintenance doses of œstrone were continued. In two such cases injections of male sex hormone three times weekly restored sexual potency while the œstrone treatment was continuing. There is thus no reason to believe that even after years of œstrone treatment sexual power is permanently abolished.

Following this work it was decided to offer such treatment to persons convicted of sexual crimes who by their own wish (expressed in writing) applied for treatment. The full implication of the treatment [p. 313] was explained and it was made clear that this is not castration but the modification of sexual impulses by suppressing pituitary function and that the effect can be graded and is reversible. In all fifteen cases have been, or are, under treatment.

Examples are as follows:

. . .

Male aged 37 years, who came voluntarily, being concerned by his definite homosexual bias which, though it had not since puberty found overt expression, was nevertheless a source of great worry and unhappiness. The man was of superior intelligence and culturally and ethically of good status. He appeared after close investigation to be a homosexual deviant of very long standing—an "innate" pattern rather than one of traumatic "level fixation"—and he failed to respond to psychotherapy. Treatment by œstrone has been followed by loss of sexual libido and he reports considerable "mental relief".

Male aged 51 years with two convictions for "indecent assault" on male persons. An intelligent man in good position he was considered to be a schizoid personality and that his deviation was of long standing and he himself not amenable to psychotherapy. Œstrone treatment abolished sexual libido and he was discharged to ordinary life within two months of the institution of treatment. Eighteen months later he is reported to be in full gainful occupation and no longer "concerned with sexual feeling".

. . .

[p. 315] It is submitted herein that there is now available a method of selective suppression of a basic instinctual drive—to wit the diminution or suppression of sexual feeling in man . . . Controlled dosage has so far averted feminisation and in fact one case employed as a miner (underground) has continued in this heavy employment with full satisfaction and has, since the institution of treatment, grown an adequate moustache. The suppression of feeling, where desired, has however been "total" and with it has gone "appreciation" of sex so that the individual's attitude can perhaps be compared with that of the "tone-deaf" in the presence of classical music or of the recently full-fed in the presence of a further offering of food...

...

*

ii. Mental Health

(a) HO 345/7: Memorandum from Paddington Green Children's Hospital Psychology Department on Homosexuality and the Law, March 1955[29]

...

Psychological Theory Relevant to the Problem of Homosexuality

It is now well-known that the sexual instincts which develop into biological operative functions at puberty derive from a gradual gathering together of all instincts, including that of appetite. This gathering together occurs in the first years, so that by the age of five a child is psychologically or emotionally ready for a full sexual experience, although biologically immature. In sexual development puberty follows on the development that has already been made by the beginning of school age. A great deal is thus crowded in to the early years, and the pattern of future sexual trends is already set by the age of 5. This is the surprising discovery of modern psychology, and it has a direct bearing on the subject of normal homosexuality.

Home Setting

When the family setting is good, small children can afford to explore all trends whether active or passive, male or female, sadistic or masochistic, etc., etc. They can identify with either parent in the common game of "families" or of "fathers and mothers". A child develops one aspect of his nature in relation to one sibling or some other relative and another aspect in relation to others. In this way the whole range of human experience is explored. In the majority of cases the main trend is along the line of the anatomical provision—that is to say boys want to be boys and girls want to be girls, but this is by no means something to be assumed. A great deal depends on the attitudes of parents and also on the provision society makes at any one time for expression of aggressive and motility urges.

A break up of the family while a child is under 5 can be disastrous in any respect, but such a break up may easily lead to a disturbance of the growing tendency of a child to become predominantly his own (anatomical) sex. In so far as emotional development is healthy, however, it is not expected that homosexual trends will bring the child into a clash with the law, either as a child or later as a man.

Psychiatric Illness and Homosexual Compulsion.

There are certain <u>illnesses</u> that quite clearly lead <u>to</u> compulsive homosexual prac-
<u>tice</u>, that which society must take into account. These illnesses include the
following two broad types:

(a) A distorted development resulting from <u>the combination of a dominating</u>
<u>mother and a dependent, compliant father.</u> The child remains permanently
attached to the mother but develops a relationship to the world through the
father, and through brothers and men friends. Sexual practice in relation to
woman (in this case the mother herself) is unthinkable, and the only chance
for sexual experience is along homosexual lines. This illness can be detected
early, and it is very difficult to cure. Psycho-therapy must be started early, and
the mother is unlikely to be co-operative.

(b) A <u>psychiatric illness</u> of well-known type with a strong <u>"paranoid quality"</u>,
that is to say, with a delusion of persecution as a dominant theme. This
illness is not uncommon, can be clearly recognised in childhood, even in
early childhood, and it can be severe enough to lead to a severe disturbance of capacity for relationships and perhaps to breakdown into paranoid illness in later childhood or in adult life. The "paranoid" (expected
persecution) anxieties can be intolerable. One of the ways these patients
seek relief is through sodomy. In sodomy the persecutory element (always
expecting to be catching the patient up from behind) is humanised into
a love partner, <u>and the sexual practice that results is part of an attempt</u> (albeit
an abortive attempt) <u>at self cure.</u>

The active partner in these practices is originally the persecuted accepting
partner, but gradually or suddenly he reverses the role thus preserving his sexual masculinity while living the passive role through the partner. The love
and trust that can exist between such partners reflects the <u>utter loneliness</u> that
belongs to the paranoid illness.

The treatment is of the paranoid illness, not of the homosexuality, and it can
always be said that treatment ought to have been instituted near the inception
of the paranoid illness, which means in early childhood.

These two types of illness, along with genetic predisposition, cover a great deal
of the subject of that compulsive homosexuality which is antisocial. The rest of
the vast general subject of homosexual trends is absorbed in that of the general
cultural plan, and calls for no legal attention.

. . .

*

(b) HO 345/8: Memorandum of the Royal Medico-Psychological Association,
June 1955[30]

. . .

4. INCIDENCE OF HOMOSEXUALITY.

There are no reliable statistics of the number of men in Great Britain who indulge in homosexual practices. Statistical surveys in some American and European countries have estimated that between 3% and 4% of their male populations are exclusively homosexual in their sexual impulses. It is probable that a much higher percentage of men indulge in some homosexual activity at some time in their lives. Latent homosexual tendencies, more or less unconscious, are found in a large proportion of men undergoing psychological treatment, but unless there were some overt activity which would at present be criminal, these would not come directly within the scope of the present inquiry.

Psychiatrists conducting out-patient clinics see very many practising homosexuals who are never charged, and who probably never become known to the Police. The statistics of convictions for homosexual offences are unreliable as an index of homosexuality in the general population.

5. HOMOSEXUALITY AND MENTAL ILLNESS.

Homosexual patients are more commonly seen as out-patients than as in-patients. They may be referred to out-patient clinics either as primary homosexual problems or because of some concurrent psychological abnormality.

(a) Patients referred primarily for Homosexuality

  (i) Men awaiting trial charged with some homosexual offence.
      Accused persons are sometimes referred by the Magistrate's Court, by Probation Officers, or by the man's general practitioner or solicitor, for a psychiatric report—particularly as to whether there is concurrent psychiatric abnormality, or whether treatment would be effective in preventing a repetition of the offence. We feel that all offenders should be referred for psychiatric opinion before sentence is passed.

  (ii) Patients, not charged with any offence, seeking psychiatric help in dealing with their homosexual impulses.
      The statement sometimes made that homosexuals never wish to be treated is not true. Not all homosexual patients are curable in the sense of becoming normally heterosexual, but the great majority could be helped to some extent in coping with their homosexual impulses if there were sufficient doctors with sufficient time to undertake the task. It is noteworthy that patients are sometimes seen whose impulses are predominantly or exclusively homosexual, who have never in their lives succumbed to temptation and indulged in any overt homosexual activity.

(b) Homosexual patients referred to psychiatrists because of concurrent psychological symptoms.

  (i) Symptomatic Homosexuality.
      Homosexual behaviour may occur occasionally as a symptom of mental illness. It may be that in these cases homosexual impulses have remained

latent until moral control is diminished by mental disease. There is also a group of patients who seem never to have developed moral control. Some of these patients are particularly prone to homosexual behaviour.

(ii) Patients with psychological illness resulting from homosexual impulses.
Men who are predominantly homosexual may be faced with more stresses and conflicts than their heterosexual brethren, and in consequence are more vulnerable to neurotic and emotional upsets. These stresses result in part from the conflict between the man's sexual desires and his moral sense, but chiefly from the fear of shame and social ostracism, the fear of committing a criminal offence with possible detection and imprisonment, and, by no means infrequently, the anxiety resulting from being blackmailed.

6.   FACULTATIVE OR SUBSTITUTIVE HOMOSEXUALITY.

Many of the Members of the Royal Medico-Psychological Association served in H.M. Forces and have had experience of homosexual problems in the Services. It is doubtful whether the incidence of homosexual activity is higher than in civilian life, but occasionally where numbers of men are kept in close contact for long periods with no opportunity for heterosexual outlet there may be a temporary increase in homosexuality. This facultative homosexuality soon ceases, however, when heterosexual contacts again become possible. There is little evidence of fixation resulting or of the men having become homosexual by contagion. The homosexuality is substitutive and does not persist.

7.   HOMOSEXUAL PHASE OF NORMAL DEVELOPMENT.

It is now widely accepted that for a time most normal boys pass through a phase in which their interests are predominantly directed to others of their own sex. This is regarded as an integral part of normal development. Sexual activity, or awareness of sexual desire may be exceptional, but the strong attachments which spring up between adolescent boys probably have an unconscious sexual basis. Sometimes overt sexual activity occurs between boys at this stage of development, but it seldom leads to any permanent fixation. The majority of boys who have had such experience during this phase ultimately develop into normally heterosexual adults. It seems likely that in those cases of apparent fixation there was already some pre-existing latent abnormality: the majority of homosexuals seeking treatment remember that they were aware of being in some way sexually different from other boys at a much younger age.

Though homosexual impulses are so common in normal sexual development, homosexual activity in adults is not frequent. Normally a man outgrows his adolescent homosexual tendencies and becomes heterosexual, any residual immature trends being restrained by his moral sense or by fear of the sanctions of society.

8.   CAUSATION.

There is not one single cause of adult homosexual behaviour, and there are usually several contributing causes in each individual case. Anything interfering with normal sexual development might predispose to homosexuality. Possible causative factors fall into the following three groups:–

(a) Genetic.

Incapacity to develop to full sexual maturity may sometimes be genetically determined. A number of workers have studied the incidence of homosexuality in the twins of known homosexuals. They have all found that the incidence of homosexuality is much higher in uniovular, than in binovular, twins. Binovular twins have similar inborn tendencies (as have brothers), but in uniovular twins the hereditary endowment is identical. Over a large series of cases environmental influences would be similar in uniovular and binovular twins. The much higher incidence of homosexuality in uniovular twins can be explained only on the basis of inborn characteristics, not of any environmental factors.

(b) Endocrine.

In a small percentage of cases the error of sexual development may be due to some endocrine abnormality. There is no convincing evidence that there is endocrine imbalance in the majority of practising homosexuals. There is undoubtedly an endocrine factor for the production of sexual desire, but nothing conclusive is known of the effect of endocrines in determining the direction of sex drives.

Some homosexuals differ from the normal in physical as well as mental attributes, e.g., deficient growth of hair on face and body, a more feminine distribution of body fat, a deviation from the norm in skeletal measurements. There may possibly be endocrine or other physical causes for these deviations which have not yet been discovered.

(c) Psychological.

There seems no doubt that the majority of cases are dependent upon psychological influences, whether or not there are also physical causes operating. There is not just one psychological situation, or mechanism leading to homosexuality, and in each individual case there is usually a number of contributing psychological determinants.

Psychoanalytic evidence suggests that the psychological causes are of a special kind in those cases in which an adult male's sex impulses are directed towards boys rather than adult males. It seems probable that these are two quite different conditions. It [is] seldom that the repeated offender against boys indulges at times in homosexual activity with adults, and the ordinary homosexual seems no more disposed to commit sexual offences against children than is the heterosexual.

There is another current belief that homosexuals as they grow older tend to prefer younger and younger partners. Experience suggests that this theory is erroneous. The homosexual's preference for a partner of one age-group usually remains fixed. We have observed this in homosexuals as old as eighty. Sometimes old men as a result of senile changes commit offences against boys, (or against girls), but it is unusual in these cases to find a history of previous homosexual practices with adults.

9. TREATMENT.

(a) Psychoanalysis.
It is seldom possible so to change the sex impulses of a man who is exclusively homosexual that he becomes heterosexual. Though psychoanalysis has much to tell us about the unconscious mechanisms causing men to become homosexual, as a method of treatment for homosexual patients its results are often disappointing. Moreover it is time consuming and costly, and it is impossible to provide psychoanalysis for the large majority of homosexuals seeking treatment.

(b) Shorter methods of Psychotherapy.
Much can be done by shorter methods of psychotherapy to help patients to cope with their homosexual impulses, to adapt themselves to fit into society harmoniously, and to relieve secondary neurotic symptoms. One cannot hope by these methods to convert many patients to complete heterosexuality, but one may succeed in reducing their homosexual activities. On the whole the earlier that treatment can be given the better the results are likely to be.

(c) Endocrine treatment.
Sexual drives in the male can be diminished or eradicated by taking endocrine preparations regularly over a long period. The effect is to diminish all sexual desires, heterosexual as well as homosexual, and enable many homosexuals to stop their activities with a minimum of tension or stress. It is necessary for this treatment to be continued permanently. Relapse occurs if for any reason it is stopped.

10. REASONS WHY CHANGES IN LAW ARE DESIRABLE.

  (i) The present methods of dealing with homosexual offenders fail in reforming homosexuals, and even fail in preventing the continuance of homosexual practices whilst the offenders are in prison.

 (ii) Little is done in prison to reform homosexual offenders. Many factors militate against the success of psychological treatment whilst a man is in prison, but if no attempt at treatment is made it will be much more difficult to treat him after he is discharged. Each year about 700 men are given fairly long sentences for homosexual offences but less than 30 receive psychological treatment in prison. Many repeat their offences after discharge from prison. Judges in passing sentence often say that the offender will receive appropriate treatment in prison, but this promise is seldom fulfilled.

(iii) There is evidence that there are opportunities for homosexual activities among men serving prison sentences. There are many homosexuals in the prison population who have been convicted of offences other than sexual ones. Boredom and inactivity, absence of heterosexual outlets, and close propinquity with other homosexuals, all tend to corrupt the first offender further.

(iv) Where a society punishes its members for some act committed in private, which is harmful, if at all, only to the offender himself, the situation is usually one of a majority imposing a law on a minority. Many homosexuals feel they are members of a persecuted minority. This consciousness of unfair discrimination sometimes leads to a feeling of martyrdom in the man who is punished, and something akin to heroworship in his undetected fellows.

(v) An offence committed in private is difficult of proof by cogent evidence other than the statement of one of the offenders. This gives opportunity for blackmail. It is for this reason that some experienced lawyers have called Section 11 of the Criminal Law Amendment Act, 1885, "The Blackmailers Charter". Lord Jowitt in his Maudsley Lecture to the Royal Medico-Psychological Association, 1953, stated that nearly 90% of the blackmail cases he has dealt with were cases in which the person blackmailed had been guilty of homosexual practices with an adult person.

(vi) It is an anomaly that the law as it now stands discriminates between male and female homosexuality. Homosexual behaviour in females, no less than in males, is socially undesirable, but it is not an offence if both the female persons are over the age of sixteen (though a woman having sexual relations with a girl under the age of consent would be guilty of "Indecent Assault").[31]

11. RECOMMENDATIONS.

1. All laws dealing with importuning and the preservation of public decency should remain in force.
2. The law respecting homosexual offences should, apart from the "age of consent", correspond closely to the present law respecting heterosexual offences.
3. All men charged with homosexual offences should be examined by a psychiatrist at some stage in the proceedings before sentence is passed.
4. There should be increased facilities for treatment of homosexual offenders whether they are serving prison sentences or are on probation.

. . .

6. Homosexual behaviour between adult male persons, voluntarily and in private, should no longer remain a criminal offence.

The large majority of Members of the Royal Medico-Psychological Association are in favour of this recommendation, though there are a few who strongly dissent. Opinion is divided whether "adult" should mean 18 or 21 years. An argument in favour of 21 years is: young men on National Service will thereby be protected from seduction by their seniors.

Arguments in favour of 18 years are:–

(a) At the age of 18 a young man is held to be fully responsible for criminal acts, and subject to maximum penalties for them. He is old enough to be able to fight for his country. Is not he then old enough to know where he stands regarding sex, and capable of protecting himself against homosexual seduction?

...

(c) If the "age of consent" is fixed at 21 years two persons between the ages of 18 and 21 would each be guilty of an offence. Being too old to be tried in Juvenile Courts, they would be submitted to the procedure at present in force for adult offenders, with all its attendant publicity. It is felt that it would be undesirable to punish by imprisonment two young men in the 18 to 21 age group, whilst adult offenders would be immune.

The general opinion of the Royal Medico-Psychological Association is that homosexual behaviour between male persons of approximately the same age should no longer remain an offence, but that seduction of the young by persons appreciably older should remain punishable by law.

*

(c)   HO 345/7: Memorandum on Homosexuality Submitted by Dr. Clifford Allen, Harley Street, London, S.W.1

Basis of Experience

I have had experience of homosexuality, amongst other forms of sexual abnormality, over the past twenty years. This commenced when working with L. R. Broster and his team at Charing Cross Hospital on the investigation of adrenal disorders and psychical abnormality[32] ...As a result of this I was impressed by the large number of people who suffered from sexual disorders, including homosexuality, and wrote a book on the subject. ("The Sexual Perversions and Abnormalities", Oxford University Press, London).[33] Through this book and some fifty other scientific papers I have written on sexual abnormality I have seen some hundreds of homosexuals of all types and have frequently been asked to give evidence in the Courts. At present I see about fifty to a hundred homosexuals and treat between ten and twenty by deep therapy annually.

Causation of Homosexuality

In my experience the main cause of homosexuality is psychological. It is usually found as the result of unhappy or broken homes and particularly when there is inadequacy (alcoholism, brutality, etc.) in the parent of the same sex as the patient. Endocrine disease, such as adreno-genital virilism in women and eunuchiodism [sic] in men, is rare but when present can act as an ancillary cause. I have not found that endocrine disease can act as a cause without psychological deviations. In my opinion all homosexuality is a form of immaturity.

Types

1 A large number of those who ask to see a consultant complaining that they are homosexual are those who are shy, inadequate, frightened people who have never learned how to deal with the opposite sex. Particularly in men such individuals tend to drift into homosexuality as a faute de mieux. This group contains many who would be classed as bisexuals and these are immature people who have never attained adult emotional development.

2 There is a second group who are orientated towards homosexuality but who do not wish to be homosexual and are often disgusted and horrified at their abnormal impulses. They will often go to considerable lengths to be cured. Obviously group 1 shades off into group 2.

3 There are confirmed homosexuals who do not seek help unless they are in trouble. They feel that there is nothing wrong about their behaviour and often insist that homosexuals are superior persons and the soil for genius, etc.[34] Those comprising this group build their lives on a homosexual pattern, they seek the homosexual professions—the Stage, the Ballet, interior decorating,etc. They are also to be found as homosexual prostitutes, and those who seek to gain advancement from their abnormality.

Treatment

Patients in the first group recorded above are often easily treated and can be helped to re-orientate themselves in a few sessions. If they are not treated, they may drift into confirmed homosexuality as they grow older.

Those of the second group need prolonged psychotherapy which may last from a few weeks to years. They are curable with skill and patience. If not treated they may accept homosexuality as the only outlet for their sexuality.

Those in group one and two are curable just as any other neurotics are curable and give similar results.

Those in group 3 do not wish to be cured, will not cooperate and treatment is useless.

Treatment by medicinal means—for example the so-called chemical castration— such as Stilboestrol is often useless and gives the patient a false sense of security. It may shorten the patient's life and should not be used except on older patients, if then.

Prison

Amongst the many hundreds of homosexual patients I have seen I have never encountered anyone who has stated that prison had any beneficial effect and the change produced by it is often for the worse. On the other hand I have seen many patients who have served two or even three sentences who have got into further trouble.

In my opinion all homosexuals who appear curable should be given the chance of proper treatment (not just a few visits to a clinic which is useless) but those who do not respond and form a danger to society (as those who interfere with children) should be confined to an institution (not necessarily a prison) for life.

If it is accepted that homosexual behaviour between adults should not be punishable by law then the age of consent should be placed high—as at thirty years—so that young men who may be set on the right road to recovery should not be deviated before they have had a chance to mature.

*

(d) HO 345/8: Memorandum submitted by Dr. Eustace Chesser, 92 Harley Street, W.1[35]

The following observations are largely based on a knowledge (social and professional) of homosexuals gained during the past fifteen years.

. . .

1   Those who indulge in homosexual practices can broadly speaking be divided into three groups.

  (i) The genuine homosexual—whom I would define as

      (a) one whose homosexuality is inborn—(if there be such a thing), and
      (b) one whose homosexuality is resultant on the interplay between heredity and early environment. In both, for all practical purposes, the condition may be regarded as irreversible by any present known means.

 (ii) The bi-sexual—one whose sexuality can be expressed both heterosexually and homosexually, and who must therefore be deemed to have freedom of choice.
(iii) The heterosexual—who may indulge in homosexual practises for one of many reasons.

2   A distinction must therefore be drawn between the genuine homosexual and the others. It could be argued that the genuine homosexual must be held responsible for the overt expression of his homosexuality just as the aggressive psychopath must be held responsible for his aggression, if this be criminal. Nonetheless, an informed society should appreciate that, to all intents and purposes, the genuine homosexual is a creature of a different species from the rest of us—what Carpenter has called the 'Intermediate Sex'.[36] A clear recognition of this fact would enable a more objective and dispassionate approach to be made on the subject of homosexuality.

. . .

4   Public and Private Conduct

"The Englishman's home is his castle"—and it is usually assumed that practises between consenting adults in heterosexual relationships—at all events from a criminal point of view—must be regarded as their own affair. It would not therefore be unreasonable for the same to obtain between consenting homosexuals. The removal of the stigma of criminality from such relationships might do much to lessen the sense of guilt and grievance from which the homosexual suffers so badly.

The genuine homosexual is as condemnatory in his attitude towards the homosexual who offends, for example, by importuning or any form of public indecency as is the heterosexual.

5 "Age of Consent"

This poses the question—at what age should homosexual relationships be permitted? I would suggest that 21 should be accepted as the 'age of consent' and responsibility ...

6 These age divisions are admittedly somewhat arbitrary, but they would go some way towards doing justice to the genuine homosexual as well as satisfying society's legitimate concern for propriety and for the protection of the young.

7 While it might be difficult if not impossible to persuade society to like its homosexuals it might at least be gradually induced to allow the genuine homosexuals some small measure of justice in humane compensation for his state, one for which he can hardly be held responsible.

8 The removal of criminality from consenting adult relationships—accompanied by the enforcement of existing laws for the preservation of public decency—seems the most feasible answer to an admittedly controversial issue.

9 Treatment of Offenders

. . .

16 Until such time as we have more insight into the causation of homosexual tendencies, we are in the main limited to dealing with the end-produce of deviated sexual growth. Psychiatry's efforts should be applied to preventing, or at least modifying this deviation. The age at which the best help could be given would probably be that of round about puberty. Such help could be made more available, if, by removing the criminal stigma from homosexuality, we were able to encourage those who were aware of divergencies in their sexual leanings to approach the doctor or psychiatrist for help. Just so long as they are allowed to struggle with their own difficulties, unable through anxiety, fear or shame to discuss their tragic situation, then the problem of homosexuality will remain with us. The 'genuine' homosexual, as I have already said, may for the present be beyond redemption. But many border-line cases could be helped—if no more than that—by the removal of the taboo that surrounds homosexuality and by a change in society's general attitude towards it. We must not forget what happened in the field of venereal disease, when a sinful label was attached to it.

*

(e) <u>HO 345/8: Memorandum of Evidence from the Tavistock Clinic,</u>[37] July 1955

...

II. Homosexuality.

(12)   (a) <u>The Nature of Homosexuality.</u>

The staff of this Clinic are unanimously of the opinion that homosexuality is a disorder of personality and as such to be regarded as an illness affecting <u>both</u> sexes and resulting from many and complex factors. Two main spheres of causative factors may be discerned which in their interaction account for this form of behaviour disorder, as for other deviations from sexual normality. In general they may be characterised as <u>early environment</u>—in the sense of parent–child relationships—interacting with <u>innate factors,</u> and these together result in deviations from the norm of the <u>emotional predisposition</u> of the individual. Homosexuality is only one among many other possible deviations resulting from this combination of factors.

(13)   We wish to emphasize especially that the interaction between these major determinants results in a considerable number of <u>types</u> of homosexuality or perverse behaviour in both sexes. The essential <u>feature</u> which all cases of homosexuality have in common is an inability to form adequate and lasting affectionate and sexual relationships with partners of the opposite sex.

(14)   That homosexuality is an illness[38] whose impact falls chiefly on the mind and emotions can be shown by close study of the personalities of homosexual persons of both sexes in spheres other than that of their narrower sexual behaviour. Thus there are frequently disturbances in the degrees of development of the conscience in the shape of either exaggerated guilt feelings or their lack (moral indifference). Secondly there is discernible in a large proportion of homosexual persons a lack of capacity to form lasting affectionate relationships (of a non-sexual character) towards any other persons, with a correlated morbid degree of self-centredness which may take the form of self-admiration or self-abasement. Thirdly, all psychiatrists are aware of the close connection that exists between homosexuality and certain well-recognized forms of mental disorder. We refer to alcoholism, paranoid states, and (less frequently) true schizophrenic psychosis.

(15)   (b) <u>Adolescent Homosexuality.</u>

The picture of homosexuality in the public mind is complicated by the widespread recognition that during the years of immaturity homosexual relationships are a "normal" phenomenon of an almost universal nature in both sexes in our society. Whilst in a large majority of such adolescent relationships the homosexual manifestations are limited to such phenomena as "close friendships", "crushes" on admired figures of the same sex, and affectionate play, every gradation between such common "normal" manifestations and active homosexual practices in the ordinary sense of the term may be found to exist, as a transition phase in the development towards full and normal heterosexuality.

(16)   It is perhaps in the failure to overcome, and develop successfully beyond, this phase that the difference between the healthy and the ill personality may be found. That such a division is not, however, absolute may be demonstrated by the frequently observed facts of regression towards homosexual behaviour under conditions of stress or isolation from suitable partnerships with members of the opposite sex.

(17)   (c) The Morality of Homosexuals.

A homosexual mental constitution is compatible with a high and sometimes over-developed sense of morality, conscientiousness and social obligation. Paradoxically, to the lay mind, such homosexuals may be models of duty and circumspection, and it is usual for this type of homosexual with a high moral sense to control his impulses, or, if he fails, to confine his activities to consenting and willing adults of the same sex similarly constituted to himself. Not infrequently homosexuals, aware of their disorder and the unhappiness which it causes them, voluntarily seek medical or psychological treatment in order to be relieved of it and made into normal heterosexual individuals.

(18)   On the other hand, as indicated, there is at the other end of the scale a type of homosexual who, either blatantly or by means of specious rationalizations of loving intent, ruthlessly exposes other human beings to corruption by importuning or seduction, using various methods to entice and deprave vulnerable subjects, often the young, and denying his or her own guilt. These basic differences between moral attitudes of homosexuals seem to us to bear a relationship to the handling of the problem both in legislation and in Court practice.

(19)   (d) Rationale of Treatment.

It will be apparent that on the basis of manifest behaviour we distinguish at least two major groups of homosexuals who differ radically in their attitudes towards their own and other people's sexual drives, even though we regard all forms of homosexuality as a species of psychological illness.

(20)   We feel that regard should be paid in public policy to the difference between the homosexual whose psychological development makes him or her want only willing partners of approximately equal age and type as against the type of individual whose homosexual behaviour is an expression of his need to hurt and deprave, howsoever speciously rationalized.

(21)   With regard to the former type we are of unanimous opinion that the practice in strict privacy of homosexuality between adult consenting parties, that is, persons over the age of 21, should cease to be a criminal offence. We are of this opinion for the following reasons:–

(a) It would do away with a great deal of potential blackmail and exploitation, e.g. by homosexual prostitutes, of this type of individual.

(b) Many homosexuals who at present feel attracted to secret relationships with one another would feel less impelled to romanticize these if this type of relationship no longer had the glamour of being forbidden by society.

(c) Lest apprehension be felt about the likelihood of the spread of homosexuality in the community as the result of relaxation of the law in this respect, we should like to point out that the statistical incidence of homosexuality appears to be constant, and that nobody not now so afflicted would wish to change from the more mature (heterosexual) to the less mature (homosexual) life merely as the result of a change in the law. On the contrary, some homosexuals now deterred by the state of the law might feel safe to declare themselves for the purpose of medical help.

(22)   We are of opinion, however, that the severity of the law should not be relaxed towards any act which tends to spread homosexuality among potential victims, especially the young. We hold the same view in regard to any other sexual act which offends public decency or decorum in this respect, or which tends to flaunt or glorify this mental illness as if it were a superior social cult. For this reason we are in favour of strict legislation in relation to the offences of importuning, corrupting, soliciting, or the establishment and maintenance of clubs or "maisons de rendezvous" for homosexual purposes.

(23)   In regard to these matters we wish to stress that we should like the two sexes to be treated on a basis of equality since similar considerations of corruption of the young and the importuning of innocent but vulnerable persons enter into the behaviour of both. In Roman law the sexes are regarded as equal in this respect. There is probably just as much mental unhappiness caused by the activities of female homosexuals ("Lesbians"), which in this country has been tacitly ignored but whose effects we as psychiatrists have had many opportunities of observing.

(24)   We consider that public policy has nothing to gain from the maintenance of the somewhat artificial distinction between "the abominable crime of buggery", i.e. intercourse in ano, and other forms of homosexual behaviour, since it is well-known to clinicians that all manner of perverted practices are easily interchangeable in this form of illness, while on the other hand the same "abominable practice" is not infrequently carried out by otherwise happily married couples, by mutual consent. Here, also, the operative distinction is between consent and the one-sided infliction of undesired indignity or depravity, among which buggery is only one phenomenon.[39]

(25)   In sum, we are of the opinion that the considerations on which legislation should be based are those of preventing harm or offence to innocent persons, the defence of public decorum and decency, and on the consideration that abnormal sex behaviour should be regarded basically as a public health problem. The homosexual should be thought of and proclaimed in the public mind as an immature, sick and potentially "infectious" person, and the whole subject divested of the glamour of wickedness as well as aesthetic superiority. We believe that if

this were the attitude of authorities, some of the attraction of it for immature minds would be taken out of the whole subject and some of the sobriety of the medical-preventive attitude infused.

(26) We feel that this view of the subject would be in line with the advance of public opinion from an attitude of disgust and horror towards one of scientific investigation, understanding and prevention. In prostitution and in homosexuality as in other behaviour disorders the most hopeful possibilities of prevention lie in the increased awareness of the mental hygiene aspect of child and maternity welfare and family care, with facilities for early competent diagnosis and therapeutic intervention, long before the emotional dispositions of potential homosexuals have crystallized into set characteristics.

(27) (e) <u>Treatment</u>.

. . .

(28) It has been the practice of Courts to sentence homosexuals to probation "on condition that they undergo treatment." We feel that this "automatic", if humanely inspired, practice has not worked well. Many ingrained homosexuals who would not have had any wish to submit themselves to treatment have accepted the condition as an alternative to imprisonment, and, irrespective of the actual progress of treatment, have abandoned it as soon as the fixed time limits necessarily imposed have expired. That is to say, they "went through the motions" of being treated without the intention of change, which is a very difficult state of mind in which to help psychiatric patients.

(29) We feel that prior to such expensive treatment being ordered by the Court, there should in all cases be instituted a system of remand for medical and psychiatric investigation of the accused . . .

. . .

(31) . . . [W]e are not in favour of the limited sentence in which psychiatric treatment is a condition, without prior expert assessment as to whether such a sentence would be likely to ameliorate or cure the homosexual. Still less do we favour the automatic imprisonment of homosexuals without adequate diagnosis, as it is only too likely that in prison they would not only have opportunities to practice homosexuality but might also strengthen such antisocial attitudes as they already have, with subsequent recidivism on discharge.

. . .

*

(f) <u>HO 345/8: Supplementary Memorandum from the Tavistock Clinic, March 1956</u>

In accordance with the request of the Chairman, we are submitting the case materials of a number of patients who have been treated by members of the staff of the Tavistock Clinic and which, in our view, illustrate...that in certain selected cases psychological treatment can be successful in changing apparently deeply entrenched homosexual attitudes and behaviour in the direction of normal heterosexuality: (cases 1, 2 and 3).

...

Case No. 2. (Dr. L. B.)

This man came for treatment in 1940. He was aged 27 and a graduate of Cambridge University. During the war he was a conscientious objector doing refugee work. He came for treatment on account of homosexual tendencies which took the form of a great interest in men and a wish to have relationships with them. He was particularly preoccupied with the desire to see if men were circumcized or uncircumcized. For this purpose he would frequent public houses, where he would "pick up" men and indulge in sexual relationships with them. He had never been in the hands of the police, but always feared it. Frequently he would loiter in public lavatories and try to pick up men with whom he would indulge in mutual masturbation.

...

In treatment he proved very co-operative and the material soon disclosed a very close identification with and fixation on his mother...

As treatment progressed further, material was elicited in relation to his strong suppressed aggression, again particularly in relation to his parents. He felt himself completely overshadowed by his father, which he resented deeply but which he would never dare to express. As a reaction to this, he would find himself obsessed by a desire to be "wicked."

...

As treatment progressed the improvement in his condition was evidenced in the change in the content of his dreams...

It was significant that in the progress of the treatment it was invariably after experiencing some powerful emotional frustration or disappointment that there would be a return to the more pathological dreams and preoccupations with lavatories...

After about a year's treatment the patient became interested in a girl. The friendship developed and they became engaged. In April 1941 they were married. Since then everything has gone happily. He has now four children, and is working successfully in a good position with the United Nations.

Case No. 3. (Dr. C. B.)

The patient was a young man of 20, a student of Oriental languages. He came for treatment because he had for a period of more than six or seven years noticed a growing sexual interest in other men. He was the only child of a good family, and was apparently a normal, healthy active child. At school he had no difficulties in learning, and was interested in Rugger, Squash, reading and music. In spite of that he was oppressed by strong feelings of inferiority most of the time. After military service, he obtained a scholarship to a University.

When he came for treatment he described his troubles as essentially a sexual preoccupation with men and particularly a strong desire to see them naked. He had for many years indulged in mutual masturbation with men of his own age, and this was invariably accompanied by strong feelings of guilt and self-reproach. During the war he was a midshipman in the Navy and was stationed in the Far East for 8 months. It was there that his interest in Oriental languages developed, but during this time he found himself developing homosexual interests and before long found himself in regular homosexual relationships. He could never find any interest in women, had no desire for them and never thought about them...

...

His treatment started in 1948...

...

With the progress of treatment he brought up material in which he saw how much his mother acted as a model for his attitude to women. There were many dreams in which he saw her making love to him and from time to time he was very disturbed to find himself imagining sexual relations with her. Throughout this early period of the treatment there was a complete absence of any thoughts or fantasies of women other than his mother.

...

After about three years of treatment he reported that to his surprise he had found himself drawn to a girl student at the University whom he had found quite attractive. He was pleased to find that she showed great understanding of his problems, of which he was finding himself talking quite freely and sincerely to her. From this point the girl played an increasingly important role in his outlook and feelings. She fell in love with him and for the first time in his life he became aware of strong physical feelings towards a woman. Fantasies about men, the desire to see men's genitals and to be observed by them became less and less frequent. He felt more and more sure that he was genuinely in love, and although he was still continuing with treatment he felt confident that he could make a success of the relationship. In 1953 they were married.

Since then everything has gone well, and there has been no return of his symptoms and no disturbance in the relationship.

...

*

(g) HO 345/8: Memorandum from the Institute of Psychiatry (University of London), Maudsley Hospital, Denmark Hill, S.E.5, 13 May 1955[40]

...

Homosexual Offences.

2. We are of the opinion that the law should not concern itself with homosexuality in private between consenting adults, not should it differentiate between the form of gratification adopted (manual, intercrural, anal, oral, etc.) by consenting adults. The reasons for this contention are that:

(1) the law usually cannot be enforced.
(2) that it is a matter better left to conscience.
(3) that in many cases such activity is the product of damaged personalities rather than damaging in itself.
(4) that it may conduce to or facilitate blackmail.
(5) that to some extent (but not in every case) sanctions and public attention may make the problem worse.

...

3.   (a) ... Broadly speaking, a youth is set in his sexual predilections at latest by 17 years ... [T]hough not unanimous, we are inclined to place the age of consent at 17 years, this incidentally being the statutory age at which the 'young person' becomes an 'adult' (Chdn. & Y. P. Act 1933, secn. 107).[41]

4.   (b) "In private": would exclude activities in places to which the public have access—bars, cafes, lavatories but could hardly include the private clubs which might well arise or increase ...

...

A Note on the Seduction of Young People.

...

8.   ...
   A common sort of story is the following. A man of immature personality opens a shop opposite to a boys' junior school and soon encourages the boys to come in. He earns the reputation of being 'queer' and those boys who, consciously or not, are interested in this sort of relationship tend to frequent the shop. Eventually one of them either offers himself or responds to a suggestion and is likely to accept money or sweets for his services.

9.   We are strongly impressed with the regular finding that those children who become involved in this sort of way with adults are already damaged by long-standing and often gross emotional maladjustment at home, or in their early lives. We find little or no convincing support for the theory that homosexuals are predominantly created by other homosexuals.

10.   With this sort of experience in mind we are against any move towards increasing the penalties against 'child-seducers', and consider that the present sanctions are sufficient.

...

Exploitation of Homosexuals.

. . .

12.   It is exceptional for our patients to admit to having been blackmailed. More common however are certain other forms of exploitation. The homosexual may be quietly robbed of his wallet, pen or other possessions especially if intoxicated at the time. He may be led to a room or quiet place only to find one or more other men present who thereupon strip him of everything valuable which he possesses. Threat of exposure does not depend only upon legal sanctions but also upon social and family reputation. It is very impressive how much some homosexuals fear the publicity of the local press rather than legal sanctions.

Penalties.

13.   The penalties for some offences seem to us to be grossly out of proportion to the social or personal damage done. This applies especially to buggery and attempted buggery (life and 10 years respectively) ...

. . .

15.   The availability, efficacy, nature and optimum duration of psychiatric treatment within prisons as they now exist is so uncertain as to offer no basis for rational sentencing, i.e. we do not think that the duration of sentences should be influenced at present by the hope of obtaining a cure through concurrent psychiatric treatment. We think that punishment of homosexual offences must, as it were, stand or fall on its own merits, and not be confused with a totally different treatment method. We hope however that continued efforts at treatment will be made during prison sentences and if necessary continued by the same therapist afterwards ...

. . .

Treatment of Homosexuality.

. . .

(2)   Remedial:

27.   1. The various analytic therapies: (abbreviated ANALYT)
      While the light which these methods shed on the origin of homosexuality is great, their efficacy as a method of treatment is still a matter of controversy.
      The main difficulties opposing effective application of these methods are as follows:
      The patient must be intelligent, young and able to express himself well.
      He must have sufficient existing strength of character to enable him to go through with a time-consuming and often trying experience.
      There are very few trained analysts and because of the time factor they can only deal with a few cases; it therefore tends to be a very expensive form of treatment.

28. 2. <u>Psychiatric team work</u> (abbreviated PSYCH)

...

29. 3. <u>Sexual Sedative Medicine</u> (SED) always in conjunction with psychiatric surveillance and support has a useful place ...

30. 4. <u>Social Worker's Supportive Measures</u> (PROB)—the 'advise, assist and befriend' clause of the Probation Act,[42]—a method which may also be applied by other persons who are mature and understanding, e.g. the clergy.

31. 5. <u>Penal</u> (PEN) including all the constructive measures that courts (especially juvenile courts) can order.

The following types of 'homosexual persons' are briefly described to illustrate the heterogeneity of the group and to illustrate the variation in treatment method. This is not a classification of homosexuality but rather representative selections from a very long spectrum.

32. (a) <u>Adolescents and Mentally Immature Adults Disliking their Perversion</u> ...

The treatment indicated is as follows (3+ scale

> + = useful
> ++ = very useful
> +++ = essential)

PSYCH +++
ANALYT ++   in some cases which fulfill the necessary criteria
SED +   for brief period to give confidence and reduce anxiety
PROB   not very useful
PEN   usually contraindicated, though appearance in court is often quite helpful.

33. (b) <u>Severely damaged personalities.</u>

(i) The very effeminate, self-advertising, female-impersonating individual, who talks in an affected manner, walks with a mincing gait, wears frilly underclothes, make-up, etc. A much misunderstood group. Most of them are essentially self-admiring and self-sufficient but delight in attracting males—they like the chase, though they may accept some physical homosexual advances, often for money; they do not derive much sexual gratification and (like some hysterical women) would gladly forego the sexual act. Most homosexuals avoid these persons unless for want of more satisfying partners. Not socially dangerous.

PROB ++
PSYCH ++   doubtfully more effective than probation for very long periods
ANALYT   not tolerated
SED   makes little difference
PEN +   they are afraid of penal action and it helps them to modify their activity; approved schools and Borstals contraindicated; sometimes small hostels are effective.

34. (ii) The inadequate, down trodden, dull (whatever their I. Q.) very passive individuals, who are unable to make affectionate relationships. They make fleeting contacts in lavatories, sometimes for payment, and this seems to represent a compensation or substitution for affectionate relationships. Not socially dangerous.

PROB +/ PSYCH +	very difficult to help; usually no family with which to work
ANALYT	incapable of co-operation
SED	they accept it for a while and then lapse.
PEN	only temporarily effective; if young enough approved schools and Borstal Institutions may help considerably.

35. (iii) The deeply resentful, anti-socially inclined individual usually with a long record including 'beyond control' charges and often non-sexual offences, coming from a neglectful and hostile home. It is difficult to determine how strongly they are really homosexual; often they have an active heterosexual life; they tend to exploit homosexuals. Socially they are dangerous especially to other homosexuals, and through their non-sexual offences. They seem to disappear from out-patient practice after the 20s, perhaps to prison.

PROB +	
PSYCH	they cannot co-operate very well and are estranged from their families.
ANALYT	inapplicable.
SED	inapplicable unless they recognize in themselves strong homosexual desires and are frightened of punishment
PEN +	probably not very effective; punishment hardens them further, but approved schools and Borstals may help them through to some sort of maturity.

36. (c) <u>Homosexuality in relatively intact personalities, otherwise well socialized.</u>

These can be divided into the young and comely who have digested any scruples and are having a good time often at the expense of wealthy homosexuals of this same sort, and into older homosexuals who are experienced, know the dangers, rarely approach anyone before they are sure that it is safe. These people appear happy and are able to hold useful jobs successfully. They are subject from time to time (especially later in life) to depression but in general do not want to change. Only if they turn towards children are they socially dangerous.

PEN ++	they have a healthy respect for the law which keeps their behaviour within bounds.
PROB + / PSYCH +	may help a little with the young.
ANALYT / SED	not accepted.

37. (d) <u>Latent and Well Compensated Homosexuals.</u>

These are men of considerable strength of personality who may not appear before the courts until middle-life or much later. They either genuinely do not know what their real difficulty is or else have struggled against it. Very often they are intelligent, perhaps married with healthy children, and may be professional men—teachers, actors, youth workers, etc. They are particularly likely to commit their offence at the time of some additional stress and once started may continue until caught. They are also particularly likely to choose one of the symbolic or 'incomplete' offences—touching, fondling, exhibiting themselves in lavatories (often with doubtful veracity, called "soliciting" or "importuning"). Unfortunately children are often their objects of choice but even so they approach them in a gentle manner and the amount of harm they do is probably greatly over-rated. They are, of course, very vulnerable to court proceedings so that a newspaper report may spell ruin for them and their families. Moderately socially dangerous to children.

PSYCH ++ (+++ if they have been in prison, when rehabilitation is usually essential).
SED ++
PROB +
ANALYT   usually too late for such methods.
PEN   no more effective than court appearance.

38.   (e) <u>Definite Homosexual Predisposition Co-existing with other serious mental disability.</u>

This is an ill defined group in which homosexuality exists in persons who are seriously handicapped in other directions. Examples of these other conditions are:

> intellectual defect.
> brain damage with or without epilepsy.
> psychoses.
> very gross personality defects, especially sadistic tendencies, callousness with extreme self-indulgence.

Unfortunately their social handicap often predisposes them to seek children as their sexual objectives.

These individuals are of course highly dangerous. They are apt to injure children or adults in a way which those in the preceding group could never do. They do not necessarily only choose 'predisposed' children.

East	PEN +++	They need to be segregated until rendered harmless.
Hubert		Actual punishment however is probably less
establishment[43]		effective here than in any of the groups. Secn. 8 of M.D. Act[44] or admission to a mental hospital may be indicated.
	PSYCH ++	
	SED +++	but their co-operation cannot be counted upon.
	ANAL [*sic*]	unsuitable.
	PROB	unsuitable.

39. Brief Note on use of Sex Hormones (oestrogens)

These are usually administered in tablet form by mouth and have the general effect of diminishing sexual activity in males. It seems to be at its best in those rather rare individuals who, though dreading imprisonment and wanting help, are truly (biologically) as opposed to neurotically highly sexed and who cannot resist their impulses. Sometimes it is possible to limit sexual activity without abolishing it. It tends to be better tolerated later in life and it is rarely indicated in youth unless without it the individual is certain to be in trouble again, and unless other methods are ineffective. In recidivists who are frightened and discouraged by punishment, it may break the vicious circle of reconviction.

In some homosexuals who are distressed and preoccupied by their impulses it brings great relief, but in others it doesn't.

It is very useful, given in large doses for a short time, in young people who thereafter know that there is this method to fall back upon. It also demonstrates what some of them find hard to believe, that it does really leave the sexual function unaltered. It may also help in the breaking of habit and the breaking of homosexual associations.

40. Possible Deleterious Effect of Psychiatric Treatment.

It has been reasonably asked if treatment ever makes patients worse. It is thought that unwise removal of the patient's resistance without putting anything in its place and without relieving the basic problem might conceivably make him more active or promiscuous, but we have not observed this to happen. A more usual difficulty is increasing depression as the patient realizes his real predicament. Apart from this we do not feel that there is any risk of deleterious consequences.

...

APPENDIX 'A'

Homosexuals 1951–1952.

All the known cases of the Joint Bethlem Royal & Maudsley Hospital in which homosexuality was diagnosed in 1951 and 1952 have been analysed as follows. Each consultant reported on his own cases. There were 72 cases:– 63 males, 9 females.

...

Preferred Sex Object:

		M	F
Child	...	11	–
Y.P.	...	10	1
Adult	...	32	7
Variable	...	5	–
N.K.	...	5	1

Preferred Homosexual Activity:

Nil	−4	Fantasies of sado-masochistic or bizarre behaviour with no desire to put them into practice.
Masturbation	−27	In 21 cases this was mutual masturbation but in 3 cases activity consisted in being masturbated only; in one it was exposure and self-masturbation only; one, touching. One denied all 'inclination' but only "did it for money". Two had once experienced sodomy, and one intercrural intercourse.
Sodomy	−18	In 3 it was active, in 5 passive. In four there was also oral activity or flagellation. Most of the sodomists were not court cases.
Intercrural intercourse	−2	
N.K.	−12	

Four women had no sex activity; 3 preferred masturbation & N.K. 2.

...

Seduction:

Seduced by older people	−14
Not seduced	−31
N.K.	−18

No women were seduced.

...

Diagnosis:

	M	F
A. Adolescent conflict	17	3
B. Severely damaged personalities:		
1. Very effeminate, self-advertising	3	
2. Inadequate "dull", passive	15	1
3. Deeply resentful antisocial	6	1
C. Well socialised, accepting homosexuality	13	1
D. Latent and well compensated	3	1
E. Combined with serious mental disablement	6	2
(Doubtful)	(7)	

Kinsey:

1. Heterosexual with incidental homosexuality	4
2. Heterosexual with more than incidental homosexuality	4
3. Equally homosexual and heterosexual	2

4. Homosexual but more than incidental heterosexuality	8	2
5. Homosexual but incidentally heterosexual	21	2
6. Exclusively homosexual in interest & activity	22	4
N.K.	2	

The Social Threat:

	M	F
Harmless	37	7
Nuisance	14	2
Dangerous	9	–
N.K.	3	–

...

Results of Treatment or State on Disposal

Much improved	... 4
Improved *	... 16
No Change	... 33
N.K.	... 10

* Often only with regard to coexisting anxiety or depressions, etc.

Prognosis:

Good	... 2
Fair	... 16
Doubtful	... 8
Poor	... 22
V. Poor	... 10
Not stated	... 5

Treatment Recommended:

Psychotherapy	... 18
Psychoanalysis	... 3
Supportive	... 18
Other psychiatric treatment	... 11
Oestrogens in addition	... (12)
Nil	... 8
Not stated	... 5

Penal Sanctions:

Helped	13
Hindered	4
Neutral, not applicable or N.K.	46

Other points:

Six men and one woman attempted suicide before or after being seen, in one case successfully. There were two cases of paranoid schizophrenia; three men and one woman were heroin addicts and three chronic alcoholics. Seven others were severely psychopathic, being possibly pre-psychotic or with severe multiple perversion, pathological liars, etc.

<u>Comments on Appendix 'A'.</u>

...

2.   A surprising number (about half) presented themselves voluntarily and without court direction of any sort which is an indication of the suffering which the condition may cause.

...

5.   Most homosexuals are satisfied with some form of masturbation. Actual sodomy is comparatively unpopular, and most of those who practised it were not court cases. This may be because sodomy cases are referred to a higher court and presumably remanded in custody, though a number of them may be suitable for bail and examination in an out-patient department. This is an instance of the artificial barrier created by differentiating between homosexual acts.

6.   Assessment of the duration of the condition is overwhelmingly in favour of homosexuality being an ingrained condition, established in childhood, even though the first overt manifestations may not appear until later. This is compatible with both main theories of the origin of homosexuality—inborn and acquired through early family influences. There is little support for the "seduction" theory.

...

8.   61% of the group are considered socially innocuous and only 12.5% dangerous. The danger is almost entirely to children and young persons. But all who are involved homosexually with young people are not necessarily dangerous, largely because these young people may already be confirmed in homosexual ways and maybe the actively seducing party.

9.   It will be noted that in-patient treatment is rarely considered the most appropriate treatment for homosexuals. The impression is that those who become in-patients are admitted rather for intercurrent psychiatric conditions rather than on account of the homosexuality itself. A large number of cases do not continue treatment. This together with the poor results of treatment seems a fair comment on the inadvisability of basing legal action on any assumption that the problem can be settled by medical means.

10.   In general we found the impact of penal sanctions on our patients too difficult to assess reliably. Subjectively speaking, however, we can recall many cases who have been helped in self-control by the fear of legal action, as well as some cases who have been hindered by the same.

...

APPENDIX 'E'

Prison cases.

The following lists 20 consecutive convicted homosexual offenders examined psychiatrically in a short-term prison for males, and gives brief particulars of each case, with the object of illustrating the sort of problems met with and the possibilities of helping them.

1.   Age 36. Offence 'importuning'. A highly intelligent school master of good personality who only gave way to overt homosexual behaviour under the stress of conflict concerning his engagement. He was seen twice in prison and thirteen times subsequently in the out-patient department of a psychiatric hospital. His fiancée was also seen several times. He made a satisfactory adjustment, eventually married, and regained his position of school teacher. He still keeps in contact with the hospital. The outlook is good.

. . .

12.   Age 28. Indecent behaviour with a man (several mixed offences over the past 10 years). He comes from an unhappy home with much parental discord and with a terrifying father. A man of very passive, inadequate, narcissistic personality who rejects treatment.

13.   Age 29. Importuning (1st offence). A man of passive, immature, resentful personality. His mother was a dominating woman and his father inferior and belittled by her. It was only possible to see him once and sufficient contact with him was not made.

. . .

15.   Age 22. Importuning. An established homosexual prostitute, the son of a prostitute who left her husband when this patient was 1. It was not thought possible to influence him at all. He did not want to change.

. . .

20.   Age 55. Indecent assault of boy (1st). A married man whose homosexuality is of secondary importance to wider abnormalities of personality. He had a despotic father and was spoiled and fussed by his mother. He had been seen regularly in prison and his wife referred to the psychiatric social worker. It may be possible to help them to a limited extent.

*

(h) HO 345/8: Memorandum of the Institute of Psycho-Analysis,[45] March 1955

. . .

SECTION II—SUMMARY OF MAIN POINTS

4.   Three important facts relevant to the control of homosexuality emerge from psycho-analytic investigations:

(a) homosexuality develops in early childhood;
(b) homosexuality is a reflection of profound psychological disturbance in the individual;
(c) most normal personalities are found, on analysis, to contain in a suppressed form minor traits bordering on homosexuality.

5.   From these three findings, and the evidence on which they are based, we consider that:

(a) the existing law does little to reduce homosexual practices, or to prevent individuals from becoming homosexual;
(b) the existing law may contain unrecognised elements derived from the unconscious anxiety about homosexuality of the individuals who constitute the public;
(c) a change in the law such that sexual relations in private between consenting adults would no longer be a basis for a criminal charge would be most unlikely to increase homosexual behaviour and might even lead to a reduction of its more antisocial manifestations, partly because it would also reduce homosexuals' feeling of being persecuted by the law;
(d) legal restraint should remain with regard to homosexual offences against minors. Psychiatric examination and treatment should be made available for such offenders, but segregation may be unavoidable in some cases. It should not, however, be in an ordinary prison, and the aim should be prevention, not retributive punishment;
(e) the severity of some of the legal penalties is excessive, particularly for buggery and indecent assault; this is illustrated by the disproportion between the penalties where the victim is a man and where it is a woman.

SECTION III—GENERAL STATEMENT

6.   It is the main purpose of this memorandum to draw attention to three essential facts which emerge from the psycho-analytic investigation not only of homosexual patients but also of other persons with or without neurotic symptoms; and to show that these three facts have important implications for legislation concerning homosexual practices. These three facts are the following:

(a) Homosexuality originates in early childhood.
(b) Homosexuals are not in general people who have deliberately chosen a vicious mode of life; nor freaks of nature, unrelated to normal humanity. They show a gross exaggeration of attitudes that can be found in the normal and this exaggeration is a symptom of a deeply rooted disturbance in personality development.
(c) The other fact is the reverse side of the same coin. Most normal personalities are found, on analysis, to contain in a suppressed form minor traits bordering on homosexuality. This leads in many cases to some anxiety about the suppressed homosexual feelings, though the person may be quite unaware of their presence.

7.   Legislation against homosexual activity may be regarded as the reaction of the people in the second, majority group against those in the first group, and it is liable to be influenced by the anxieties just mentioned, since homosexual activity in others is felt as a threat to the society with which the normal person identifies himself, ultimately as a threat to his own heterosexual adjustment. According to the present law of England, any kind of sexual activity between males, under any circumstances, is a criminal offence, and where assault is involved the maximum penalty is 10 years imprisonment in contrast to 2 years for a similar offence against a female. These facts show clearly how much more the anxiety of the public is aroused by homosexual than by heterosexual activity.

8.   We feel that this legislative anxiety has obscured the essential issue of distinguishing between those homosexual activities which constitute a real menace to society or to certain of its members, and other homosexual activities which may raise no important social issues, however significant they may be as symptoms of disturbance in the individuals concerned. The position is analogous to that in cases of mental illness, which is regarded as a medical, not a legal, problem except insofar as the patient is a menace to himself or to others.

9.   These considerations lead us to draw a sharp distinction between homosexual activity carried out in private between consenting adults on the one hand, and on the other hand (a) activity involving the homosexual seduction of minors, (b) actual sexual violence against a non-consenting partner, and (c) acts involving public indecency.

10.   Short of creating a police state, we doubt if legal sanctions can do much to prevent private homosexual activity between consenting adults. What they can and do achieve is to foster a feeling of persecution and injustice amongst homosexuals, which will increase any antisocial tendencies they may have; and to create ample opportunity for blackmail.

11.   It is sometimes argued that to alter the law so as to exclude such homosexual activities would mean that society was expressing its approval, or at least its sanction of them. We do not agree with this view. Although society does not approve of drunkenness, it is not considered necessary to legislate against it as such, but only against its antisocial consequences. It is public opinion, not legislation, that has diminished drunkenness, and public opinion will always be a powerful force acting against homosexual behaviour. Other such forces are the potential homosexual's knowledge that he will not have children, and that he is condemned to a lonely life because an affair with another man is never permanent, so that he can never have a settled life in the community, where his relationship to women will always be socially ambiguous. Thus no one who has a real choice in the matter is likely to choose homosexuality, but only a man with a deep-going personality disturbance, and this cannot be altered by legislation.

12.   The homosexual seduction of children and minors involves different considerations, and here legal sanctions are clearly necessary for the protection of the

public. The importance of seduction in childhood or adolescence as a cause of lasting homosexuality has, we believe, been considerably exaggerated, and most boys who have this experience are probably not permanently affected by it; those who are so affected have usually a strong predisposition to homosexuality from other causes.

13.   Nevertheless, children and young people require protection by the law from such emotionally disturbing interference. In view of the later sexual maturing of boys as compared with girls, it may be advisable to consider raising the "age of consent" for boys from 16 years to a later age. The principles we would suggest in the case of such offenders against minors are, first, to prevent repetition of the offence (if all else fails by some form of segregation, though not in an ordinary prison, which is notoriously a breeding-ground for homosexuality) and secondly to provide for the psychiatric examination and, if applicable, treatment of the offender.

14.   Another aspect of the law to which we would direct attention is the seemingly disproportionate penalties for the offences legally known as buggery and attempted buggery. Since this offence is not necessarily homosexual, though doubtless usually so (it refers to anal intercourse with a woman or animal as well as with another man), it is evident that another psychological factor has been at work here in the public mind, namely intense abhorrence related to the function of excretion. This type of strong reaction against erotic pleasure connected with excretory functions is very familiar to psycho-analysts; and everyone with intimate knowledge of small children knows how normal such pleasure is to them. The normal adult has erected a strong barrier of disgust which makes him feel horrified that any adult should take pleasure in such things; and this, we believe partly explains the irrational violence underlying previous legislation on this matter. In part, however, it is truly a reaction against homosexuality, motivated by anxiety about the idea that a man can be treated sexually just like a woman.

15.   We consider that there are certain respects in which the present law is too severe, and as a result lends colour to the view the homosexual is apt to take of himself, that he is a persecuted victim of society. Such an impression is given further support by some of the police methods which are found necessary to ensure conviction. Psycho-analytic investigation has demonstrated a close intrinsic connection between homosexuality and feelings or delusions of being persecuted, and the severity of the criminal law unfortunately confirms such feelings.

SECTION IV—THE PSYCHOLOGY OF HOMOSEXUAL AND NORMAL SEXUAL DEVELOPMENT

. . .

17.   Sexual feelings and behaviour begin in early childhood, not at puberty. Furthermore, this "infantile sexuality", as Freud called it,[46] is something that has innate, instinctual roots, rather than something that needs seduction from an outside source to bring it into being, although such seduction may certainly foster it

and warp it in various directions. Study of infantile sexuality has made it clear that the essence of sexuality in general is not an irresistible attraction exerted on members of one sex by the other, but rather that normal heterosexuality comes about only as the end result of complicated developments. Many of the forms taken by sexuality in childhood resemble adult perversions, for example sexual looking and exhibiting, and sexual pleasure associated with the excretory functions. Sexual activity with a partner of the same sex is likewise common in childhood.

18. Another fundamental view point [*sic*] of psycho-analysis is the notion of bisexuality. Anatomically each sex shows certain vestiges of the reproductive organs of the opposite sex, and also in the psychological field we find regularly that an individual has at least traces of the mental qualities which we are accustomed to associate with the opposite sex. Thus we may say that everyone is potentially bisexual in a psychological sense, but in most cases the psycho-sexuality is predominantly either masculine or feminine, and this predominance usually corresponds to the physical sex of the individual. Nevertheless, the latent homosexual tendencies of apparently normal individuals are shown by the high incidence of homosexual behaviour that occurs where the opposite sex is not available—in boarding-schools, ships, prisons, etc. It is a fact of experience, not simply a matter of theory, that prison is a breeding-ground for homosexuality, and nothing could be less likely to cure it.

19. It should be noted at this point that the adjective "homosexual" is currently used to denote two different aspects of sexual behaviour, which may or may not go together—first, the choice of sexual object of the same sex as the subject; and secondly sexual activity or attitudes appropriate to the opposite sex, i.e. a female type of sexual behaviour in the male. By no means all homosexual men are predominantly feminine in type, and these masculine types may choose sexual objects who, although technically male, i.e. possessed of male genitals, are yet as like females as possible, e.g. young boys. This peculiar compromise has been shown by analysis to depend essentially on psychological factors operating during childhood. In the other type, which shows feminine behaviour, we also find psychological determinants for this attitude; it is likely, however, that an endocrine factor may be operative in some of these cases.

20. It will be clear from the foregoing remarks that psycho-analytic work with heterosexual as well as homosexual persons has shown that homosexual desires and attitudes are normal components of sexual development. They become superseded by heterosexuality only when adult sexuality is fully established after adolescence. When overt homosexual behaviour or attitudes are retained in adult life it is a sign of a neurotic disturbance of the whole personality and of a failure to achieve a satisfactory solution of the problem with which every individual is faced, the problem of reaching an adjustment between the demands made upon him by his innate instinctual forces and by the forces of the social environment.

21. The continued existence of potential homosexual desires in the normal person is manifested chiefly in negative ways, that is, by reactions against

homosexuality. Thus in social life one can frequently observe signs of anxiety about the subject. Various social mechanisms operate to reinforce the individual's efforts to overcome his anxiety about his latent homosexual feelings and to deny them. One of these is the overemphasis on the virtues of toughness and aggressive heartiness which is characteristic of the atmosphere of some boys' schools.

22.   But the mechanism most relevant to our present discussion consists in achieving complete blindness to one's own latent homosexuality by developing an intense preoccupation with homosexual manifestations in others—saying, in effect, "It is not I who am homosexual, he is". The feelings of guilt and anxiety which are properly related to the individual's own homosexual impulses then become converted into condemnation and persecution of the other homosexual. We regard the operation of this mechanism in so many overtly heterosexual men as the most important source of the emotional heat and controversy surrounding the subject of homosexuality and the problem of legislating for it.

SECTION V—PREVENTION AND TREATMENT OF HOMOSEXUALITY

23.   Because of the origins of sexuality in early family relationships, it will be apparent that its prevention can only indirectly be helped by legislative action. It is a matter for mental hygiene and for the social influences which may favour the development of happy homes where the parents themselves have a good emotional relationship and a healthy attitude towards sex. We therefore do not intend to discuss prevention in this memorandum.

24.   Nor would it be appropriate to discuss treatment in any detail. Psychiatric and psycho-analytic treatment has its greatest chances of success with adolescents, who are in a more plastic state; the die is perhaps not yet cast and the outcome may be either permanent homosexuality, or a turning to heterosexuality after passing through a temporary homosexual phase. Under these circumstances, psychological treatment may tip the balance in a favourable direction. Such youths may need help and encouragement to seek treatment, and the courts might co-operate in this by the use of the probation service, which has already proved its value in this connection.

25.   As regards older men who are confirmed homosexuals, the outlook for treatment is less favourable, but in the presence of a genuine desire for change, i.e. real unhappiness and feelings of guilt and not merely concern about the legal consequences of the homosexual activities, the prospects are far from hopeless. A psychiatric examination and assessment of treatment possibilities is needed. Where practicable, treatment is particularly urgently called for in the case of men who make a practice of seducing young boys. There are often specific psychological reasons for this predilection, the simplest being fear of approaching an older man, but commonly the boy represents a mixed female–male idea in the man's imagination. In the case of persistent offenders of this type, treatment facilities should be combined with segregation if that should prove inevitable.

26.   What form of segregation is suitable for these cases merits serious study and presents a difficult problem. On general principles we do not favour prison sentences for homosexual offenders because the exclusively male environment is more likely to accentuate than to check a homosexual predisposition; and further-more they are likely to spread homosexual practices amongst other inmates of the prison. However, we confess it is difficult to envisage a really suitable institution for the segregation of such people.

27.   Psycho-analytic treatment is costly and time-consuming and requires much co-operation from the patient; it can therefore be applied only in a limited num-ber of selected cases. Its importance is not that it provides an immediate large-scale solution to the social problem of homosexuality, but that it is a method of investi-gation and research which has already contributed much to the understanding of the subject and whose continued application promises still further illumination.

*

(i) HO 345/9: Memorandum submitted by the British Psychological Society[47]

. . .

HOMOSEXUALITY

A.   General Psychological considerations

(1) So far as we are able to judge at present, the principle of multiple aetiology applies to sexually-coloured relations between two persons of the same sex as it does to other problems of human behaviour. The main aetiological factors are commonly discussed under the following headings which are not mutually exclusive:

  (i) Biological (e.g. most investigators agree that about 4% of adult males are biological variants,[48] i.e. true inverts, in that they never experience heterosexual desires throughout their lives).
  (ii) Social and Cultural.
  (iii) Constitutional (e.g. homosexuality in males may possibly be shown to be associated with certain body-types).
  (iv) Psychological (e.g. unsatisfactory relations with other people, particularly parents, occurring early in childhood may, it would seem, lead to an arrest of psychosexual development.)
  (v) Biochemical and endocrinological.

Any one or more of these factors may appear to be significant when an indi-vidual case is investigated by physicians and psychologists. However, we have at our disposal insufficient knowledge to be able to arrange these factors in order of aetiological importance.

(2) It has been assumed by some that a determinant of subsequent homosexual development is to be sought for and found in early seduction. In our view,

the evidence in favour of this view is far from conclusive. On the other hand, most psychologists would agree that violence or undue persuasion in a sexual situation of any kind may be very traumatic at any age, but especially in the case of the immature.

(3) The concept of homosexuality appears to embrace widely diverse phenomena which may require different methods of study. Thus, considerations affecting pederasty, occurring throughout the great part of Hellas and Magna Graecia, where it was socially accepted and indeed encouraged for several centuries, differ markedly from those affecting homosexual love between consenting adults occurring in a social framework, such as our own, which forbids homo-erotic experience of any kind. Different methods and indeed different specialists might be needed to investigate these divergent types of sexual behaviour.

(4) If the principle of multiple aetiology, as propounded in (1) be conceded, it stands to reason that no one therapeutic approach can be generally applied. It seems desirable, therefore, that every homosexual male who wishes to become heterosexually orientated should have his case carefully assessed by a competent physician (with the aid, if desired, of a qualified psychologist) before the appropriate therapeutic method can be determined. In our view, too, any male charged with homosexual offences should be offered the opportunity of an investigation of this kind, though it should not be made compulsory.

(5) (4) does not imply that such a re-orientation is always possible, even if it be considered desirable.

(6) According to Kinsey and his research-team, 4% of adult white Americans are exclusively homosexually orientated throughout their lives. 37% of males have experienced orgasm in a homosexual situation; and this percentage rises to 50% in the cases of males who remain unmarried until the age of thirty-five. It may be that this statistical survey does not accurately apply to Great Britain; but a large-scale research has not been carried out in this country. We are, however, satisfied with Kinsey's main contention, namely that men should be assessed in accordance with a heterosexual–homosexual rating scale, whose polar opposites are exclusively heterosexual—exclusively homosexual, rather than be allotted arbitrarily to one or other of these two categories. The widespread incidence of homosexual behaviour-patterns that appears to obtain inevitably in most societies does moreover suggest that it is the duty of the state to protect the sexually immature from premature sexual experience, insofar as it is able, by penalising gross physical relations between boys below the age of sexual discretion and adult males.

(7) In view of the paucity of reliable scientific data on the phenomena connected with homosexuality, we should like to stress the need for systematic research into the various aspects of the problem. Such research should be the concern of a team of workers from different disciplines—e.g. psychology, sociology, medicine and biology. The findings of such a team should be carefully coordinated.

B. Some considerations relevant to the law relating to Homosexual offences

(1) If the Departmental Committee decide to recommend the abolition of the Criminal Law Amendment Act, 1885, which abolition would give a certain freedom of sexual expression to consenting adults of the same sex, an age of consent will have to be established. The age of sexual discretion recognised by the Common Law is 16, since a male may contract a valid marriage at that age. It must therefore be a matter for careful consideration as to whether the age of sexual discretion should be different in a homosexual context.

(2) It should be borne in mind that in a relation between an older man and a youth there may be much of educational and emotional value, and that a tendency to over-stress possible sexual factors complicating such a relationship is to be deprecated since it might discourage certain human contacts of undoubted value.

(3) It must be appreciated that there are many forms of physical contact in homosexual relations, as indeed there are in heterosexual congress. Certain homosexual practices are more repellent to many heterosexual people than are others; but such sentiments should not be allowed to influence the decisions of those who are considering the amendment of the laws covering homosexual offences with judicial detachment.

APPENDIX I

RELEVANCE OF ANTHROPOLOGICAL OBSERVATIONS ON HOMOSEXUALITY AND PROSTITUTION[49]

At the present juncture, when our own social attitudes towards prostitution and homosexuality are under review by the Departmental Committee, it may be useful to recall the diversity of attitudes shown towards the same phenomena by other societies, and at other times.

...

Homosexuality

A review of data in the Yale University index[50] of descriptions of seventy-six contemporary societies has shown that in the majority of cultures homosexuality is condoned or encouraged for at least some members of the population. An extreme example is that of the Siwans of Africa[51] who expect homosexual behaviour of all men and boys: both married and unmarried men are expected to have both homosexual and heterosexual affairs. This shows that homosexuality can co-exist with a tolerant opportunity for heterosexual activities; but the former does appear to be actively encouraged by such communities as the Moslem Arabs, whose women are secluded while men and boys are engaged in common pursuits for long periods of time.

Men of the Keraki tribe in New Guinea enforce passive homosexuality on all boys approaching puberty, as do the Kiwai. In a number of communities

(American Indian, Chuckchee of Siberia[52]) there exists an institutionalised role, the berdache or transvestite, in which a man adopts women's clothes and women's ways and lives as the wife of another man of the tribe. In some cases (Comanche, Chuckchee, Koniag) such persons are regarded as shamans and are believed to be endowed with supernatural powers. This is not always the case. In North India, for example transvestites known as hinjras [*sic*: hijras] subsist by begging and male prostitution; they apparently cater for a concealed demand, but they are a despised and outcast section of the community.

Where institutionalised homosexuality has been reported, sodomy is the rule. In a few societies (Hopi, Wogeo,[53] Dahomey, Crow) other practices are also, or alternatively, carried on. In the Trobriands sodomy is known to occur, as there are stereotyped expressions to describe it, but the practice is scorned rather than abhorred vehemently.

The widespread occurrence of homosexual practice appears to confirm the belief that most normal individuals have potentialities for developing homosexual interests which may or may not be brought out by their exposure to social learning. This belief is further strengthened by the observations which have been reported on "inter-sex" cases. These are persons whom a developmental anomaly has endowed with external genitalia of indeterminate appearance. Their physiological sexual identification depends upon whether they are born with testicles or with ovaries; but it has been found that these individuals generally show a strong preference for continuing in the sex role in which they have been brought up, regardless of whether it corresponds to their gonadal equipment or not. This emphasises the great importance to early experience and social conditioning upon human sexuality.

...

Evidence of animal observations

Apes, monkeys and baboons have all been observed to show active homosexual behaviour, co-existing with heterosexual mating. This tendency seems stronger in immature than in mature members of the species.

Some naturalists maintain that mounting and adoption of the passive sexual role demonstrate the animals' dominant or submissive relationship, regardless of their sex. Subordinate individuals of either sex frequently make the female sexual presentation as a demonstration of submission; they may then be permitted by the dominant partner to take food, or carry out some otherwise forbidden activities.

Homosexual arousal is much less frequent in female primates than in males; but has been observed in chimpanzees and in monkeys.

Lower animals may show homosexual mounting but this may be due to misidentification. It occurs most readily when sexual drive is heightened by abstinence. The passive partner does not play the appropriate receptive role, as does the primate. Female sub-primates mount each other when on heat: passive partner shows sexual arousal. Females on heat will often mount a sluggish male and thereby stimulate him to satisfy them with an exercise of his male function.

In summary, it can be said that a biological tendency for inversion of sexual behaviour is inherent in most if not all mammals including the human species.

Homosexual behaviour occurs in many human societies, is more common in adolescence than in adulthood, and in men than in women. In human societies some homosexual behaviour persists in spite of stringent taboos. Where it is condoned, such behaviour is common, if not universal.

*

(j) HO 345/8: Memorandum of a Joint Group appointed by the National Council of Social Service and the National Association for Mental Health[54]

. . .

APPRECIATION OF THE PSYCHOLOGICAL ASPECTS.

. . .

3. In the case of the very large group of homosexuals whose condition is not directly attributable to physical causes it is generally accepted that they show a sexual abnormality which is due either to early psychological disturbances or to an arrest of sexual maturation at a pre-adolescent stage. With the prostitute, on the other hand, it has been widely assumed that she indulges in an excess of normal sexual activity for pleasure as well as gain. Psychological investigation, however, has shown that many prostitutes are sexually frigid and this not only secondarily as a result of their life, but rather primarily as a result of deep-seated difficulties in their psycho-sexual development and in their child–parent relationship. It is highly significant that they may reveal a good deal of overt or latent homosexuality.

4. There is yet another point where the two problems meet: a considerable number of male homosexuals appear to be more promiscuous than heterosexual men. They find it impossible to have sexual relations with their male friends, but resort habitually to contacts with casual "pick-ups". This is to some extent due to the social difficulties in the way of a more permanent homosexual union, and the fact that the social stigma is greatly aggravated by the legal one.

5. The more fundamental cause, however, is to be found in a marked split between the sexual desires and the tender loving ones, the latter being directed exclusively towards the idealized parents and their later substitutes, the former towards strange and often despised partners. This dissociation between sexuality and love accounts also for the desire of many otherwise normal men, in addition to sexual perverts of various forms and degree, to seek out prostitutes; and it can be seen to operate in the prostitutes themselves.

. . .

7. The psycho-sexual development is a very gradual and complex process which can be held up or disturbed by a variety of inner conflicts and environmental

factors. We are mainly concerned with the influence on the children's mind of the parents' attitude and conduct ...

8.   The result of adverse environmental factors depends on their varying degree and combination, the stage of the child's development at which they occur, and on individual pre-disposition. It is not possible to attribute any given effect, e.g. homosexuality, to specific environmental factors. But there is one fact very commonly found in the history of passive homosexuals, namely the parents' wish for a girl and their persistent treatment of the boy as if he was a girl or as if he would have received more affection as a girl. In general however the pre-requisites of a normal sexual development are the same as those of a good emotional and character development: a secure home atmosphere where the parents themselves have a good emotional relationship and a healthy attitude towards instinctive demands, which enables them to respond to the children's needs without being over-stimulating and over-indulgent, or so forbidding as to prevent their natural emotional and social growth.

9.   The importance of a good mother–child relationship has become so well known that it seems relevant to stress that the father too has a considerable role to play in the children's life; and this not only in the later stages of childhood and throughout adolescence but from fairly early on.

10.   The adequate planning and application of effective prevention is the more important and urgent task since the treatment possibilities are severely limited, particularly those for sexual deviations and anti-social character cases ...

12.   The treatment prospects of male homosexuals depend largely on the category to which they belong. For a broad classification one may use the following three categories, whilst realising that they are to some extent overlapping:

(a)   The "true invert" who, on account of an abnormal physical constitution, cannot develop normal heterosexual interest. There is no possibility of physiological cure. Medical treatment by endocrine preparations can diminish or even abolish all sexual desire, but very prolonged treatment can have undesirable physical and mental side-effects. Psychotherapy can be used to enable him to make a more satisfactory adjustment to his disability and thereby to achieve better control over his impulses.

(b)   The psychogenic homosexuals, whose condition is mainly due to early psychological disturbances, have better treatment prospects in theory: certain groups of this category can respond to psycho-analysis or to analytic psychotherapy, but these forms of treatment are necessarily so time-consuming as to be unavailable to more than a small minority of cases. In practice, the majority cannot expect more than the limited help available to those in the first category.

(c)   The "late developer" whose emotional development may not be seriously disturbed, but only retarded and arrested at the homosexual stage of adolescence, has the best prognosis provided that treatment is not delayed too long.

Many youths in this category may not require intensive psycho-therapy but could respond to guidance from psychiatric social workers whose training is adequate for this purpose.

...

APPRECIATION OF THE SOCIAL SITUATION.

14. ... We believe that it is essential to deal with the social aspects of the problems with a corresponding understanding and detachment if society is to grapple more successfully with the issues than it has succeeded in doing in the past.

...

16. The homosexual is a person who, for one reason or another, innate or acquired, has become a deviant from the normal personality pattern. His form of deviation, when actively practised, is one which society is bound to find unacceptable, since it strikes at fundamental relationships between the sexes and at normal family life, and these form the basis of much of our social structure and outlook. It is necessary at the outset to emphasise that, although changes in the law will be suggested later in this memorandum, these are not meant to imply that homosexual practices should be looked upon as any less undesirable than they have ever been; no condonation is intended. Indeed, changes in the law are proposed mainly because it is believed that they would prepare the way for a better approach towards lessening the incidence of these practices.

17. In dealing with its deviants, society must have in mind several considerations. Though it may condemn the form of deviation, it does not follow that the deviant himself should be damned out of hand.

...

19. ... [G]iven that homosexuality actually exists as a social problem, the main considerations appear to us to be the following:

(a) Since the homosexual cannot exist out of contact with other people, society has a right and duty to give protection against homosexual practices to those of its members, especially the young, who may be most susceptible to seduction and who may themselves suffer permanently from such approaches.
(b) In its own interests, society must seek to rehabilitate the homosexual, both by its outlook and approach to the problem, and also by providing such treatment as is possible, under conditions which will encourage and enable homosexuals to take the best advantage of it.
(c) The homosexual deviant who seeks to come to terms with his condition, is able to avoid homosexual practices and to divert his emotional life into generally acceptable channels, ought not to have to feel that he suffers social disapproval.

20. At present neither the attitudes of the community as a whole, nor the existing legal provisions, succeed in achieving a great deal of this. Society in general, little

aware of the psychological and social facts of the situation, appears to demand simply the permanent and total suppression of the homosexual's sex activities. Suppression alone, however, cannot solve the problem; it adds to the strain and will probably make the individual even more unstable. He is likely to need help if he is to succeed in re-canalizing his emotional life. Unfortunately, however, it appears to be a matter of chance whether those he is most likely to approach (doctors, clergy, etc.) will be able to give appropriate advice and, if wrong advice is given, the man's condition may easily be made worse both from his own point of view and from that of society as a whole. It is clear that a much better-informed and therefore more understanding public opinion is essential to progress, together with a fuller realisation of the factors involved, and of the nature and limitations of treatment. The present discussions and the eventual Report of the Departmental Committee may well provide the basis for such a re-orientation of public thought on the subject.

...

GENERAL EDUCATION FOR PREVENTION.

...

32.   The causes of many forms of homosexuality and prostitution appear to lie in strains and stresses which hinder and divert the growth of the personality, seen both in its individual aspects and also in its relationships with others. Much of the pattern of personality is formed during infancy and childhood, and the importance of the emotional and social attitudes of the family, and the inter-relationships of its members, is therefore obvious. The primary emphasis then should be on promoting, especially in the family, those conditions likely to develop a stable and well-rounded personality in the child. It is our present social insufficiency in this respect which creates so many later social problems and delinquencies, of which homosexuality and prostitution are simply particular examples.

...

THE LAW—POSSIBLE CHANGES AND THEIR SOCIAL IMPLICATIONS.

...

43.   We recommend that the law should be modified so as to take no cognizance of homosexual practices undertaken in private by two consenting partners who are both over the age of 21.

44.   In support of this change, we mention the following matters. Firstly, we believe that there is probably a greater social readiness now to extend to these cases the general contention that the State should not normally be competent to interfere with the private actions of consenting adults, a situation already recognised in the statutes of many other European countries. Secondly, the opportunities for blackmail would be much reduced. Thirdly, some diminution in the making

of casual contacts with strangers and other offensive behaviour in public places might result. Fourthly, it would be easier for conscientious homosexuals to seek treatment without any fear that they might be laying themselves open to police action. Finally, with an easement of the social attitude towards this particular group of offenders, the tendency of some of them to feel persecuted and to seek compensation in anti-social or ostentatious conduct might be lessened.

45. We have selected the age of 21 after careful consideration … [W]e have been influenced not to suggest a lower age by the fact that public anxiety would certainly be stirred if the protective arm of the law were unable to reach the young man during the time of his National Service.

46. If a change in the law on the lines of this recommendation were introduced, it would be essential not to give the impression that conduct which was wrong before 21 suddenly became acceptable after that age.

47. We recommend that in order to deal with persons under the age of 21 who are considered to be in moral danger through contact with homosexual offenders or because they themselves have been indulging in homosexual practices, the relevant provisions of the Children and Young Persons Acts should be extended to the age of 21 …[55]

48. We feel strongly that it is in the age-group 17–21 that most can be done. The jolt which could be expected from appearance before a court might serve to arrest a developing homosexual tendency and prove a useful deterrent. At the same time it is with this age-range that the help of psychiatrists and adequately trained social workers, when it is available, can be expected to be most efficacious.

. . .

50. We recommend that it continue to be an offence for a person who has reached the age of 21 to engage in homosexual practices with any person under that age in any circumstances or with any person against his (or her) will, or to behave indecently in public.

51. The experience of psychiatrists in dealing with such offenders seems to indicate that there is little hope of eradicating homosexual tendencies over the age of about 25. An attempt should be made to help them by means of treatment to live with their disability so that they may cease to injure either themselves or society, though it has to be borne in mind that the older a person gets the more difficult treatment becomes. A prison sentence, may not contribute to this end and is not generally conducive to change; indeed it may serve to aggravate the condition. Consideration should be given to treatment in centres on the lines of the projected East-Hubert Institution,[56] or by means of an extended use of probation, combined with out-patient treatment.

. . .

*

(k) HO 345/7: Evidence Submitted by Dr. Winifred Rushforth, Hon. Medical Director of the Davidson Clinic, Edinburgh[57]

...

HOMOSEXUALITY

I should like to offer briefly conclusions drawn from my own personal experience on some aspects of this problem. I propose to deal with causation, treatment and prevention.

Two main groups of homosexuals are recognised.

1 The overt homosexual. This group includes all who are aware of their homosexuality, including those who are apprehended by the police ...
2 The crypto-homosexual. This group includes many patients who come suffering from anxiety and other neurotic symptoms. In the course of analysis, their essential homosexuality becomes evident.

From a psychological point of view, adult homosexuality is regarded as immature sexuality. It is essential to realise that each human being passes through a homosexual phase in adolescence, but the normal individual leaves this behind him in the ordinary process of emotional development when he is able to fall in love with women of his own age. In both men and women, full maturity involves hetero-sexual capacity.

CAUSATION

Common factors in its production are

1 The segregation of the sexes which limits opportunity of developing relationship [sic] between them. Such segregation occurs in schools and colleges, in the army and navy, in mines and other occupations, as well as in prisons and reformatories.
2 Family situations where the parents are not adequately related. The mother tends to relate herself too closely to her son and the father to his daughter. Such possessiveness hinders growth of the child's personality by denying freedom, imposing adult approval or disapproval instead of allowing freedom of choice. A dominating parent keeps the offspring immature. Typically, very devoted sons and daughters are likely to be homosexuals. A misogynist or woman-hater is a man who claims his freedom and independence by avoiding women—essentially he is avoiding a dominating mother.
3 The absence of the mother in early childhood may occur through death, or because she works, or because she is socially occupied and uninterested in her children. A similar state of affairs occurs if the child is institutionalized or hospitalized for a period. The necessary love, that is power to relate to another, simply is not elicited in the child and he remains immature, narcissistic or homosexual.

4 The absence of the father operates by denying the boy a figure with whom he can identify, a pattern of behaviour and a natural support for his masculinity. Very passive and irresponsible fathers act in the same way.

5 Harsh, tyrannical, ill-tempered fathers create fear in their offspring and an unwillingness to identify with them. Whenever the son identifies with the mother, either because she is more lovable, or more adventurous, or more interesting, or for any other reason, he is likely to be in difficulty over sexual development.[58]

6 Rejection by the parents, one or both, because of illegitimacy, because he/she is not of the desired sex, or for any other of the many reasons that make him unwanted and unvalued, handicaps him in the same way.

7 Inadequate housing and indiscriminate sleeping arrangements are serious factors. When children have to sleep with adults, it inevitably creates difficulties. If a room of one's own, so desperately wanted by the adolescent, is not a possibility yet a bed of one's own might be a more immediate goal and would do something to ease the almost intolerable problems of incest and its allied miseries in the under-privileged classes.

It must be recognised that in our present-day society many of these factors are operative and that homosexuality is therefore an extremely common condition which merits investigation on a national level.

TREATMENT

Since the condition is psychologically determined, treatment by drugs is unlikely to be effective. Imprisonment creates further problems as prison conditions predispose to further homosexuality and, in my opinion, is only of value in segregating hardened offenders who are a potential danger to young people.

Psycho-analytic therapy in certain cases at least is effective in bringing the homosexual into a more mature state in which he can relate to women. It operates by enabling the patient to understand his problem, to see it objectively and by releasing him from the fears and inhibitions which have hindered him to accept his manhood and the responsibility it entails.

. . .

Success in psycho-therapy depends

a) on the desire of the patient to be treated and in his co-operation with the analyst.

b) on the devotion of the analyst, his optimism and ability to encourage the patient to undergo a thorough treatment. These qualities, together with a thorough training in his profession, make for a successful analyst.

. . .

PREVENTION

Prevention lies in better preparation for marriage and in a social service (such as the Davidson Clinics) where aid in psychological adjustment can be provided for families in difficulty.

It lies also in better housing, affording common decency to boys and girls as they develop.

I would suggest that throughout adolescence mixed activity in clubs should be encouraged. Scouts and boys brigades, guides and girls guildries should be encouraged to amalgamate and share certain activities.

. . .

Public opinion, if educated to regard homosexuality as a problem of present day society in which all are involved, might be willing to work towards the removal of causes rather than punishing the victims.

. . .

SUMMARY

Parental attitudes in the developmental period of their sons' lives have determining influences in their sexual development.

All human beings pass through phases between birth and maturity when their relationships tend to alternate between homo- and hetero-sexuality.

Analytic therapy has as its aim the liberation of the creative spirit so that the individual takes responsibility for the full development of self into husband and father.

Homosexual cases of all ages who profess a willingness to co-operate may well be given the chance to obtain analytic help.

The most hopeful cases are of men under 30 who have not been deeply involved in homosexual practice, although at the Davidson Clinic good results have been obtained even in older men with much experience of homosexuality.

*

(l) HO 345/9: Memorandum by a Joint Committee Representing the Institute for the Study and Treatment of Delinquency and the Portman Clinic (I.S.T.D.), London W1.[59]

. . .

I. STATUS OF INFORMANTS.

The I.S.T.D. was founded in 1931 for the study and treatment of delinquency. Its first concern was to organise a Psychopathic Clinic for the examination and treatment of delinquents, which at the same time provided a service of Court Reports on recommended cases. In 1948 the Psychopathic Clinic was taken over by the

National Health Service under the title of the Portman Clinic, and the I.S.T.D. continued amongst other and educational functions the organisation of research into various problems of delinquency. Both in the Portman Clinic and in the I.S.T.D., a multi-disciplined approach has been followed involving the co-operation of psychiatrists, psychologists, organic physicians, social workers, sociologists and statisticians.[60]

. . .

II. PRELIMINARY COMMENTARY ON THE STANDPOINTS OF (A) THE LAW, (B) CLINICAL PSYCHOLOGY, (C) RELIGION AND MORALS, AND (D) PUBLIC PREJUDICE, ON THE PROBLEM OF HOMOSEXUALITY.

. . .

B. The Standpoint of Clinical Psychology.

To the psychiatrist the problem of homosexuality raises no question of criminality unless the sexual deviation is associated with acts of violence, assault or seduction of minors. It is regarded quite simply as one of a number of deviations from the biological aims of heterosexuality, for which three main factors are responsible: (a) constitutional or innate factors, (b) developmental factors operating in early childhood and again at puberty and (c) immediate or precipitating factors promoting sexual tension and encouraging a homosexual form of discharge. Both the psychiatrist and the social psychologist agree on the existence of constitutional factors. Man is constitutionally a bisexual animal. The main difference between these two groups of specialists is that whereas the former stresses individual developmental factors, both conscious and unconscious, leading to the organisation of a homosexual system, and regards many forms of homosexuality as forms of mental disorder (disorders of instinctual expression), the latter is under no obligation to recognise the 'pathological' nature of some forms of homosexuality and stresses mainly the individual and social factors which promote a homosexual organisation. The psychiatrist recognises in fact that, whilst for constitutional and developmental reasons homosexuality is in many instances a natural form of sexual deviation which cannot be described as a 'disease', in many other instances it is a sign of mental disorder obtaining mainly sexual expression and having the same dynamic significance as the psycho-neuroses and other classical forms of mental disorder. In certain instances both sexual and non-sexual forms of mental disorder may exist in conjunction. For these reasons the psychiatrist is, subject to the exceptions indicated above, unable to accept the existing legal assessment of homosexuality as a criminal manifestation. It is either a natural deviation or a mental disorder.

The psychiatrist further recognises that, contrary to both legal and popular opinion, sexuality does not originate at puberty but exists from birth ...

The psychiatrist recognises also that the sexual impulses of small children are bisexual in nature, i.e. can apply to objects of either sex. When the heterosexual elements, either for constitutional or developmental reasons, are inhibited and the impulses remain directed to an object of the same sex, a state of homosexuality exists. This may be either transient or become organised at any age from early childhood onwards. Small children of from three to five years frequently practise homosexuality of a simple type and it is notorious that during the later school periods from 9 to 18 homosexuality is rife in both sexes.

This developmental approach serves to correct the general impression that homosexuality is an isolated, adult manifestation of a perverse and criminal nature. Not only does it develop extensive ramifications from childhood onwards, but the effective sublimation of these impulses contributes greatly to the social cohesion and potential friendliness existing between persons of the same sex without which no society would remain stable. Should these sublimations be disturbed or fail to develop effectively, some form of mental disorder is likely to ensue; and, as has been indicated, this may take either non-sexual or sexual forms. It is from this latter group that the 'pathological' types of homosexuality are largely recruited.

Moreover, homosexuality varies from extremes of active masculine to passive feminine types (in which latter, constitutional factors play a more important role) and in fact 'pathological' types are more frequently encountered in the active group ...

Two of the simplest examples must suffice to illustrate the developmental (pathological) type of organised homosexuality. In one case the boy with a rather passive constitutional disposition is treated by his father as if he were a girl, unconsciously identifies with his mother and seeks for love-objects of an active masculine type to whom he reacts as if they were father-lovers. In the second case an otherwise active type identifies himself with his father's masculine role but being also mother-fixated is unable to form love attachments to women and seeks for objects of the same sex who seem to have feminine characteristics or attitudes.

...

C. The Standpoint of Religion and Morals

...

... To the priest and minister of religion all sexual licence and unchastity are sins, and homosexuality in all its forms is regarded by them as an immoral sexual activity. But over the years it has steadily been borne in on priests dealing with this particular form of sexual abnormality that there are many offending members of their congregations who are mentally sick, and that though what they do is still a sin it cannot be considered apart from their mental abnormality. And they note that this sinful and abnormal behaviour does not respond to penances.

Moreover, it has become increasingly recognised that, sexual behaviour apart, many confirmed homosexuals manifest the greatest moral and ethical integrity in their private and social lives. Others again, particularly those with strong religious convictions, continue to seek help and guidance for their abnormality but without relief. It is now a common practice amongst ministers of religion to refer many of these individuals to psychiatrists for treatment. Like the psychiatrist the priest forms his conclusions from intimate personal knowledge of and relationship to his parishioner; and he recognizes that there are certain homosexuals with genuine tendencies and predisposition towards their choice of a sexual object, whom it is beyond their power to change. But whilst the priest has developed this modified and tolerant attitude to these sinners no such tolerance would for a moment be entertained for homosexual acts that were criminal, such as offences against youth or any age involving antisocial acts that are the equivalent of rape.

...

D. The Standpoint of Public Opinion

In view of the opinion frequently expressed by legal authorities, viz., that the Law should give some expression to popular feeling regarding alleged criminal offences, it is essential to point out that, however much or little the general public may condemn homosexuality on religious or biological grounds, its attitude of condemnation is mainly influenced by what can only be described as profound emotional prejudices which are neither moral nor rational in nature. Accurate statistics on the subject are not available but roughly there are two main varieties of public prejudice. The first and more vociferous takes the form of angry disapproval together with an openly expressed desire for condign punishment of buggers, who are regarded as degraded and debauched persons. A second group is on the whole inclined, for equally irrational reasons, to condone homosexual practices; and between these extremes exists a more or less neutral group, indisposed to pay much attention to the subject except in the form of obscene wit for which, significantly enough, they evince considerable relish, and a genial contempt for the pansy.

In the popular imagination homosexuality is taken to be synonymous with buggery and most popular reactions of disgust are due to the anal and excretory associations of this practice. No doubt a sharp reaction would also exist to homosexual fellatio if the widespread existence of this practice were more generally known. Incidentally much more tolerance is exhibited, certainly towards fellatio and to a certain extent towards buggery when practised between heterosexual partners. And the general attitude of the male towards female homosexuality is still more tolerant in spite of the fact that, allowing for the absence of a penis, female homosexuals are just as uninhibited as male homosexuals, if not more so. That the commonest forms of male homosexuality are 'petting', exhibitionism and mutual masturbation does not prevent all practices that go by the name of homosexuality being regarded as equally heinous in the eyes of the more hostile prejudiced group. In short there is neither rhyme nor reason but a good deal of prejudice behind most actively expressed popular reactions.

The sources of this prejudice lie ... in the fact that strong unconscious and conscious reactions against any form of sexuality are established in the early childhood of every individual at a time when the forms of sexual impulse are 'polymorphous'. Specially strong reactions of disgust and guilt are built up against anal and sadistic sexual impulses, and, though to a lesser extent, to oral practices (fellatio). In some instances these reactions are later expressed in sharp disapproval of such conduct in others.

As has been pointed out, man is constitutionally a bisexual animal and in the early years of life the distinction between male and female function is not as fully developed as it is in later life. Every individual has to cope either consciously or unconsciously with bisexual problems and here again it is characteristic of persons who are not manifestly homosexual but whose own defences against homosexuality are not very securely established to disapprove strongly of homosexual practices in others.

One of the consequences of the existence of earlier phases of and defences against polymorphous sexuality (which in course of later childhood are usually forgotten) is that the adolescent enters adult life with the most naïve conceptions of the nature and function of sexuality, as witness the common superstition that heterosexual coitus is the only 'normal' form of adult sexuality—this in spite of the fact that heterosexual activity amongst adults includes a wide range of polymorphous practices mostly in the form of fore-pleasure which, if existing in isolation, would be regarded as perversions or deviations.

With all this prejudice against abnormal forms of sexual activity, it is remarkable that the public are not more concerned about what happens to homosexuals sentenced to terms of imprisonment, and what effect the homosexual environment of a prison has on them, and they on others. Dr. Mackwood informs us that during ten years experience of psychiatric treatment of convicted offenders of all kinds in Wormwood Scrubs Prison, including many homosexuals, in all the voluminous correspondence from the relatives of these offenders he never had a single enquiry as to the effect of homosexual prisoners on their relatives and friends in prison, nor of the effect of their homosexual relatives in prison on other prisoners. Senior Prison Medical Officers confirm this lack of interest in homosexuality in prisons.

In short it can be maintained with a good show of reason that many of the emotional reactions of the general public to homosexuality and in particular the reactions of the 'antagonistic' group are neither rational nor particularly moral in the adult sense of the latter term. To give undue weight to the existence of public prejudice would merely tend to perpetuate in the criminal code obscurantist and emotional attitudes which run counter to modern opinion, both humane and scientific. Enlightened opinion recognises that amongst a minority of the population homosexual practices are widespread and vary in degree of organisation, from transient erotic manifestations to a degree of attachment both physical and mental which, save for the difference in sexual object, cannot be distinguished from the corresponding manifestations of heterosexual love. It can also be shown that whereas in some instances homosexuality is a sexual deviation of a constitutional nature which is not answerable to treatment, in others the condition can be recognised as a disorder of the sexual instincts differing only in form from other mental disorders such as the neuroses. In neither instance can it be regarded as a crime.

Conclusions

From the general considerations set out above, the following conclusions can be formulated. Bisexuality and homosexuality are not crimes, although in certain cases they can be associated with criminal acts. They are remainders of or regressions to infantile forms of sexuality which deviate from adult biological heterosexual aims and objects. From the point of view of treatment, they can be roughly divided into two main groups: (a) cases in which the constitutional and developmental factors combine to form a fixed deviation which is usually refractory to treatment and (b) cases in which pathological factors in development give rise to symptomatic homosexual disorder of the adult sexual impulses, but which are to a varying extent amenable to treatment ...

The concern of the psychiatrist as psycho-therapeutist is primarily with the treatment of pathological types. He is also concerned to help fixed homosexuals who desire to control their deviated impulses to secure this control. Under existing laws which stigmatise and punish homosexual conduct as criminal, his task is frequently restricted to securing this control. As a forensic psychiatrist his concern is primarily with cases of homosexuality in which the condition is associated with criminal conduct of a pathological type—e.g. offences against public decency, sexual assault or violence, or seduction of adults or minors. In this way he is able to serve the social purposes of the community by recognising and attempting to reduce the anti-social and dangerous proclivities of some homosexuals. But he cannot assent to the proposition that homosexuality between consenting persons of the age of sexual discretion is per se criminal.

...

V. ON THE PROGNOSIS AND TREATMENT OF HOMOSEXUALITY

...

D. PROVISIONAL CONCLUSIONS.

... [I]n a so-far unascertained but certainly large proportion of cases, there is no answer to homosexuality save tolerance on the part of the intolerant anti-homosexual groups in the community ... The group of homosexual cases that is dealt with at present at a delinquency clinic is constituted of homosexuals detected by or reported to the police, or, in simpler terms, the group whose homosexual activities are so indiscreet or compulsive as to lead to exposure. Amongst these too a large proportion, possibly over one-half, are cases in which there is no answer to homosexuality and, in the sense of criminal conduct, no need for an answer. The cases in which psychological treatment is appropriate are ... e.g. cases in which conflict exists and a desire to be freed from the deviation, cases of pathological homosexuality, cases in which owing to age, seduction, temptation, and other factors, a person who might otherwise have developed in a heterosexual direction has become temporarily homosexual or has developed a homosexual organisation, and cases in which the homosexual urge leads to criminal conduct of a pathological type (violence, rape, seduction). The effect of treatment should be assessed exclusively on the results obtained with these 'appropriate' cases.

... [M]ost psychiatrists who are experienced in the handling of homosexual cases preserve an attitude of expectant reserve as to the outcome of their treatment

...

... [I]n most of the cases dealt with it is easy to establish the factor of inadequate or faulty sexual education and there is no question that appropriate educational measures are of the greatest service in promoting homosexual control. Should those measures be applied not merely to cases actually charged with homosexuality but during the early phases of sexual development from childhood to puberty, the possibility of reducing the number of adult pathological sexual deviations would be increased to a considerable extent. It is in fact premature to regard the limits and limitations of psychological, educational and social measures as fixed. The results already obtained by those means should, on the contrary, provide an incentive to renewed and expanded effort. The application of penal law to sexual deviations (save in such cases as offend against the laws governing violent conduct, seduction, etc.) is in fact a retrogressive policy running clean against the whole trend of modern scientific thought and experience.

...

*

(m) HO 345/8: Appendix: Statistical Analysis of 113 Homosexual Offenders Discharged from the Portman Clinic (I.S.T.D.) During the Two Years 1952–3[61]

...

VIII. Summary and Conclusions.

1. An analysis has been made of the case histories and treatment records of 113 cases who were referred to the Portman Clinic for homosexuality and discharged in 1952 and 1953.

    ...

3. The last conviction was for importuning in 45% of the cases, for gross indecency in 20% and indecent assault in 27%. 11% had not appeared before a Court for homosexual offences when referred ...

4. When rated for degree of homosexual interest and behaviour, heterosexuality was more dominant in 25% of the cases, and homosexuality was more dominant in 40%; 35% were equally attracted to both sexes. Only 15% of the adults had been exclusively homosexual since adolescence ...

    ...

6. 23% of the group were neurotic and 8% were psychopathic; there were 2 cases of psychosis and 3 of organic deterioration. 54% had no marked mental disorder or defect.

7. 5 of the juveniles were having homosexual relations with other juveniles, 4 with adults and 1 with both. 70% of the adults currently had had adult

partners, 27% juveniles and 3% both. Only 2 had changed from juvenile to adult partners.

8. The prognosis was good in 31% of the cases, fair in 32%, doubtful in 23% and poor in 14%. Prognosis tended to be worse the greater the degree of homosexual interest and activity, and also worse where the aim was anal intercourse

...

9. 81 cases were treated at the Portman Clinic for varying lengths of time. The chief method was psycho-therapy, hormone treatment being used in 7 cases. At the end of the treatment 36 no longer had homosexual impulses and 21 who still had homosexual urges had achieved discretion or conscious control. 6 were definitely unchanged. None of those who had been exclusively homosexual since adolescence had lost their homosexual impulses, compared with 51% of the bisexuals; 3 of the unchanged were exclusively homosexual.

...

10. The follow-up of cases about whom information could be obtained revealed that 19 were known to have made a satisfactory adjustment and 34 were presumed to have done so. 19 cases were convicted again after referral, 12 for homosexual offences. The lack of recent contact in a large number of cases, and of the relevant information in others, means that it is not known whether those patients who had lost their homosexual impulse at the end of treatment had any recurrence of homosexual interest or activity.

*

(n) HO 345/9: Memorandum from Drs. Curran and Whitby[62]

...

INTRODUCTION.

This is not a full dress memorandum, but notes and points ...

The memoranda are referred to as M and the transcripts referred to as T with the appropriate numbers following.

1.　Aspects and viewpoints.

Different approaches emphasise different aspects. The viewpoints must be clearly distinguishable and the most [*sic*] in practice are:

A. Biological or medical (one question being what is disease?)
B. Moral (one question being what is sin?) and
C. Social (one question being what is crime?)

...

Definition.

The concepts of "unnatural" and "sinful" are neither useful nor applicable in the definition of biological phenomena. Again, what social consequences may follow from the presence of a condition should be separated from its definition.

A Note on the "Unnatural".

"In all the criminal law, there is practically no other behaviour which is forbidden on the grounds that nature may be offended, and that nature must be protected from such an offence. This is the unique aspect of our sex codes." (Kinsey 1949[63]).

Cultural Variations.

Convincing evidence is available that what is socially acceptable or ethically permissible has varied and still varies enormously in different cultures—a point well documented in the appendix of the British Psychological Society memorandum.

2.   Definitions.

. . .

Homosexuality

The condition of homosexuality must be distinguished from homosexual acts.

   The condition of homosexuality refers to the direction or preference of sex object; homosexual acts refer to activity motivated by this direction or preference. (Homosexual acts may not be homosexual offences, e.g. masturbation with homosexual phantasies, by far the commonest specifically homosexual act, is not an offence).

   The condition of homosexuality does not always lead to overt homosexual behaviour or to homosexual offences. The results of homosexual preferences may be socially useful or expressed in socially acceptable behaviour—the classical example being the well sublimated schoolmaster.

   As in other fields of behaviour, the subject may not be aware of his motivation; or may reject and then project these motivations—an explanation several times given for special repugnance to homosexuality. (M.42, M.76, M.90[64]).

. . .

   Homosexual acts are not always performed by individuals who are primarily homosexuals as is clearly seen in situational homosexuality and in adolescence and also in certain primitive types who want sex and are indifferent as to whether the partner is male or female.

. . .

3.   Classifications.

Wide variations in subject, object, accompanying disorders, cause or books and prognosis have led to numerous qualifications.

   For descriptive purposes the Kinsey Scale has the advantages of (a) objectivity, (b) that it does not import moral judgment, (e.g. "perverts" as opposed to "inverts" or "true inverts"), (c) nor does it assume any special theory of aetiology (e.g. "congenital" or "constitutional" as opposed to "acquired").

Illustrative Examples.

In line with the above the indulgence in homosexual practices is not necessarily highly correlated with the place an individual may occupy on the Kinsey Scale. Moreover, the "Kinsey rating" of an individual can and does vary a good deal with age; and other factors (including Providence and Psychotherapy) play their part too. Thus many pass through a transitional homosexual phase in development but lead satisfactory heterosexual lives in maturity. Others with no obvious homosexual tendencies may indulge in homosexual practices when placed in special circumstances that prohibit contact with the opposite sex but will return to heterosexuality thereafter—thereby illustrating the often temporary effect of opportunity, influence or seduction. Conversely, the fact of being married and a father (as was Oscar Wilde) does not prevent a person from having homosexual preferences whether he indulges them or not. Similarly the refusal to indulge in such practices is no guarantee of a heterosexual "constitution"; a number of predominantly homosexual individuals refrain from fear of legal and social consequences or because of their ethical standards.

4. Origins.

...

2. Physical pathology.

No <u>physical</u> pathology for the <u>condition</u> of homosexuality has been demonstrated. Biochemical and endocrine studies have so far been essentially negative (M.95)[65]; and anthropomorphic investigations of body build and the like are as yet quite inconclusive (T.40).[66] Kalmann's [*sic*] twin studies, on the other hand, if they can be accepted, do however suggest a high degree of genetic disposition—and should be given somewhere in extenso.

...

Symptomatic Homosexuality.

All the medical witnesses were unanimous in holding that homosexual acts and homosexual offences could be symptomatic of other definite diseases, e.g. schizophrenia or senile deterioration or decay (it must also be remembered that heterosexual offences can also be symptomatic of disease in the same way). It is proper for medicine to stress this possibility; but little evidence was adduced that the association with definite mental disease was at all common ...

...

3. Psychopathology.

In the absence of a physical pathology, a number of psychiatrists have tried to justify the view that homosexuality is a disease on psychopathological grounds,

(M.76, P.36[67]). But the psychopathological factors adduced also occur in the heterosexual (M.36, T.37, T.34[68]), possibly however with less frequency.

Several medical witnesses maintain that in addition to the homosexuality (whether as a condition or in acts) other evidence of disease occurred in the form of other psychological disturbances or other evidence of "personality damage", (M.90, M.76, M.42, T.35, T.36[69]). But psychiatrists tend to see just such cases (T.35, M.57[70]). Nor is evidence of other psychological disturbances or personality damage obvious in all cases (M.90).

It has also been argued that the psychological disturbances adduced are secondary to the strain and conflict imposed by homosexual preferences rather than concomitant or causal and that they are far less prominent or even absent in societies and cultures in which homosexuality is less taboo. (Personal communications from psychiatrists in Latin and Eastern countries and also from Kinsey).

4.   Homosexuality and "Immaturity".

A common variant or amplification of the psychopathological theme is seen in those psychiatrists who regard homosexuality as a condition of arrested development (or near disease) resulting in "immaturity". (M.57, M.42, M.76).

Any theory of psychosexual development that regards homosexuality as an invariable phase in the course of development makes the immaturity concept almost a diagnosis by definition; evidence of immaturity other than the homosexuality (and by definition homosexual acts are immature acts) is often not apparent or may be quite absent—except in the very important group of transitional homosexuality in adolescence, which often seems little more than experimental mucking about.

. . .

B. Natural deviation.

Homosexuality as a "Natural Deviation".

This possibility has also its advocates, e.g. "it is either a natural deviation or a mental disorder" (M.54,[71] Professor Penrose, M.95[72]), i.e. in different cases.

A common view is that the sexual urge is at first undifferentiated. A homosexual orientation results from the variable importance of (a) genetic predisposition, and (b) external environmental factors in the development of the individual. The Freudian view that everybody passes inevitably (M.42) through a homosexual phase in development, or that there is anything like a rigid or fixed pattern of psychosexual development of this kind is disputed by Kinsey (1949).[73]

The occurrence of homosexuality in all known cultures is also adduced in favour of homosexuality being a biological potentiality of universal incidence.

. . .

D. Practical issues.

. . .

(1) Age of Fixation of Sexual Pattern.

In assessing the relative importance of nature and nurture the emphasis was laid heavily by most medical witnesses upon the latter (M.57, M.76)—although some degree of "predisposition", i.e. varying potentialities, was always ultimately invoked (T.34).

The interesting practical point on which all medical witnesses were unanimous was, that however it may have come about, the main sexual pattern is laid down in the early years of life, and the majority held it was more or less fixed in main outline by the age of sixteen many holding even earlier. (M.90, T.54,[74] M.57).

Kinsey is quite specific on this point: "The data we have already published on social levels show that by fourteen perhaps as many as 85% of all boys have acquired the patterns of sexual behaviour which will characterise them as adults, and something like nine out of ten do not materially modify their basic pattern after sixteen years of age." (Kinsey 1949, and M.93[75]).

No medical witness differed substantially from this view which, if correct, has clearly an important bearing on any "age of consent", for a decreasing minority will be permanently affected by homosexual contacts or experiences as the years pass (M.93) ...

(2) Seduction.

Medical opinion was unanimous that there was no single simple cause for either the condition of homosexuality or for homosexual acts or behaviour (just as there is no single simple cause for the condition of heterosexuality or heterosexual behaviour). In brief many factors must be taken into account.

The medical witnesses were unanimous in holding that the effect of seduction in the production of homosexuality had been greatly exaggerated; held that in general seduction had little effect; and pointed out that the younger partner was quite often the seducer (M.42, M.57, M.86,[76] T.35, and T.36).

Seduction might, however, produce general emotional damage (M.92, T.36 and T.54) rather than special damage in the production of homosexual deviation; but the occurrence of the latter in a minority of cases could not be gain said, (M.76 and T.36).

Medical opinion was, therefore, in strong contradiction to the commonly held lay and legal opinion of the devastating and all important effects of isolated acts of seduction in the production of homosexuality. Some stressed that court procedure could be harmful (M.91[77] and M.95).

It may be considered that the seduction theory is yet another example of the desire to find and the emotional satisfaction in finding a single simple cause that can be blamed.

Parents it would seem can be genuinely reassured that in the vast majority of cases isolated acts of seduction do little harm.

On the other hand repeated acts of seduction skilfully managed over a fairly prolonged period at a susceptible age and in a predisposed individual may have a profound effect in tipping the scale towards homosexuality in a small minority. This is especially the case if carried out by a member of the family.

Kinsey (M.93) stresses the great importance of a psychosexual arousal of this kind occurring with the first orgasm.

5.   Manifestations.

(a) Latent homosexuality of some degree is claimed to be found with great frequency in the course of analysis or other psychiatric treatment—some say universally (M.36, M.58,[78] M.42, M.90 and T.54).

...

... Latency may be apparent in poor relations with wife or completely unsuccessful love affairs, or neurosis of various kinds (M.36), or inferable in psychopathic manifestation, e.g. the well-known case of Lord Castlereigh [*sic*] who developed a paranoid melancholia in which he believed he would be caught and imprisoned for homosexual offences and which led to his suicide.[79] We have recently had personal experience of a case of melancholia with the symptomatic obsessional thought of a penis in a man's mouth and many psychiatrists would infer from this a special interest of fellatio which, however, the patient completely denied.

(b) Overt ...

Mutual masturbation is the most common activity followed by fellatio (but according to Kinsey this is as common as mutual masturbation—not I think the general experience in this country), followed by sodomy; but most practising homosexuals have indulged at some time or other in all three both actively and passively.

...

Practical Issues and Paedophiliacs.

Only a very small percentage of homosexuals—according to Kinsey 1% to 2%—are paedophiliacs, although these bulk very large in criminal practice.

...

Medical opinion was unanimous and was backed up by the experience of prison medical officers. Those who are attracted by pre-pubertal age groups and post-pubertal age groups falling to different populations with very little overlap.

...

Rake's Progress.[80]

It has been suggested that an alteration in the law might result in a Rake's Progress so that individuals become insatiated with adults, run down the scale of their loved object, and will end by seducing little boys. Medical opinion was unanimous including prison medical officers, and this Rake's Progress does not in fact occur ... except with extreme rarity ...

...

The Distinction between Sodomy and other Offences.

Medical witnesses (with one possible exception, namely Dr. Matheson,[81] with whom the Director of Prison Medical Services did not agree) no medical witnesses saw any justification.

As regards the prognosis of sodomy, opinion differed, some (T.41,[82] T.54) holding the prognosis was worse and others (T.35) that it was not worse. Nor was there agreement as to whether or not sodomy was associated with more anti social tendencies or more aggressive personalities (T.39,[83] T.41). Kinsey (M.93) held it more popular amongst lower social classes.

...

Lavatory offences.

We think we should point out that the Institute of Psychiatry was in favour of warning more often. We should also mention the question of the police being agent provocateur. The shocking story of our police witness who lured a man into Marlborough Police Station before arresting him and did not seem to realise the implications of this behaviour should, I think, be quoted.[84]

The position in Scotland with its one case of importuning last year is convincing evidence I think of the effect of legal procedures, with special reference to the Scottish need for corroboration and their different methods of taking statements.[85]

C. Lesbianism.

Lesbianism is uncommon as a medical problem according to most medical witnesses, but opinion is greatly varied as to its frequency of occurrence. The Tavistock Clinic seemed to be the only medical body who considered lesbianism as common as homosexuality between males. No medical witness was in favour of changing the law on lesbianism.

...

6. Incidence and alleged increase.

...

Incidence and the general population.

According to Kinsey it will be remembered about 4% of the white males are exclusively homosexuals all their lives and over a third of all males admitted at least some adult homosexual experience. Further, 10% of males were more or less exclusively homosexual in their outlets for at least three years consecutively (the above figures apply to adults).[86]

Incidence in Great Britain.

The only figures given to the Committee in a "normal" population were those provided by Miss Davidson (T.48[87]) who stated that of a hundred undergraduates who were not patients, five still had homosexual whims and fantasies, whereas 30 had them at an early period. Those 25 had shown a change in pattern since the age of 16 or 17 and the age of 21 plus.

. . .

Difficulties in estimating.

There are two main difficulties in ascertainment (i) what are the criteria for homosexuality—how wide should the net be cast? Thus, according to the psychoanalysts (T.54), a homosexual component is present in everybody and in this sense homosexuality is universal. (ii) The natural reluctance of individuals to admit to a component or preference that is socially condemned or to acts that are illegal and liable to a very heavy penalty . . .

General Figures.

Havelock Ellis (if I understood him correctly) estimated that in the professional and middle classes about 5% were predominantly homosexual in their orientation, i.e. presumably Kinsey rating 4 or more.[88]

The Kinsey figures are given above in the report. Some medical witnesses (T.40) considered them possibly applicable in Great Britain, others (T.51[89]) regarded them as too high for this country.

Incidence and general practice.

As a problem this seems surprisingly low. A perceptive and experienced general practitioner in Chelsea told us he had only met one homosexual as a problem in an adult and Dr. Whitby's experience in general practice is much the same, but both knew of homosexuality in their patients as an occurrence, but not as a problem.

Incidence in Psychiatric Practice.

This is surprisingly low—about 10% of the 800 new patients seen annually at the Tavistock Clinic—and about the same percentage of the 100 new patients seen annually at the Davidson Clinic (T.36).

. . .

. . . The above figures suggest that Kinsey's findings, however correct they may be, for North American males (white) may not be correct for Great Britain, but even with a figure as low as one or two percent with a Kinsey rating of three upwards the number of homosexuals in the United Kingdom is very large indeed, and the problem is certainly not "negligible."

. . .

Fallacies in figures.

Selected groups are seen both by the law and medicine. Psychiatrists see homosexuals with a high proportion of psychiatric abnormality (M.57). The law probably sees an undue proportion of careless and indiscreet homosexual types and a totally misleading proportion, taking homosexuality as a whole, of paederasts, but incidence of homosexuality cannot reasonably be best seen by the doctors or the courts. Many homosexuals are well adjusted and never reach either. A further fallacy lies with fluctuations in police activities both in time and place, which are apt to lead to erroneous conclusions. This is well exemplified in the New York police figures (M.93).

Alleged Increase in Homosexuality.

More widespread recognition is not evidence of any real increase. The increased conviction rate can be a reflection of increased police activity, a striking example being quoted from New York (M.93).

Only a very small fraction of homosexual offences are ever found out. Such evidence as we have from other countries gives no evidence of either increase or decrease whether changes in the law have occurred or not.

7. Effect of law.

The Effect of the Law on Opening the Flood Gates.

In general, medical opinion was that the legal position did not prevent the development of homosexuality (M.42), was not a great deterrent (T.34, M.40[90]), although some adolescents might be deterred (T.40).

. . .

Deterrence. If, as Romilly and others have maintained, the deterrent value of the law derives from the probability of conviction rather than from the severity of the penalty inflicted if convicted,[91] and if it is correct, that only a very small fraction of homosexual offenders are convicted, the deterrent effect of the law must be small.

The "Flood Gate" argument of what would happen with any relaxation of the law does not, on the information available, seem to be supported by experience in other countries.

It may be considered that to uphold strongly the "flood gate" argument suggests a high homosexual potential in the upholder of the argument; for most normally heterosexual individuals have little temptation to homosexual acts.

8. Treatment.

A. Natural history of the homosexual

According to Kinsey the percentages on his scale are briefly as follows:–

0–	50%
1–3	34%
4–5	12%
6–	4%

and in his experience those rated in 5 and 6 almost never develop a heterosexual pattern (exclusive) after some years of experience and very few rated 4 do so either; the possibilities of change seem to depend upon (i) the years of experience and (ii) the exclusiveness of homosexual interests or practices. Kinsey stated he had never seen an individual rated 5 or 6 developing an exclusive heterosexual pattern as the result of penal or psychotherapeutic efforts.

. . .

Treatment.

As regards the medical treatment of homosexuality, the possibility of medical help is more important than an academic discussion as to whether the condition—or the acts—should be called a disease.

It is not for the psychiatrist in his professional capacity to decide what form society should take but to try to help people in a better adaptation to society as it is. Even if the homosexual orientation is not changed, nor the number of homosexual acts, it is something to be able to help individuals to be happier in themselves, more efficient at their work or to be more discreet in their sexual activities.

B. The Objects of Treatment can be arbitrarily divided under four main headings—(i) change of direction in the sex urge, (ii) greater continence, (iii) greater discretion and (iv) a better adaptation to life in general. The question arises as to whether it is always desirable to try to treat the well-adjusted homosexual.

C. Methods.

Psychiatric Treatment does not consist of either pills and potions, i.e. physical methods of treatment, or psychotherapy, but consists of a mixture of physical, psychological, social and environmental measures in varying proportions according to the case.

. . .

Physical Treatments: Oestrogens.

Castration would improbably be tolerated; and it is not successful (T.35).

Oestrogens do not affect the direction of the sexual impulse but reduce sexual desire or libido.

All the medical witnesses who have had personal experience considered oestrogens had a place in treatment although they varied somewhat as to its importance. They did not consider it dangerous or had really harmful side effects and they were in favour of allowing its use—with suitable safeguards including the patient's consent—in English prisons, (it is now used in Scottish prisons). (M. 95, T.51, M.84,[92] T.39, M.90, T.54, M.57, T.35).

. . .

D. Results.

The Results of Treatment.

In trying to assess this the same problems arise as in ascertainment, with special reference to the age of the patient and to his "Kinsey rating". The best results in medicine are always obtained in those who don't need treatment.

...

Change in Sexual Orientation.

It can, of course, always be argued that when this occurs the patient was not 100% homosexual or a Kinsey 5 or 6.

It is a striking fact that none of the medical witnesses were able to provide any reference in the literature to cure by psychotherapy what might be called "established" cases of homosexuality; but both the Psychoanalysts and the Tavistock Clinic have now sent in a few examples.

The great difficulty in assessing results is well illustrated by an elderly married man with children who sought advice on his wife's instigation for impotence. It transpired that he had apparently always been a "Kinsey 6", had only married in the hope of cure and had only achieved intercourse with the aid of homosexual fantasies—an interesting example of socially successful masturbation per vaginam. In almost any 30 year follow-up without personal interview and willingness to be frank, this would have seemed a striking example of change in sexual orientation or of the success of treatment.

Psychotherapy.

There is general agreement that some cases can be greatly helped, disagreement as to how many, but agreement in general that the proportion is small ... From the Institute of Psychiatry approximately 40% were classed as improved and much improved.

E. Practical difficulties.

The Criteria for Psychotherapy were usually given as (i) relative youth, (ii) a fair intelligence and (iii) a genuine desire to co-operate (M.76, T.36, M.57, T.35, M.90, T.54, M.58, T.40).

...

According to the Prison Commissioners 90% of the cases in prison were unsuitable for psychotherapy, failing in one or all of the criteria mentioned above.

...

Psychotherapy, Time and Personnel Requirements.

According to the psycho-analysts (T.34) owing to the time and cost involved—the time running into hours weekly over years, psycho-analysis is not a very practical proposition except for very few. According to the I.S.T.D., however, (T.54) no clear relationship existed between the time expended and the results obtained.

F. Place of prison in treatment.

The Value of Prison in Treatment. Only one doctor denied that it ever could have any value (M.15[93]). Other doctors considered most cases unsuitable for treatment in prison (M.90, T.39 and T.41). It was plain that no sexual reorientation occurred (M.93), or that cases were made worse ceasing to care what happened (T.39), or that their "neurosis" became aggravated (T.35). On the other hand the prison doctors considered that prison could be helpful because of the healthy tone that could be created, and because successful psychotherapy could be and was being carried out (M.86). Also because the recidivism rate was low (M.86 and M.93) because a prison sentence pulled some up sharply and beneficially, making them re-assess their conduct (T.54) and that they could be even grateful for this (M.86). Better control could also be taught (T.35) and a prison sentence could help to start after care and after treatment (T.40).

...

*

iii. Scientists

(a) HO 345/8: Memorandum submitted by the Institute of Biology[94]

The Council of the Institute of Biology supports the general tenor of the memorandum submitted by Professor C. D. Darlington, Sir Ronald Fisher and Dr. Julian Huxley ... [T]wo sets of factors, inheritable and environmental, are involved in the development of homosexual behaviour: while the inheritable factors cannot easily be changed, the environmental factors are more subject to control.

A detailed survey of patterns of sexual behaviour in mammals as well as in human societies has been made by C. S. Ford, an anthropologist, and F. A. Beach, a psychologist and animal behaviourist, both of professorial standing, both of Yale University, in their book: Patterns of Sexual Behaviour (1952). Professor F. A. E. Crew, F.R.S.,[95] supplied a preface to the English edition. This book summarises work by recognised authorities on the sexual behaviour and physiology of man, apes, monkeys and lower mammals, over the last thirty years, together with studies of human societies by reputable anthropologists. It is clear from this survey that the behaviour of no single human society can be regarded as typical of the human race as a whole; it is also clear that there are certain common elements in the sexual behaviour of all mammals—including man.

In most mammals, the activities of courtship and mating are closely linked with reproduction. In all human societies, however, sexual relations serve a variety of non-reproductive functions. The difference between the lower mammals and man with respect to sexual behaviour is not, however, as great as has been supposed; those mammals closest to man in evolutionary status also resemble him most closely in sexual behaviour ...

Homosexuality occurs not only in infra-human primates but in lower mammals as well, though it is more marked in primates. The mounting of one female by another is common in cats, dogs, sheep, cattle, horses, pigs, rabbits, guinea-pigs, hamsters, rats and mice. Feminine and masculine reactions may occur in the same female in rapid succession, and such inversion of mating behaviour is not uncommon. In apes and monkeys, masculine homosexuality is not solely a substitute for heterosexual coitus; some adult monkeys maintain homosexual alliances concurrently with heterosexual activities.

Ford and Beach's survey of human societies is confined mainly to existing, pre-literate, communities. Their data, which form a valuable supplement to the evidence of the Kinsey reports, indicate that while homosexual behaviour is nowhere the predominant type of adult activity, it occurs in nearly all human societies of the present day and is generally more common in men than women. It seems probable that all men and women have an inherited capacity for erotic responsiveness to a wide range of stimuli (that is, to a wide range of partners of different age, sex and even species), and that all societies enforce some modification of this generalised capacity, so that the type of behaviour preferred by any one society exerts a normalizing influence on children born into that society.

The evidence available shows that man closely resembles other mammals in the general pattern of his sexual behaviour. Nevertheless, with the development of the brain, learning plays in human sexual patterns, and to a lesser extent in those of apes, a more important role than it does in those of the lower mammals. The fact that a majority of the population, in any given culture, does not overtly display the complete range of possible types of sexual behaviour common to mammals and to human societies, but only those which are socially acceptable in that society, is probably due to learning and to the exercise of moral control.

... Homosexuality is often regarded as a pathological or abnormal tendency to be extirpated; an alternative attitude would be to accept a certain incidence of homosexual activity in a community as normal and to recognise that the actual incidence depends on genetical, environmental and, in man, educational factors that are as yet almost undefined. Without prejudging the results of the inquiry that we advocate, it may be legitimate to suggest imprisonment, often in the company of other homosexuals, may not be an effective treatment even if the community decides to treat overt homosexuality as a criminal matter.

*

(b) HO 345/8: Memorandum Submitted by Professor C. D. Darlington, Sir Ronald Fisher and Dr. Julian Huxley[96]

I General

1 Homosexuality occurs in parallel circumstances, in all gradations and probably in similar frequencies in the two sexes. In both sexes and in both married and unmarried persons the frequency is higher than is usually believed since only its exceptional public manifestation attracts notice.[97] It has occurred in all civilisations and in all races of men whether it has been socially approved or disapproved.

2 The documentary and literary evidence of homosexual behaviour in notable men and women is necessarily confined to past times and is necessarily therefore circumstantial but it is very extensive. And it agrees with what we expect from modern scientific studies. It indicates that among homosexuals there are individuals with characteristic gifts as well as characteristic defects. Such individuals have made varied and often important social contributions in art, science, politics and war.

3 The harm done to society by homosexual acts (as by heterosexual acts) arises from exceptional manifestations, <u>viz</u>. seduction, rape, incest, public indecency, soliciting, prostitution and conflict with the duties of marriage and parenthood.

4 Persecution of homosexuals has occurred intermittently throughout history, partly because these anti-social manifestations attract attention and perhaps partly because even in Christian countries persecution of minorities gives emotional satisfaction to the majority. It has applied to male rather than to the less obvious female homosexuals. In certain societies at certain periods male homosexuality has however been openly approved.

II Genetic and Environmental Components

1 Homosexuality (like heterosexuality) is immensely diverse in its manifestations. It exists in all gradations of intensity and durability. It is not sufficiently or properly describable as abnormal or unhealthy or immature or intersexual since no individual is perfectly normal or healthy or mature ...

...

3 In regard to the reaction of heredity and environment, a helpful analogy with homosexuality is found in left-handedness. A disposition to left-handedness occurs in various degrees in a small minority of children. This disposition is genetically determined in the sense that it is characteristic of the individual and cannot be changed without external compulsion and greater or less danger to the mental character of the individual. We used to make the left-handed child conform. Now we have learnt that we must tolerate his abnormality. The effects of homosexuality are more complex and more serious but some of its causes are clearly of the same kind.

4 Certain social conditions can undoubtedly encourage homosexual behaviour in apparently normal people in whom it would otherwise be suppressed. In many kinds of public institution (e.g. public schools, colleges, prisons) the use of segregation and the fear of conception and disease as the means of preventing intercourse between the sexes has in the past (to a slight but deplorable extent) encouraged homosexual intercourse. Education should aim at reforming customs which encourage homosexuality in adolescents and adults, especially the deprivation of the company of the opposite sex.

5 Extreme social discouragement, on the other hand, has never succeeded in eliminating homosexual behaviour in extreme cases. The royal examples are most fully documented. Parental, political and dynastic pressure failed to discourage the symptoms of homosexuality in five Kings of England, in Henry

III of France or Frederick the Great of Prussia, most of whom lost their lives on account of it.[98] These individuals were moreover unique in their own families.

6 Again the experience of begetting, bearing and bringing up children has been found in men and also in women to have no effect on a strong homosexual disposition.[99] The present law sometimes prompts homosexuals to marry in order to divert public suspicion from their activities and thereby ruins the life of an innocent partner.[100] Incidentally such marriages may also serve to propagate the genetic disposition for homosexuality in the progeny.

7 Such evidence leads us to suppose that the genetic as opposed to the environmental component of homosexual behaviour is sometimes very high. This view is confirmed by the evidence of Galton's method of twin study.[101] As the following table shows, one-egg or identical twins behave in the same way; two-egg or fraternal twins show a merely family resemblance as regards homosexual behaviour.

Table: Difference in frequency of homosexual behaviour of Co-twins of like sex of 71 male homosexuals according to whether they are One-egg or Two-egg Twins; (after F. J. Callmann [*sic*], Amer. J. of Human Genetics, 4, p. 142).[102]

Type of Twin	Degree of Homosexual Behaviour				
	High	Medium	Low	Nil	Total
One-egg	28	9	–	–	40 +
Two-egg	1	2	8	15	31 e

+   3 unclassified;   e   5 unclassified and 14 females in addition.

The contrast between the two groups shows how high the genetic component in sexual behaviour can be. But the two-egg group alone shows something more. Here there are 15 pairs of individuals who stand at the opposite extreme of behaviour. Thus the example of each twin has had no effect whatever in leading or misleading his twin brother.

8   We have to conclude that society has no more hope of compelling the healthy individual of an extreme homosexual disposition to renounce homosexual life than it has of compelling the healthy heterosexual individual to renounce heterosexual life; lifelong seclusion from the object of sexual attraction is the only means of prevention in either case.

III Recommendations

1 Homosexual acts between consenting adults committed in private should not be classed as criminal offences.

2 Homosexuality of either sex which involves such offences as rape, incest, public indecency, prostitution, soliciting, etc., and particularly offences against minors or seduction of minors should be punishable by law on the principles applied to heterosexual offences, the two groups of offence being legally assimilated.

3 Homosexual acts committed by either sex should be recognised as a breach of the obligations of marriage and a ground of divorce on a similar footing to adultery.

These changes would have the following effects:

(i) They would remove persecution from individuals in other respects useful and sometimes outstandingly valuable members of society.
(ii) They would equalise the legal position of the two sexes by removing the stigma of criminality from male as contrasted with female homosexuality.
(iii) They would stop blackmail and other corrupt practices associated with the present enforcement of the law.
(iv) They would remove the present inducement to marriage for homosexuals with its deplorable effects for the partner and the offspring.
(v) They would make possible the scientific investigation of the problem of variation in sexual behaviour.

All of these changes seem to be desirable in an enlightened society.

*

(c) HO 345/9: Notes of a meeting with Dr. Alfred C. Kinsey at 53 Drayton Gardens, SW10, Saturday, 29 October 1955[103]

1 ...

(DR. KINSEY): I think the difficulty in estimating the levels of sex crime depends upon the fact that under American law at least it is only a minute fraction of one per cent of the illicit behaviour that is ever apprehended and charged before a Court and brought to conviction. Our laws are of such a nature that they call criminal a great deal of sexual behaviour which is common in the total population ...

The question concerns the difficulty in estimating whether there are increases or decreases in sex crime ... I feel very sure that your incidence of a thing like homosexual activity in this country cannot be too radically different from what it is in the United States and again it would be a minute fraction of one per cent of such conduct ever apprehended.

2 ...

It fluctuates tremendously, depending upon police activity. In one year in New York City there were four thousand arrests on a misdemeanour level and in the next year 137 arrests. There is no indication whatsoever that that represented any change in the behaviour of the population. It was a new Police Commissioner who had a different policy and was putting the whole of his police force into holding down other types of crime.

3   Q. [CURRAN]. One of the arguments put forward to us against any relaxation of the law is what you might call the "floodgate" argument that once you change

it you would increase activity ... A. ... [I]n New York City the law was reduced so that there is no penalty for homosexual relations between consenting adults in private, only five years ago.[104] There is no one who has suggested that there has been any modification in the actual behaviour of the total population. Sweden and Denmark dropped their penalties for adult relations some years further back. Once again, no one has suggested that dropping the penalty has modified the behaviour in those countries.

4   Q. Another argument put forward to us which we have often heard is that what the Committee has often called the "Rake's Progress" argument that people will slide down the scale, or might do so, from preferring adults to preferring children ... A. It would be a very small percentage of persons with a homosexual history who are interested in contacts with minors ...

I have carefully examined the histories of over six thousand males who have homosexual histories and many hundreds of females, and off-hand I can recall very few instances, perhaps three or four or half a dozen of males who had once had a preference for older males and became interested in younger males. I have seen many instances of males who had once upon a time a preference for younger males but gradually built up their preference for older males and dropped that activity.

...

7.   DR. CURRAN: ... In your experience, and I think this may be of some importance in connection with the assessment of therapeutic claims, how much do people move up and down the Kinsey scale ...?

...

A.   ... An individual who has become a five or six almost never develops any kind of heterosexual pattern after he has had some years of experience. If they are ones or twos there is still considerable choice, or even some of the threes may be redirected by a clinician still one way or the other: so it depends upon the number of years of experience and the exclusiveness of their homosexual experience. There are only fifty per cent of the males who have exclusively heterosexual histories, only four per cent who have exclusively homosexual histories, so the forty-six per cent of males have some combination of homosexual and heterosexual history. Three-quarters of those would rate as ones, twos or threes, so there is a very considerable percentage of those with some experience who may be redirected by the clinician into more exclusively heterosexual patterns.

...

13.   ... We have seen persons who have never had homosexual experience until they are in their mid-fifties or even later and who become exclusively homosexual at that late age; that is one reason why we object to the notion that this is an innate matter which predetermines an individual for a lifetime.

14.   ... [W]e find that for most males the general orientation is pretty well fixed by sixteen years of age. We can predict the general pattern of behaviour for more

than 90 per cent of all our males if you give us just that portion of the history which has been completed at sixteen years of age ...

...

17. ... [W]e have never seen a person who is a five or six, and relatively few persons who are fours, who have been affected by penal punishment or clinical treatment to further development of an exclusively or primarily heterosexual pattern. What we have seen accomplished by clinical service, but not by mere imprisonment, is a change of mode of behaviour so that they no longer come in conflict with public reaction.

...

23. ... Juveniles who grow up in average homes in our social organisation ultimately end up with about a third of the population with some homosexual history.

Juveniles who go through State institutions, live in institutions at the age at which they turn adolescent, and particularly in their middle and later teens, have about some 70 to 80 per cent ending up with homosexual histories. I think there is no doubt that is primarily a product of institutional life. There may be some selection in the persons who get into such institutions, but it is inevitable when they are at the peak of their sexual activity and are shut up in an exclusively male community, whether that be a penal institution or the best boarding school or the Foreign Legion, they are going to have to find sexual outlet with other men.

24 ... Certainly your English public school pattern demonstrates that a great many of them who have some such experience in these early ages do not turn out primarily homosexual. The vast majority of them do not. It depends upon the exclusiveness of the pattern built up and how many years it is maintained.

25 DR. CURRAN: A thing which has concerned us, naturally enough, is the question of the harm done by relatively isolated acts of seduction on minors or juveniles; firstly, what harm does it do, and secondly its importance in connection with the reassurance that can often be given to parents about the harm done to their children?—A. This is one point at which our findings are quite in accord with Freudian theory; the nature of the first experience is of exceeding importance. The first experience in sexual contact, the point of satisfaction for the boy or girl, begins to condition the individual and such early experience will have more force if it is unsatisfactory, or if it is satisfactory it will shape the later patterns very often more than later experience will ...

26 ... If it is the very first experience in social-sexual conduct to the point of orgasm, it is of very considerable significance; or if it is very early in the experience of the individual in socio-sexual conduct, I would imply at this point that the surest antidote is early heterosexual experience to the point of orgasm. That being taboo in our Anglo-American culture, it offers us the problem of

how to orientate the individual heterosexually so that he is safeguarded against developing a homosexual pattern.

. . .

45   DR. WHITBY: In the course of your histories, did you find that people change in the modes of homosexual intercourse, going for buggery, sometimes, or mutual masturbation at other times, or do people largely stick to one? Is there interchange of methods?—A. It depends on the age of the individual and the amount of experience they have. There is usually a steady progress. The first homosexual contacts are usually manual, they subsequently arrive at oral contacts, and in many histories do not get anal contacts until later. Anal contacts are more frequent at rather low social levels, and oral contacts are more frequent at our upper social levels. There would, however, be at least 90 per cent, with any extensive homosexual experience, who had all three techniques, and still others, in their histories, who I think would use them quite indiscriminately.

   . . . The European literature, including the Viennese sexual analytical literature, has suggested to us that anal relationship is more common, both in homosexual and in heterosexual experience, in Europe.

46   DR. CURRAN: My impression for what it is worth—I spent a year in the States a year ago—was that oral performances were more common than they are in my experience in this country . . . There is one final thing I would like to ask, and that is what are we to make of the value of Kallman's [sic] twin studies.

(Dr. Kinsey answered off the record)[105]

47   Q. As regards Lang's work on Siblings,[106] what would you say, Professor Kinsey?—A. I would say another great defect in Lang's work—this I can put on record—and I mean upon the twin work or any other investigation of inheritance, is the fact that they depend upon public knowledge concerning the histories of relatives. If you actually got complete sex histories from all these persons whom you are comparing, that would be one thing, but to depend upon merely socially demonstrated homosexual or heterosexual history is not a reliable source of information. Q. If I understood you rightly, from the point of view of the note, you think that Lang has not made out his case?—A. That is so.

DR. WHITBY: That is a very important point.

. . .

3
Homosexuals

Introduction

At the outset, committee members whimsically took to calling homosexuals and prostitutes 'Huntleys' and 'Palmers', after the brand of biscuits. This practice, apparently to save the blushes of the female stenographers, was relatively short-lived; the graphic nature of the witness statements would have made it redundant. But in the early months the committee had to decide whether it wanted to interview any of these Huntleys, in full committee or smaller gatherings, and—if so—what type of Huntley.[1] The correspondence between Wolfenden and the secretary, Conwy Roberts, indicates a willingness to entertain the idea but no great enthusiasm. A number of Huntleys wrote offering their services; correspondence and informal meetings took place.[2] Some could be easily dismissed—there was no evidence that one R. Devereux Shirley, for example, was qualified to 'represent the beliefs and needs of the big majority of the 500,000 homosexuals in Great Britain'[3]—while others were more promising. Roberts thought that 'an invert in a responsible University post' who came to see him was just the 'very decent sort of chap' who would be suitable: unlike many with his 'particular disability' he had no axe to grind and was 'the personification, in fact, of the D. P. P.'s "genuine" homosexual!'[4] Nevertheless, when it came time to organize some full interviews with homosexuals in the summer of 1955, Wolfenden's priorities lay with the other scheduled witnesses; the homosexuals could be offered whatever time might be left over. Roberts even suggested that, 'If the idea is merely to let the Committee see what a few Huntleys look and behave like, then the proceedings could be informally conducted over a cup of tea.' Wolfenden demurred at this: since two of the men were rather distinguished blokes, if the committee were going to put itself to the trouble of seeing them at all, it ought to be reasonably thorough about it.[5]

In the end, just three self-identifying homosexual men appeared before the committee (and another two sent in memoranda). The first was the journalist Peter Wildeblood, not long after his release from prison. He had volunteered for permission to do so in a letter to the Home Secretary from gaol, 'because I thought there were probably very few other men who were able or willing to put forward the viewpoint of an admitted homosexual'.[6] ('I confess I am not looking forward very much to our interview with Mr. Wildeblood', Wolfenden wrote to Roberts, 'but I guess that once he gets going he will do most of the talking.'[7] The chairman's antipathy to the special pleading of a convicted offender was abundantly clear in the hostile comments he scribbled on Wildeblood's memorandum.) The other two were the noted ophthalmologist Patrick Trevor-Roper (brother of the historian Hugh Trevor-Roper) and Carl Winter, the Australian-born Director of the Fitzwilliam Museum and a Fellow of Trinity College, Cambridge. In an interview recorded in 1990 Trevor-Roper explained his motivation. He and a number of other homosexuals of his acquaintance were apprehensive that the committee would only hear from the readily identifiable—those caught and imprisoned, exhibitionists or men from the 'ballet-ish, transvestite world'—and not the considered opinions of homosexuals 'in fairly established jobs'. So he and his friend the novelist Angus Wilson contacted Wolfenden through Goronwy Rees, whom they both knew slightly, to suggest a possible joint interview. Wolfenden invited them, along with William Wells, to a rather awkward dinner at his club, the University Club. Wells was very sympathetic; 'Wolfenden was slightly holding his nose in the air.' The subject at hand was not broached until the last course, when Wolfenden broke the ice by raising the case in the news about the drummer boys at Windsor (see II: 1, n. 112). He then intimated that he and the committee were on the same page as his dinner guests, in that they preferred to listen to them rather than the 'manifestly gay' (Trevor-Roper's phrase) they had heard from, who were neither very articulate nor balanced.[8]

Carl Winter, a mutual friend, had already been in contact with Wolfenden,[9] and arrangements were made for all three of them to be interviewed together. 'I would not put it past them to develop cold feet', Wolfenden wrote, 'but I hope that their exhibitionist impulses are stronger than their fears.'[10] In the event, only Wilson dropped out ('Angus was going to be away on some tour'), so it was Trevor-Roper and Winter, fortified by whisky, and appearing before the committee as 'Doctor' and 'Mr. White' respectively to preserve a semblance of anonymity, who faced a studiously polite chairman and committee (only James Adair was hostile in his questioning) in the intimidating grandeur of the Home Office. They emerged after an hour, shaken but—Trevor-Roper thought—having achieved something.[11]

This was progress, of a sort. As Goronwy Rees pointed out [(d)], prewar organs of government would not have allowed an unfiltered voice to such men. But therein lay a problem: 'such men' were remarkably similar to 'chaps like us'—respectable, well-educated, well-connected, masculine-presenting, professional men, alike in every respect except their sexual orientation, men who conducted their sexual affairs discreetly and in private. The committee did not solicit the point of view of the flamboyant quean, working-class trade, bisexuals and promiscuous cottagers, let alone paederasts and paedophiles: Wolfenden was never going to make the world safe for them. The committee largely chose to avert its gaze from the sociable queer world of pubs and private clubs as well.[12] The right type of homosexual in the right type of domestic social environment was going to get the nod, and no one else.

As an ex-con, Wildeblood was an ambiguous figure here. Trevor-Roper and his circle wished to distance themselves from him,[13] but Wildeblood plainly positioned himself as the acceptable version of homosexual in demeanour and purported practice.[14] And it is Wildeblood, in particular, who has divided opinion between those who see him as courageously proclaiming his homosexuality and battling for the cause and those who accuse him of selfishly fighting for people like himself and throwing the rest of the queer world under the Piccadilly line train [(a)].[15] His contempt for effeminate men, for example, was striking, and his plea for tolerance and understanding for 'real' homosexuals bordered on the maudlin. Still, his exclusionary tactics perhaps indicated a calculated assessment of what the committee was willing to hear and what was feasible in the prevailing political and social climate. His own life-experience and some of his writings suggest that he was rather less judgemental and intent on silencing alternative voices than he let on to the committee.[16] And, even within his limits, he had some sharply pointed things to say in his memorandum: that homosexuality was neither a crime nor a disease; that police tactics against suspected homosexuals were scandalous; that the laws against homosexual sex in private were far from being a dead letter, as he could attest from bitter personal experience; that homosexuality was largely tolerated in prison; and that attempts at 'cure' were farcical.

Wolfenden never wavered in his civility in public,[17] but in private he was not much more impressed with Trevor-Roper's memorandum than with Wildeblood's, describing it as a 'screed' (presumably because of its allegations of widespread police corruption). In contrast, Roberts found Winter's memorandum 'very sound and sensible'.[18] Regardless of the impression they made, in their memoranda [(b–c)] and joint interview [(d)] both of the witnesses bolstered the notion of the respectable, discreet homosexual. Like Wildeblood, they stressed that homosexuality was not an illness. Winter thought heredity or a combination of heredity and early environment accounted for the vast majority of genuine homosexuals; Trevor-Roper divided homosexuals into 'genetic' and 'sporadic' (the product of divorced or alienated parents). His theory about homosexuality and family senescence was bizarre, and for a highly educated man he proved himself to be surprisingly ill informed about the law regarding homosexual offences. The two witnesses agreed on the need to protect the young and public decency and insisted that homosexuals and paederasts were different types—that the seduction thesis was bogus. They also gave short shrift to the notion that those who enjoyed sodomy were somehow more depraved than masturbators and fellators—though Trevor-Roper expressed an opinion that anal sex appealed only to a minority of homosexuals.[19]

Both highlighted how the many upstanding and successful homosexuals in Church, state, the armed forces, the professions and in business were vulnerable to the potential exposure of the most private of acts. They disagreed over the prevalence of blackmail, but their combined list of foreseeable improvements if law reform were enacted included: an end to *agents provocateurs* and other disreputable police stratagems; the mending of embittered and solitary lives; a reduction of male prostitution; and a decrease in cottaging and other public displays. Much of this was 'on message', but there were also fractures in Winter's discourse where alternative discourses threatened to escape. One was his assertion that real cures for homosexuality were rare and ephemeral but that psychoanalysis could enable homosexuals to accept their own natures with equanimity and courage. This making of more self-confident queers was not, as he pointed out, what the magistrates who advocated psychotherapy had in mind. A second was his suggestion that homosexuals did not pay much attention to social status—that a peer and a farm labourer, a professor and a seaman, an eminent novelist and a policeman might pair up and find love.[20] This transgression of class—possibly a worse sin than the sex itself, as Oscar Wilde had discovered—threatened to destabilize the comforting notion that, with decriminalization, only social equals would couple together.[21] And, thirdly, Winter's confession that when he was a boy he seduced the family gardener complicated the developing wisdom that homosexuality and paedophilia were quite distinct.

If Wolfenden turned out to be an exercise in marking out boundaries and firming up binaries—establishing which emerging construction of a homosexual type should be released from the law's grasp—any suggestion of a fungibility of inter-generational tastes and desires for and by teenagers was inconvenient. It is perhaps because of the danger to the reform project of any such damaging questions that many of the witnesses felt the need to draw a very thick line at age 18 or 21 or higher. The memorandum from an 'ORDINARY and COMMON male homosexual' [(e)] was no different. The seduction thesis was greatly exaggerated, he claimed, but in his plea for reform he still felt the need to reaffirm that not only should the law against minors (he suggested below the age of 18) be upheld, it should be strengthened. The final statement in this section, submitted by a homosexual medical practitioner [(f)], had a more reasoned take, suggesting that any age of consent above 16—by which time sexuality was fixed, he claimed—was emotional rather than logical. If young men who were predominantly heterosexual practised homosexual acts during National Service they would revert thereafter, and no damage would be done. Picking up on Kinsey, he was rather contemptuous of the search for aetiologies, since sexuality, whether biological, hormonal or environmental, was beyond the control of the individual—and, as he astutely remarked, nobody felt any need to explain the provenance of heterosexuality. His memorandum, albeit coloured by the prevailing distaste among Wolfenden's witnesses for effeminates and for the visible portion of the homosexual iceberg, was nevertheless a robust, no-nonsense appeal for sanity.

*

(a) HO 345/8: Statement submitted by Mr. Peter Wildeblood[22]

(Age 32—Sentenced to 18 months imprisonment, March 1954. Served sentence at H.M. Prison Winchester and Wormwood Scrubs. Released March 1955.)

The misconceptions and prejudices which surround the question of homosexuality make some kind of clarification necessary before individual problems can be discussed. For the purpose of this statement, I propose to divide the homosexual population into three main groups:

A. Those men who, through glandular or psychological maladjustment, regard themselves as women and behave accordingly. They attract rather more than their share of attention and are, therefore, the type most readily associated in the public mind with the word "homosexual". Since they are not responsible for their physical or mental make-up, it seems unjust to treat them as a social menace, although they may admittedly be a social nuisance.

B. Pederasts, in whom the sexual impulse is directed towards young boys. Presumably everyone would agree that some deterrent is necessary in such cases, as it is in the case of offences against young girls. I cannot speak on behalf of this group, which I regard in the same way in which a "normal" man would regard anyone having intercourse with female children, but my experiences in prison lead me to believe that their present treatment leaves much to be desired.

C. Homosexuals within the strict meaning of the word: that is to say, being attracted to men like themselves. This group is, I believe, by far the largest, being about equally distributed among the various levels of society, but it is of necessity extremely cautious and discreet. For this reason, its members tend to deplore the behaviour of Group "A" almost as much as that of Group "B"; which is illogical, but understandable.

I do not intend to say much about Group "A", except to point out that in other countries their social nuisance-value is somewhat diminished by their tendency to congregate in exclusive meeting-places, out of sight of the general public. The popularity in Britain since the War of frankly homosexual entertainments such as the "Soldiers in Skirts" revues suggests that such men are now regarded by middle- and working-class audiences with tolerant amusement, instead of with scorn; the same impression is given by the appearance of numbers of men en travestie at "family" public-houses in the East End of London.

Members of Group "A" acquire an extraordinary amount of licence in prison, and several of them have told me that they almost prefer it to life "outside".

The problem of Group "B" seems to me essentially the same as that of their heterosexual counterparts. In many cases they are not specifically homosexual: they have chosen their sexual objective by age rather than by gender.

It would be logical to treat offences against children of either sex in the same way; but at the moment it is quite clear that sentences given for offences against boys are disproportionately severe.

. . .

If any far-reaching changes in the law were to be recommended they would, I suppose, be principally concerned with Group "C", to which I belong myself, and which is the only one for which I feel qualified to speak.

I have already said that, under present circumstances, this group is so circumspect that its full extent will probably never be known. Whatever changes may be made, I believe that the great majority of homosexuals of this kind desire to lead their lives with discretion and decency, neither corrupting others nor publicly flaunting their condition.

I have used the word "corrupting" because some people appear to fear a change in the law on the grounds that homosexuals are forever proselytizing, and that this tendency would run riot if the punishments were abolished. In my experience, this is not true. Most homosexuals would not wish, and in any case would not be able, to proselytize a "normal" man. What they wish to do is to find another man of their own kind, and if possible, form a permanent attachment.

But they are placed by the laws of their country in a position of permanent danger. This applies to all alike, the faithful and the promiscuous, the discreet and the outrageous, so that there is no particular advantage in obeying the universal moral rules and ordering one's private life with discretion and fidelity.

Indeed, there are grave disadvantages. A promiscuous and temporary liaison is far less likely to provide corroborative evidence in Court than an association in which genuine trust and affection play a part...Letters, photographs and so on, innocent enough in themselves, are sufficiently damning in the atmosphere of a Court—as my own case proves. A promiscuous homosexual takes enormous risks, but they are not as great as those which are run by a man who lives quietly and faithfully with another, with no question of corruption or of public scandal.

...

...Some months before my arrest I was walking along the Old Brompton Road. It was midnight, and outside a closed public-house I noticed two men loitering. A man aged about 70 came down the street, turned down a side-alley, and went into a lavatory beside the public-house. He was followed by the younger of the two men, and almost immediately there was a sound of scuffling and shouting.

The elder of the two men ran into the lavatory, and they dragged the old man out, crying and struggling. When I shouted at them to let him go, they told me they were Police officers. A woman who had joined us on the street corner asked what the old man had done, and one of the detectives said that he had been "making a nuisance of himself".

The old man began to struggle violently, and the detectives pushed him up against the railings of the Cancer Hospital, outside which we were standing. His head became wedged between two iron spikes, and he started to scream. The detectives asked one of us to ring up Chelsea Police Station and summon a van; the woman said "you can do your own dirty work, damn you," but I thought the man was likely to be badly hurt if the struggle went on, and going into a nearby telephone box I informed the Duty Sergeant, as requested, that his colleagues were "at the top of Dovehouse Street["]. He was evidently expecting this message, because the van arrived in a couple of minutes and the old man was removed, moaning and bleeding from the nose, having fallen on the pavement while I was telephoning.[23]

During my 12 months in Wormwood Scrubs I spoke to several ex-policemen who had been convicted of various offences, and they told me that this kind of practice was by no means unusual. One of them explained to me that promotions in the Police Force depended largely on the number of convictions obtained, and that each Police Station displayed a kind of scoreboard on which the convictions obtained by the various officers were tabulated.

He explained to me quiet frankly that since "real criminals" were difficult to catch, and homosexuals "dead easy", visits to the public lavatories were the usual path to promotion. The higher-ranking officers were of course quite aware of this, but they condoned the practice because they, too, needed a good average of convictions in order to impress their superiors. Nearly all the accused men, my informant added, could be frightened into pleading Guilty at the Magistrates' Court on the assurance (not always accurate) that by so doing they would escape publicity.

. . .

In Britain, homosexuals are prosecuted under 17th Century laws of ecclesiastical origin[24] and under the Criminal Law Amendment Act of 1885 . . .

The Criminal Law Amendment Act has been described as the Blackmailer's Charter and is widely held to be a bad law, frivolously conceived, and unsatisfactory in practice. Even the legal profession is said to be in favour of abolishing it—not, I think, on humanitarian grounds, but essentially because of its impracticability.

It would, I believe, be a tragic mistake if the Committee, persuaded by this purely professional argument, were to recommend that the 1885 Act should be abolished and the older laws left intact . . .

A large number of men would continue to live, as they do now, in fear of prosecution. Blackmail would continue to flourish . . .

It may be further argued that if the 17th Century laws were allowed to remain intact they could, in deference to public opinion, be treated in practice as a dead letter. This seems to me a most dangerous fallacy . . .

Until three or four years ago it was always supposed by homosexuals that the laws, in so far as they concerned consenting adults acting in private, were in fact a dead letter. During Sir Harold Scott's tenure of office at Scotland Yard[25] it was tacitly understood that, whatever the actual laws might be, the Code Napoleon was the basis of legal practice.[26] Prosecutions were extremely rare, and it was said that homosexuals who were blackmailed were encouraged to take their troubles to the Police, who at that time were more concerned with apprehending blackmailers than with their victims.

In about 1952 this tendency was reversed,[27] and when I was in Wormwood Scrubs I met two men who had been convicted on their own evidence after complaining of blackmail to the Police. In both cases the blackmailer went unpunished.[28]

. . .

. . . I would like to turn to the question of medical and psychiatric treatment of homosexuals in prison. There seems to be some confusion on this point, due, perhaps to the understandable tendency of doctors and psychiatrists to claim—in public—a rather greater success than they are in fact achieving. In private, I have never met a doctor or psychiatrist who claimed to be able to turn a homosexual into a heterosexual; nor, in prison, did I meet a single offender who was receiving regular treatment.

I myself applied for treatment when I arrived at Winchester Prison in March, 1954, and was interviewed by the Prison Doctor, Dr. Finton.[29] Five weeks later I was transferred to Wormwood Scrubs... Out of 1,000 men at Wormwood Scrubs, less than 20 were receiving psychiatric treatment, and only a few of these were homosexuals.

When I saw the Principal Medical Officer, Dr. Landers, who is himself a psychiatrist, he told me that Dr. Finton had expressed the opinion that I was not a suitable subject for treatment. In the course of a long conversation Dr. Landers said he did not believe there was any effective "cure" for homosexuality, but it was sometimes possible to give treatment for a resultant neurosis—that is to say, to relieve a homosexual of his anxiety about being a social misfit. I did not see how this could possibly be reconciled with the idea of punishment, and in any case it did not seem to be what I required, so I asked him about the possibility of treatment by injection of sex-hormones, which I had already discussed before my trial with a distinguished, but sceptical, endocrinologist.

Dr. Landers' view was that this experiment was fraught with unknown dangers and should be used with extreme caution; it had already been tried at Wormwood Scrubs, resulting in one case in a further conviction, and in another in distressing biological changes.

The only thing he could recommend to me was a lengthy course of psychoanalysis combined with psychotherapy, preferably undertaken after I had completed my sentence. He admitted, however, that if this were successful it might have such far-reaching effects on my mental make-up and personality as to interfere seriously with my capabilities as a writer. The matter was not, therefore, taken any further.

...

The atmosphere of a prison is unusually conducive to homosexual relationships, with or without physical expression.[30] On conviction, a homosexual thus finds himself transferred to a community which is actually more tolerant of his condition than the one in which he has previously lived. Group "A", with its capacity for outrageous gaiety, is almost too popular; it plays havoc with prison discipline and has a demoralising effect on the warders, among whom the percentage of homosexuals is at least as high as anywhere else. Group "B" comes in for a certain amount of moral disapprobation, though rather less, surprisingly, than its heterosexual counterpart. Group "C" is not only unreservedly tolerated; it is expected and even encouraged to embark on serious emotional relationships, either with fellow-members of the group or with other prisoners who, though convicted of other crimes, happen to be homosexuals.

Although the moral atmosphere of a prison is hardly typical, I think it has some bearing on the present public attitude towards homosexuals ...

My own exposure and subsequent imprisonment have not only failed to rob me of any of my friends; they have actually increased their number. I have received large numbers of letters from men and women previously unknown to me, expressing sympathy and good wishes for the future—and these have not included a single one of the obscene or condemnatory kind which might have been expected. It is, I think, of interest, and particularly to politicians, that this attitude is taken by people of every social class. My working-class friends and acquaintances have proved, if anything, to be even more liberal-minded than those of the middle-class.

It is strange that public opinion should have been moving towards greater tolerance at the same time that Authority was deciding on sterner methods of repression; but that is quite clearly what has happened, and the resultant open clash in Press and Parliament is, of course, the reason for the existence of the Committee.

In conclusion, I would say this.

I do not believe that homosexuality between adults is a crime, because it lacks the essential characteristic of a crime—the doing of harm to another person. The imprisonment of men like myself is logically indefensible and morally wicked[31]; it weakens the whole concept of Justice in our country. Furthermore, the law as it present stands [*sic*] offers the Police an incessant temptation to act in a way more appropriate to a police state than to Britain.

I am by no means sure, on the other hand, that homosexuality can fairly be considered an illness. It has existed for so long, and among communities so different, that I believe the view that it is a kind of moral sickness, symptomatic of decay, to be entirely false. Even if it were an illness, it would be unjust to prescribe compulsory treatment when the doctors themselves are divided as to its cause and cure.

Havelock Ellis once compared homosexuality to colour-blindness.[32] You do not punish people for being colour-blind,[33] and you do not force them to take medical treatment, for none exists. Is it any more logical, or just, to punish people like me?

We do not ask for any special consideration, but only for the rights which are common to all free men. We may always be looked down upon by the others, but at least we should be allowed to seek what happiness we can.

The shadow of fear is a terrible thing; it cripples a man's character and distorts his moral sense. Set us free, and we can at least try to order our lives with decency and dignity; leave us in this shadow, and we shall continue to be bitter, secretive and warped,[34] a persecuted faction incapable of good.

There are many thousands of us—how many, we do not know—and among us there are no doubt some who will never grow into good citizens. But the rest of us, and, I believe, the great majority, would be better and more useful members of society if we were allowed to live in peace, instead of being condemned to live outside the law.

*

(b) HO 345/8: Memorandum submitted by one of the witnesses to be heard at 2.15 p.m. on Thursday 28th July [1955][35]

The treatment of Homosexuals in Britain gives rise to many social evils, which would all to some extent be mitigated if private homosexual acts between consenting adult men ceased to be criminal. Among the more grave of these ill effects on society are suicide, blackmail, and the encouragement of police 'agents provocateurs'. But since the homosexual factors in a suicide are normally concealed, while few cases of blackmail and almost none of 'agents provocateurs' reach the newspapers, their extent is difficult to assess and often minimised. It therefore occurred to me that some indication of the frequency of suicides, blackmail, and 'agent provocateurs', as a direct result of the present laws regarding homosexuality, might be given if I were to list certain cases in which my necessarily limited circle of homosexual friends and acquaintances have been personally involved.

SUICIDES

Suicide among homosexuals is most commonly the result of exposure and its sequels. Such cases are much more frequent among those who live furtive lives, committing occasional homosexual acts when opportunity offers, but who have never become integrated into a group of other homosexuals from whose experience and advice they might profit. For that very reason I have never met any such person, although by secondhand I know of many; and of men who kill themselves for no apparent cause, there has always seemed to me to be a preponderance of the unmarried, whose occupation, interests, and personality are of the homosexual type, and in whom the fear of blackmail or exposure may well be responsible.

   The following three men, whom I myself knew, were all among the smaller grounds of suicides who had established an apparently satisfactory homosexual way of life, and were as free as any homosexual can hope to be from the immediate risk of exposure or blackmail.

. . .

R.—aged 19, a 'brilliant' Cambridge undergraduate, who gassed himself five weeks ago (3rd June). Although well integrated into a group of other homosexuals at the University, a lifetime of persecution and ridicule from his fellows had led to a sense of isolation, of which suicide seemed to him the natural sequel. At the inquest it was stated and confirmed that he had been driven to death because of this persecution, and he wished these reasons to be made public in the hope that his end might draw attention to, and thereby alleviate, the plight of his fellows.

. . .

BLACKMAIL

Blackmail is by far the most common social abuse due to the existing laws concerning homosexuality . . .

. . .

C.D.—a critic now in the late twenties, received the accompanying letter six months ago. It was sent by a sailor who had approached him in a Soho pub, and then asked for a bed for that night; although physically attracted by the sailor D. made no mention of matters connected with sex until the sailor announced that he would prefer to sleep in bed with D. rather than in the alternative single bed offered. The sailor's conduct subsequently suggested a considerable homosexual experience; money was not mentioned, and they parted on apparently friendly terms. This letter was received eight days later—on a Saturday. He was unable to contact his doctor or lawyer, and since the sum was not embarrassingly large, he sent the £10 forthwith and had heard nothing more. A medical examination next week showed no sign of venereal disease. It may be added that the letter, which combines many illiteracies with an unusual expressiveness, suggests that it had been used on a number of occasions or had been written with the help of a more educated accomplice.

[COPY   Annex A
Thursday 27th January, 1955   L/Seaman X.Y.Z.
                                          Chatham, Kent.

Mr. _____

I have a rather unplesant supprise for you, (I hope). You, to use, the vernacular are rotten. In case you are not aware what that means, I am telling you that you are infected with V.D.

By last wenesday I also was infected and had to report sick. That is why I was not available for the week-end.

I recived your telegram thank you, but as you can gess I was in no mood to answer verbaly or otherwise.

I can't come ashore again until I am cured and goodness knows when that will be. So I can't take any direct action, I could however take some indirect action this end.

When you go sick with V.D. in The [scribbled out] you are asked to make a statement as to when, where and with whom. I have said so far that I was drunk. It was difficult to remember and it was proberly some slut. Now if I were suddenly to remember when, where and with whom, it might be very embarasing for you.

I think you can see where this is leading. I can't have physical revenge so I will have it another way. I think you should compensate me for what has happened. Weather you were aware of your condition or not I don't know, for sure. I think you were. It must show in some way even in you. As you know I have an M/C and at the begining of each year I have to tax and insure it. That costs roughly £10 and you would not want me to pay a sum like that out of my pay, would you?

You may please your self what you do with this letter. If however you do answer it, be positive. I don't want any half-measures, and I don't like Banks. I know you will think I am mercenary and vindictive, but I can assure you it is only the latter, and after what has happened I think I have good cause to be.

                    Yours X.Y.Z.]

Comment

...

I do not, of course, defend the promiscuity that invites blackmail, however readily it may be blamed on the awareness of a hostile society, but I am only concerned here with the encouragement that the present laws give to a multitude of weak, but not yet vicious youths, mostly servicemen on meagre pay, to embark increasingly on a life of crime.

...

AGENTS PROVOCATEURS

The existence of 'agents provocateurs' is a commonplace among that small minority of homosexuals who, from a mixture of fear, ignorance, guilt and defiance, seek their sexual outlet through casual 'pick-ups' in parks, bars and public lavatories. These police 'agents' frequently combine their role with that of blackmailer as in the recent case with which the Committee will be familiar, fully reported in the Manchester Guardian of 29th March, 1955—the only occasion to my knowledge in which the police offenders have been exposed and sentenced.[36] I have heard of a number of instances, but the following two, as told to me by the 'culprit', are illustrative.

G.H.—aged thirty, an anatomist at one of the London teaching hospitals, was passing a public lavatory in Gloucester Road in May 1953 at about 7 p.m., when a goodlooking young man beckoned him with a tilt of the head. H. was not in the habit of seeking sexual partners in such places, but on the spur of the moment followed him in. A second man then entered the lavatory and stood at the far end. The first man made suggestive gestures, and when H. seemed to respond, both men announced that they were policemen, and led him to Chelsea Police Station. On the way there they sympathised when he told how this exposure would ruin him in his job, and offered to take from his pockets all evidence of his identity, so that he could pose as a clerk, plead guilty, and that 'if they benefitted' they would prevent any report from reaching the newspapers. This went off as planned, and next day he was dismissed from the court with a fine of £8. On leaving court he met the two policemen outside, received back his property and arranged to meet them in a pub later where he handed over an envelope containing £2.

J.K.—a specialist at a London Teaching Hospital, aged 30, visited a public lavatory in Bayswater in 1933, whither he was followed by two men. They made advances and the doctor states that these 'were not acquiesced in, but not utterly repulsed'. On leaving he was arrested, convicted at the police court for importuning, fined a nominal sum and released. The subsequent newspaper publicity was fatal to his career, he was struck off the register, and only readmitted in 1941 (not from humanity, but from war-shortage of doctors!) He has now no self-pity, commenting only that 'he got what his carelessness and folly merited'; but he feels that the disquieting factor was the evidence given by the two plainclothes policemen who supported their case by stating that he had been followed from one public lavatory to another (he had only visited the one in question) and that he made certain salacious remarks and actions, which I can aver, are totally out of keeping with his character and education.

Comment—that agents provocateurs should be constantly employed in a liberal country is naturally offensive, and damaging both to society and to the standing and morale of the police themselves. This is particularly distressing when the only 'crime' that they seek to uncover is one which in no way harms society and whose very 'sinfulness' is in doubt.

It may be argued that nearly all these cases concern the seekers after promiscuous sexual relationships in public places (although the actual intercourse may well be performed subsequently in private). I would only add that most homosexuals, like heterosexuals, pass through an apparently promiscuous phase, before they settle down with a chosen partner (unless their whole career is spoilt and their outlook warped by some such police exposure); and those men who continue to frequent such 'picking-up' places later in life, often do so for a spurious sense of adventure and defiance induced by the feeling that society has already branded them as criminals.

As a postscript, there is one other observation I would like to add on the larger question of the genesis of homosexuality.

On analysis of the homosexuals, I know they fall into two quite separate aetiological categories of almost equal size, which I will call the 'genetic' and the 'sporadic' homosexuals.

The genetic homosexuals are those in whose families other members are affected, and there is no history of parental disharmony. Such families often seem to be dying out, at any rate in the male line, although prolific in former generations, and many of its male members who are not homosexual seem to be relatively sterile. If this is true, it would seem as though the germ cells which are normally immortal, but which as they pass through each generation, bud off mortal body-cells with their appropriate life-span, do themselves sometimes senesce; and the moribund stock is often stigmatised by a 'crop' of homosexuals. This phenomenon is more familiar among titled families whose genealogies are more readily known or ascertained, but in whom the homosexuality is only inferred from gossip or conjecture; but this associated homosexuality is more arresting when the genealogies of homosexual friends are traced.

The Sporadic homosexuals appear as isolated members of otherwise 'healthy' stock; they seem to be so constantly the sons of divorced or alienated parents that I assume this to be the underlying cause. For the same reason, such sporadic homosexuals are particularly assailed by that feeling of insecurity which all homosexuals suffer, are generally the more neurotic, and it is thus this group that turns more readily to suicide as a gesture of defeat.

Concerning female homosexuals my knowledge is very slight, but it does suggest that the same two categories are found, the genetic group sometime associated with the male genetic homosexual in the same family, and sometimes in separate stocks.

I offer this analysis with great diffidence since my numbers are statistically insignificant, and I am largely unfamiliar with the literature on this aspect of the subject. Even if my conclusions are valid, they may have little relevance to the present enquiry; but they would at least seem to confirm that the deviation is inborn or acquired in early childhood, and psychiatric treatment can thus palliate or advise but never alter the percentage of the homosexual in each individual, while any suggestion that homosexuality is attributable to seduction after the age of puberty is manifestly incorrect.

*

(c) HO 345/8: Memorandum by Mr. C. W.[37]

... [H]uman personality—the human body and the human soul—is not divisible into two perfectly clear-cut sexes, the wholly female and the wholly male; but ... on the contrary we all embody, for better or worse, many characteristics of both sexes, sometimes more or less balanced and reconciled, and sometimes with one or the other sex predominating, but not necessarily always in a body of the same sex as the psyche that inhabits it. Upon these inborn, inherited foundations of character the structure that is later built depends, during the unconscious formative years of a child's life, not upon himself but upon his parents, his loss of parents, or other combinations of extraneous circumstance. The profound effects of these childhood influences are too well-known to need emphasis. It is my conviction that persons entirely or predominantly homosexual are largely either born so, or born and made so as the inescapable result of their heredity and their early environment; and that the number of people who deliberately adopt and persist in habits of homosexual feeling and intercourse, as a vicious preference, is an infinitesimal minority of a minority.

It follows from these considerations:

(i) that I do not hold homosexuality, among either men or women, to be per se either vicious or criminal, or good, any more than I hold heterosexuality to be per se vicious, or criminal, or good ...

(ii) that I do not hold homosexuality to be an illness, any more than a vice or a crime. A current medical fashion of referring to it publicly, in Parliament and the Courts, as a form of illness, seems to me to be founded less upon fact than upon a humane but misdirected desire to mitigate, by such an approach, the irrational rigour of a law that treats as criminals only such homosexuals as are male, but that makes up for such remarkable selectivity by treating all these males as criminals.

(iii) that, in my opinion, quite disproportionate emphasis is laid upon suicides, blackmail and scandals involving homosexuals ... On the other hand, for obvious reasons, no equivalent attention is focussed upon the successes of male homosexuals, as such ... The facts are sufficiently impressive: many homosexual men achieve the highest offices and dignities in the state, the church, the armed forces, the professions and the business world; some receive every kind of public honour and distinction; while scores of thousands of others lead the unobtrusive lives of ordinary useful citizens, unknown alike to fame or the police. And every one of them, according to his temperament, must amuse or afflict himself—to many it is a most bitter and painful reflection—with the knowledge that the Queen may be pleased to honour him, the State to entrust affairs of peace and war to him, or the Church to give him charge of souls, but that the law can ruthlessly put him in gaol and utterly destroy his reputation and career if it ever chances to learn that his sexual conduct departs from a rule of total chastity that it demands of no other body of citizens. Let him be in every other respect the ablest, most honourable, and trustworthy of human beings, he is to be totally disabled and untrusted in the one single matter which, of all others, is to all of us our most private concern. All sexual acts are apt to seem gross to all except those immediately taking part in them; few can ever possibly appear anything but discreditable when the actors are made to stand in the dock and answer publicly in cold blood to a hostile law for their most intimate private actions.

(iv) that, obviously the law inflicts a crude injustice upon many homosexual men by lumping them indiscriminately together with others who may be active or potential criminals; that it imposes upon ordinary human nature a heavier hardship than it is reasonable to expect ordinary male human nature to bear; that, in consequence, the men here concerned are left with no practical option except to break this law; and that the law, so difficult to enforce, enforced spasmodically, and sometimes with the aid of discreditable stratagems and provocation, is to that degree brought into disrepute.

The public scandal caused by certain prosecutions may possibly have the effect of deterring some persons, but it is also arguable that it may, by exciting the imaginations of others, also have the reverse effect.

(v) that, to confine a man whose weakness is for intercourse, in every sense of the word, with his own sex, in a place where he must live for years, quite cut off from all normal society and influences, and forces into the closest physical intimacy with other males, while he and society are totally deprived of all his useful activities, is surely a fatuous form of punishment, let alone of "treatment". How grotesque it is can be illustrated only by supposing that, if heterosexual immorality were also a crime, heterosexual men were sentenced to long periods of curative confinement in the cells of the females at Holloway[38] ...

(vi) that I am more than doubtful whether many mature adult homosexuals, or even youths, when of predominantly homosexual character, can be so radically changed by psychological treatment as reliably to be accounted "cured". The desire for woman and the notion of physical intimacy with her are conceptions utterly foreign and totally incomprehensible to the true male homosexual. The psychologist, it must seem to such a patient, is seeking to make him into a quite different person; but nobody can possibly conceive of himself as not himself, and very few can even wish to be a different person: most people want to remain exactly themselves, only happier, healthier, more successful, and much more beloved by the persons whose love is the only form of love they value and understand. No doubt it is a sensible precaution for any person in trouble to fetch a doctor to Court to propose that he "undergo a prolonged course of treatment"; the patient may even enter into the spirit of the thing with sanguine expectations of becoming another person, but I doubt whether real cures can be many and lasting. On the other hand I have known several homosexuals who explained that profound psycho-analysis had enabled them to accept the basic facts of their own natures with more equanimity and courage; but this can scarcely be the result contemplated by magistrates.

(vii) that, while some homosexuals may be quite as promiscuous as the generality of other males, many are extremely scrupulous and self-disciplined, for reasons that are entirely moral and that have little to do with fear of a law that they do not respect; many homosexual couples, moreover, establish unions, held together by affection and consent, that last through the lives of the two partners, and that exhibit many of the qualities that distinguish the happiest heterosexual marriages. Some such couples even legally adopt sons; while others, and many single homosexual individuals, within my personal observation, often behave towards numbers of young men with generosity, helpfulness and understanding, without the slightest implication of any kind of impropriety. The adult man who has gone through school, university, military service and the early stage of his profession, without being aware that some bachelor friend was helping him, must be both unfortunate and more than a little unperceptive; and the quality of such help is that to many men it provides the greatest satisfaction in their otherwise rather sterile and frustrated emotional lives.

(viii) that the crux of society's problem concerning male homosexuals seems to me to be, therefore, reducible to finding a means of allowing reasonable liberty to the decent and responsible among such men, while not opening the door to public disorder or to the corruption of boys and youths of tender years. That order and decency must be preserved is common ground; that the integrity of children must be protected is obvious. I most certainly think that no adult, whether man or woman, should be allowed to corrupt or seduce anyone of tender years and that protection similar to that given to girls below 16 should be enforced in the case of youths up to 17 or 18. One can no more give licence to pederasts, strictly so-called, to corrupt boys than one can allow men to seduce young girls. If acceptable age-limits for the protection of all young people could be thus established, I should then like to see perfectly equal laws, or better still no laws at all, governing immoral relationships of consenting adults, irrespective of the sexes of the partners or of their forms of sexual intercourse, so long as their acts take place in private and do not infringe the rule of public decency.

...

*

(d) <u>HO 345/14: Interview of 'Doctor' and 'Mr White',[39] Thursday, 28 July, 1955</u>

...

[Q2588] MR. REES: Mr. Chairman, I wonder if I might say something for the record which I would very much like to say. I think one of the most difficult things before this Committee is what the state of public opinion about this subject is, and it seems to me the best possible objective proof of the change in the climate of opinion that these gentlemen should have appeared before us today. Twenty years ago that would have been, I think, quite out of the question and quite impossible.[40]

...

[Q2590] [CHAIRMAN:] ... [W]hereas suicides, blackmail and similar scandals are in fact a notable part of your paper, Doctor, Mr. White's view on the other hand is that in his opinion quite disproportionate emphasis is laid upon suicide, blackmail and scandals involving homosexuals ...
   MR. WHITE: ... I know of nobody personally, I think, who has been blackmailed ...

...

[Q2591] (DOCTOR): ... [B]lackmail is in my experience most frequent, mostly covert, meaning the threat of blackmail or the fear of the threat of blackmail, or the line of theft one cannot expose because the thiever knows you are in an unsafe position and of that type; it is almost universal among those who lead promiscuous lives.

[Q2592] MR. REES: That seems to me to correspond absolutely to the facts of all homosexuals I have known who have led the kind of life you describe, and I am really surprised that one could really be homosexual and not have come across blackmail.—A. (MR. WHITE): I am homosexual and I have not come across it, nor do I have any personal knowledge of any friend of mine who ever has.

...

[Q2593] MR. MISHCON: I was really wondering whether it was not a question of the circles in which one moves ...—A. (MR. WHITE): My answer to that would be that homosexuals almost entirely do not move in circles or classes; they move in an accepted pattern through society. They do not pay very much attention to social status or where they come from or where they are going to; and I think that is one of the reasons why society is rather alarmed about them, that they do not adhere to the ordinary social prejudices and distinctions. It is quite possible that a peer may be attached to a farm labourer or an able seaman to a university professor, and I have known an eminent novelist who lived in a great state of devotion with a London policeman ...[41]

...

[Q2597] ... MR. WHITE: ... I live and have lived for a number of years in a society of which many members are aware that I am homosexual. I have many friends who are known to be so. A large number of the extremely critical people among whom I live have a very unfavourable view on this and do not wish to know me on that account, and I think it is fair to say without acrimony that I do not particularly wish to know them in return. I am content to know the sort of people I do know, which covers a very wide field, and we are all completely at ease in one another's company and the world in which we live, which is a much more extensive world, I think, than many people would suppose. We visit each other's houses, go abroad, travel, look at the sort of things which interest us, art, exhibitions, ballet, and have a satisfactory life within that sphere...

[Q2598] MR. REES: You describe this happy, successful life. Where do you think is the real instance of the injustice?—A. It seems to me unjust that society and the law should say that what I conceive may be an arresting of personal development, possibly of emotional development or may be possibly called a deformation of character—though I confess I do not know what the perfect ideal character is from which this is so marked a deformity—should be regarded as criminal, and acts not harmful to other people, not publicly indecent, which stem from that kind of personality, should be punishable, and in many cases punishable extremely severely with, it seems to me, quite a disproportionate degree of severity related to the offences with which they deal. That a man should have his entire career, the well-being of his family or his associates, the place where he works, everything smirched and destroyed because his sex life does not conform to the sex life of the majority of the population, seems to me grotesque.

...

[Q2600] [MR REES:] If you were asked to alter the law, how would you alter it?—A. [DOCTOR] Only by deleting, if I remember correctly—"or in private".[42]

Q. Just like that?—A. Just like that.

[Q2601] MR. MISHCON:...Are there many cases to [Mr. White's] knowledge where there are consenting adults who have been brought before the Courts and dealt with although they were carrying on in private?—A. (MR. WHITE): I am informed by people who do take an interest in the legal aspects of this matter...that the law does not greatly interest itself in those who are carrying on homosexual relations in private. I have however read of such cases of people who have been found out...

...

[Q2602]—A. (DOCTOR):...I happened to be visiting a country parson the other day who was very willing to come here in person—I do not think I need mention his name—who cited five people he knew personally who had committed suicide for fear of police proceedings going on in the neighbourhood or inquiries actually being made and prosecutions being made. Three, as I remember it, were discovered on the round robin business, the police overheard some conversation, grilled him, frightened him, and he then declares the names of the people he has had homosexual relations with and these people are all put in the dock, and homosexual relationships between such people generally occur in their own rooms, so those are all in fact private homosexual acts...

...

[Q2606] [DR. CURRAN:] In your experience...do you think any homosexual act with somebody who is not homosexual really has a permanent effect in switching?—A. (DOCTOR): This is always an irritating conception which is advanced, because it seems to me so totally devoid of any truth, certainly any homosexual experience after puberty has no effect on them in my experience, and I feel that particularly because at my school—it might have happened to be at a particularly busy period, going through a phase when homosexuality was general—all the boys had casual experiences of masturbation with each other with the exception of about two, and not one of them has turned out to be a homosexual. I am quite content that the damage, so to speak is done certainly before the age of puberty.—(MR. WHITE): I would entirely agree with what the Doctor has said about that and could say from my own childhood experience, although I am not of course a good illustration of the point—that I knew a neighbour's child, a boy who was one of a large family, who has since married, had a large family, and is in every respect the most normal heterosexual person I suppose I know, and he and I as children had a prolonged homosexual affair at twelve or thirteen.

In this connection I want to bring this point in, not from any exhibitionist motive but because it is a serious point—when I was a child I myself had a relationship of this sort with our gardener, against his will in the first instance. It was I who, as a homosexual boy, seduced our gardener against his will and against his better judgment...[I]t may be the duty of the adult to resist the temptation of the child who is pervertible but it is very often the child who is the determining factor in the case. I think it must be very difficult for certain people, if their interests are at all susceptible in that way, to be attacked by a persistent small boy.

...

[Q2610] CHAIRMAN: Arising out of what Mr. White said, has he any recollection of the age at which he himself came to know, or decided that he was homosexual?—A. (MR. WHITE): I think, to give the most honest answer I can to that, the pictures which I formed in my mind as a child were dominantly male pictures from as long as I can remember. During a certain period of early adolescence, twelve, thirteen and fourteen, they were interspersed with what proved to be the familiar leaning towards the opposite sex, passing inclinations and excitements with the opposite sex. As I grew into adolescence I went through a prolonged period of tension within myself over this matter and tried as hard as I could with every sort of aid and help, but privately—my parents never knew this—to overcome this, in order to conform to what I conceived to be the right and approved social pattern, which led me into a great number of minor adventures and a few major adventures with women, which all proved to be extremely unhappy, particularly for the women concerned. I think I was fifteen or sixteen when I first clearly and definitely knew why I did and wanted the things I did and wanted.

[Q2611] Q. I did hear somebody say the other day that he knew when he was eight.—A. Yes, I knew a good deal of the subject in general because I was an avid reader of such books as those on Greek mythology and things like that when I was young.—(DOCTOR): I was the reverse from Mr. White, I was not very intelligent in seeing the things going on around me. I remember being teased at preparatory school for being effeminate in certain respects. At Cambridge I had phantasies of sexual relationships with people I had known at my previous school, but again not acquainted with the people. As a medical student in London, throughout my medical student life I was unaware, except for an occasional comment in the "News of the World" of the existence of homosexuality; and the last year I lived with a woman, or had a permanent sexual relationship with another female medical student for some months, conscious that during those occasions I was frequently shutting my eyes and imagining her to be a particularly close friend, a male student. Only when the war came and I was qualified did I meet somebody who said this goes on all round you, and that was a revelation.

. . .

[Q2612] . . . (MR. WHITE): . . . I have known many, especially Englishmen who thought it best to live abroad, the kind who said: "They are too old for me when they are fifteen; on their fifteenth birthday I am not interested, I give him a large present and he can go away". It is an extreme case but I have known of such cases and I use it only to say I do not know of any single case of an adult homosexual who liked relationships with adult males who suddenly said, there is a small boy, I wish to seduce him or I wish to have him—it seems to me quite out of the question. To most homosexuals who do not like that kind of thing it is just as revoltingly strange as homosexual activities must seem to the heterosexual. It is quite incomprehensible to anyone unless he is in fact a paederast himself.

. . .

[Q2613] . . . (DOCTOR): . . . I have a very large homosexual acquaintance, 150, perhaps more, of people whose tastes I know. I know of not a single one of them who has ever had relationships, or has any desire for boys under eighteen. If I may follow that up, I have heard it said that as one gets older one likes boys younger and younger which seems to me, for all my experience, nonsense . . .

There is a broad group of homosexuals, those who like a father figure in effect, and there are those who like somebody younger, and that generally remains stationary. When he is fifty he may like someone of thirty and that sort of thing, but they are quite separate from paederasts.

. . .

[Q2615] DR. CURRAN: . . . Do you think people who indulge in sodomy differ in any way from other homosexuals?

– A. [MR. WHITE]: . . . In my experience, and I know all types, those who commit sodomy are completely indistinguishable from the others.

. . .

[Q2616] (DOCTOR):...On the question of sodomy, it is frequently suggested that it is a common form of homosexual relationship. In general, in my experience, it is relatively rare. Most homosexual acts between adults are not buggery—ten, twenty possibly, even thirty per cent., and whether it has any relevance—I agree entirely with Mr. White that it is purely a matter of taste, and that part of the anatomy is no more sinful than the mouth; indeed one has necessarily in mind—...quoting Kinsey—that 30 or 40 per cent. of heterosexual couples indulge in that same act and indeed other variations.[43]

[Q2617] Q. [Dr. WHITBY]: Do you think it would be correct to say that the so-called bisexual is really a homosexual?—A. I think so, because there is a pull to heterosexuality.—(MR. WHITE): There is a tremendous pull for the homosexual person to try to get married, even if he only wants somebody to look after his house and darn his socks and that is, I think, one of the reasons which induces quite a number of such men to do so.

[Q2618] MR. WELLS: I forget whether it was in answer to Mr. Mishcon or Mr. Rees—you said the only alteration in the law you wished to see made was the alteration to the Labouchere amendment. Would you appreciate that would leave sodomy quite untouched?—A. (DOCTOR): In that case I am lamentably misinformed. I thought sodomy in private would be covered by that...I had thought—I know very little about the legal provision—but I had thought the Labouchere amendment replaced the old laws.

...

[Q2619] MR. WELLS: I wonder if you could tell the Committee how, in your experience, you tend to find—if I may put it that way—recruits; how are your partners found?

... A....[H]ow I meet homosexuals is that one is introduced to them by other homosexuals, because my private life tends to be virtually restricted to almost the purely homosexual world and has become increasingly so.

As to recruits, one meets by ordinary introduction. How they discover the homosexual world is probably rather like myself, that one is found looking rather worried by an older doctor who asks you why you are worried and so on, and it all comes out. Two of my current students both approached me. Why did they approach me? They were frightened and felt out of society. They felt one would be sympathetic. They knew I was unmarried and, because I had dealings with them in the ordinary way they were in a position to lead up to it. I do not ask any students, but over the last ten years one or two others have done the same.

[Q2620]...A. (MR. WHITE): My own experience is that my acquaintances are made spontaneously by a kind of telepathy that you see a person or persons and you realise that they are sympathetic or likeable, and a few words of conversation will establish the fact that they are also homosexual ...

[Q2621] Q. What changes, if any, do you both think there would be in this state of affairs and in your society if the law were amended in the sense you both desire?— A. (DOCTOR): In relation to the homosexual? I do not think it would make much change really. There would be less feeling of isolation and embarrassment and unhappiness. It might just relieve the travails which go with it, or the exaggerated behaviour which may be heard in conversation, but the later effect would be that it would be a disincentive to anti-social acts by the person who hangs around public lavatories and so on, if that is what he does.—(MR. WHITE): I should hope that it would lead to a decrease of provocations and stratagems on the part of the police. I do not wish to attack the police on this matter at all, and if a thing is a crime it is their duty to find it out by all means and bring the persons to book. On the other hand it does lead to some disagreeable and, to my mind, disreputable stratagems of which I was once myself the attempted, but unsuccessful victim. I think if the law were amended as we suggested, it might help to reduce these extremely squalid and disagreeable intrigues, and I think it would tend to lessen what male prostitution there is—I do not pretend to know how many male prostitutes there are, or their activities—but it would lessen those because nobody would want to buy something if they have a better thing in reasonable privacy.

[Q2622] ... [H]omosexuals are made so young that people who are homosexual are made either before their birth or in their make-up or things which happen to them in their extreme childhood. I do not think they are made subsequently to any considerable extent by what happens later on at twelve, thirteen or fourteen and I cannot ever conceive any large number of ordinary men and women who have very strong or persistent homosexual impulses—there is a certain number and always will be—I cannot see that it will either decrease or increase the number, and I do not think it would alter my own life one jot except that the fear which is always behind one that there might be a disaster at any moment, would be removed. But one has got so habituated to that, that in my own case it would make little difference.

I think it would make a great deal of difference to the attitude of a number of homosexuals I know who are extremely embittered and rather exhibitionist, and that is merely a protest against what they feel as an injustice against their security, and you get the lunatic fringe of homosexuals, the same sort of absurd people who exist on the fringe of any group of persons.

...

[Q2623] ... —A. (DOCTOR): The point I would like to be sure I had made clear first of all in my evidence is that this question of people becoming paederasts from starting off as ordinary homosexuals is not true, that they just do not: secondly, that homosexuals are made, the percentage of homosexuality in a person is decided at some age before puberty, I suspect before five, it may well be before ten, and subsequently it makes no whit of difference to their true homosexuality what they will do at forty.

...

It would to a large extent, a change in the law, save a very large group of relatively young casual offenders who would pass probably at the age of thirty into an ordered, useful, homosexual life and get over that promiscuous age, the same as heterosexual promiscuity. It would lead to a tremendous saving of national energy as well as economy in the money wasted on useless treatments, useless imprisonments, and damaging the lives of useful citizens and so on...

. . .

[Q2625] Q. [MR. REES]: There is this theory which indicates that a lot of people believe that there is a sort of unexploited pool of homosexuality which does not come to the surface now but would if the law were changed.—A. [DOCTOR]: It is just not true.—(MR. WHITE): My view is that any severe law must have a certain amount of deterrent effect. I cannot see how you could have a severe law with severe punishment which is not to some extent a deterrent. But I do not think it is such a big deterrent as it is supposed; I think it deters a number of people who know they are homosexual from doing the remotest thing about it and they lead—I have seen it with my own eyes so often—embittered solitary and frustrated lives, unable to make friends anywhere because they feel the mere mention of the fact that they were homosexual might have shocking repercussions.

There are some, especially among elderly men who grew up in a different climate and different time than that in which others grew up, that they have remained lonely, solitary, embittered and frustrated and I think that is the result of the extremely deterrent effect the law has: but I think it has not deterred anybody from being homosexual—not even if the heavens fall would it stop any homosexual from being homosexual.—(DOCTOR): If I may, I would ask Mr. White whether the majority of those embittered, frightened men are not so for social reasons, whether in fact a change of law itself would make any difference to them.—(MR. WHITE): It would not have the remotest effect on their lives, I think. On the whole I do not think they can ever look forward to very busy or happy old ages but that is neither here nor there.

. . .

*

(e) HO 345/8: The point of view of the ORDINARY and COMMON male homosexual[44]

. . .

THE AIM:

To present to the Committee for their impartial consideration, the point of view of the ORDINARY and COMMON male homosexual. This in the hope that they may, after due deliberation, agree that in this modern and enlightened day there is required the following simple CHANGE IN THE LAW, only and simply.

That sexual relations between ADULT males in PRIVATE should NOT be against the law. It is agreed wholeheartedly that such relations should continue to be illegal if:–

(a) They take place in public,
(b) One of the persons concerned is under the "age of consent" (18 years?)
(c) One of the persons is not a consenting party.

It cannot be emphasized too strongly that in putting forward these views, I am in no way countenancing or asking the Committee to countenance any relationship between an adult male and a minor; such acts are not those of the true homosexual and the law should be strengthened rather than relaxed, so far as these offences are concerned.

A sex variant who is inclined to impose himself upon children must be segregated immediately until he has ceased to manifest these tendencies, but there is no evidence which indicates that homosexuals harbour potential tendencies to attack children any more than any other group of men. Sexual adjustment established early in life is likely to be maintained and if a man is attracted to adults, he will continue to be thus attracted.

REASONS FOR REQUIRING THIS CHANGE:

(1) History, e.g. that of Greece, tells us that homosexuality has always been a part of society: the repressive measures of this country having so spectacularly failed (as proved by the numbers of recent prosecutions), it presumably always will be. The vast majority of homosexuals grow up quite naturally homosexual, i.e., there has been no sudden crisis such as seduction, or other shock, to produce the abnormality. For example, in my own case, I only gradually realised between the ages of seventeen and eighteen, whilst at the University, that my outlook was different from that of my fellow students, and in what exactly the difference lay. That is to say, that the sexual awakening appears to the individual concerned to be equally as natural a process to the homosexual as to the heterosexual. Granted that the sexual urge is a very fundamental one, it follows that the homosexual is VERY FUNDAMENTALLY different from his normal fellows, and that his WHOLE OUTLOOK is different. Why should such unfortunate people, who have not asked to be different, have done nothing to make themselves so and do not choose to be what they are, be treated like criminals? No one asks for orgies of any sort to be made legal; but if I meet another male of similar tastes to whom I am attracted, and that attraction is mutual, can there be any harm to anyone in, e.g. kissing him? Peculiar if you like; difficult to understand, yes; should be corrected if treatment is possible, by all means; but criminal—WHY? NO.

(2) Still bearing in mind that we are discussing the more "natural" (who is by far the commoner) homosexual, rather than the more unusual effeminate type of "pansy", what is the individual who slowly discovers himself to be so afflicted to do in order that he may live within the law? At present his ONLY course is to abstain entirely from, what is to him, natural lovemaking, for not only is physical sex denied him but even making love. That is to say he must, whether he wishes to or not, live a very incomplete life. This must be compared with voluntary celibacy of either the hetero- or the homo-sexual. The strain of the unnatural abstinence

must tend to make him more abnormal than he is. The doctors tell us that cure is hopeless in the vast majority of cases. Certainly the specialists failed to change the direction of my own libido despite the great expense to which my parents were put. And indeed it would be surprising if it were not so: is it to be expected that a heterosexual would be turned into a homosexual by treatment if the boot were on the other foot? . . .

(3) What possible good can a prison sentence do? Fear of it will not stop a funda-mental biological activity like sex . . . I have long ago decided, after the long, bitter and lonely battle of adolescence that prison is easier than abstinence. And that despite the fact that, as a professional man, I should not be able to earn my living again, and should thus be punished twofold. But the constant proximity of con-viction, seeing and knowing it happen to friends, has removed the fear of prison. Lord Chief Justice Goddard has said that it is the fear of the social stygma [*sic*] rather than the sentence itself which is the greatest deterrent. I, in common with many others, no longer have that fear.

(4) That homosexuality is only driven underground, partly by the law and partly by public opinion (but after all the force of the former depends ultimately on the latter), should be stressed. The effeminate types that the man in the street regards as the only type of homosexuals are, in fact, numerically few when compared to those homosexuals who are undetectable in dress or mannerisms, likes or dislikes, except possibly by another homosexual—and not always then. Kinsey shocked America, and, as he states, shocked himself with the revelation that more than a third of the total male population had had some homosexual experience, lead-ing to orgasm, AFTER ADOLESCENCE: and that 4% of the male population are exclusively homosexual for their entire lives. Taking the male population of Great Britain as 25 million, this gives a figure of a million homosexuals. This ignores bisexuals, and males who are exclusively homosexual for only a period of their adult lives. Even so such a minority must surely deserve a better fate than to be condemned as criminals in a country which prides itself on the fair treatment of minorities. Even if the law is changed now public opinion will continue to be intolerant for my lifetime; but perhaps for those homosexuals yet unborn life will be happier.

These "ordinary" homosexuals consist of men of all ages, ranks and occupations. I personally know homosexuals in the Church, the Law, Parliament, Medicine, Veterinary Surgery, the stage, the three armed services, working down coalmines, as cowmen, on the land, etc., etc.

. . .

Further points for the consideration of the Committee:

(a)   Seduction of the young as a cause of homosexuality is grossly exaggerated.

. . .

(c)   The homosexual today lives like an escaped prisoner, waiting constantly to be discovered. Yet, search as he may, he can find nothing he has done to justify this attitude of Society against him... [T]he homosexual tends to become anti: antipolice, antisocial, antiwomen, antietc., developing an ever-growing antagonism which inevitably prevents him expressing his full usefulness to society.

...

(g)   May each member of the Committee say to himself—"There but for the grace of God go I", and remember that each day there are being born children who, in the present state of our knowledge, must inevitably grow up homosexual and who will live either to bless or to curse your findings and your recommendations.

*

(f) HO 345/8: Memorandum on certain aspects of the problem of Male Homosexuality as seen by a Homosexual Medical Practitioner[45]

...

(The writer's opinions have been formed from personal experience and from intimate knowledge of more than 200 homosexual men; they have not been formed by reading text-books written for the most part—by heterosexuals.)

[1.] Distribution of male homosexuality:

Fears have been expressed in Parliament, in Courts of Law, in the Press and in medical Journals that male inversion had assumed the characteristics of an "Epidemic" in post-war Britain.

These opinions indicate an ignorance of the true facts.

(a) Male homosexuality exists today in every country of the World, amongst peoples of every race, culture and creed. There is evidence that it has so existed since the dawn of human history.

(b) I submit that there is no evidence that there has been an increase in homosexuality in Britain, i.e. in the percentage of the total population which is so conditioned.

That there has been an increase in the amount of homosexual behaviour (an entirely different matter) is very probably [*sic*].

But such increase is comparable in every way, I submit, to the undoubted increase which has also taken place during the last half century in the total amount of heterosexual behaviour, i.e. more promiscuity amongst the young, more adultery, more fornication. Parental control has been much reduced, contraception popularised, sexual taboos discarded, and sex in all its aspects glamourised and publicised.

(c) Most of the widespread homosexuality is only visible to those who are themselves homosexual. There is no secret most homosexual men guard more assiduously than that they have homosexual desires.

Usually no man—whether his skin be white, black or brown—will reveal his homosexuality to another whom he believes to be heterosexual; not even if the latter is a medical practitioner. The violent emotional prejudice against the male invert can be as great, or greater, in the medical practitioner than the general public.[46]

This probably explains why, to the best of my recollection, no patient during my twenty years of busy general practice has ever consulted me regarding his homosexuality except those who had prior knowledge of my own inversion.

Yet, I have known personally and discussed the subject intimately with more than 200 men of all ages, of several races and creeds, and many nationalities, of all whom [*sic*] had had active homosexual experience.

(d) Statistics of police prosecutions are indications of the extent of police activity in this regard, not of the number of men with homosexual instincts who exist in the total population.

2.   Homosexuality an inherent biological characteristic?

I submit that no evidence exists which can refute the following submissions:

(a) No man, whether he be normal or abnormal, chooses the direction his sexual desires will take.
(b) A man has control over his sexual activities but not over his sexual instincts.
(c) No sane man, endowed with normal heterosexual instincts, would choose to become a homosexual. Why should he? In what way would he gain?
(d) Nobody seeks an aetiology for heterosexuality because it has been accepted as an inherent biological instinct.

Is there any evidence that homosexuality and bisexuality are not also inherent biological characteristics?

Whether all three types of instinct are genetically determined, the result of hormonal influences or environment, etc., etc. is immaterial in so far as they are all equally beyond the control of the individual.

No man has yet scientifically established a cause for homosexuality which can refute the argument advanced in this paragraph.

(e) There is no proof that the direction of the sexual instinct can be permanently altered by any known means—not even by homosexual seduction in childhood.

3. Prevalence of male homosexuality:

It is my considered opinion that Kinsey's figures in this regard are approximately correct for most countries, not merely the white population of the U.S.A.

For thirty years I have been in the habit of making estimates of the extent of homosexuality in small communities and groups of people of whose sexual habits I have had knowledge.

. . .

In the town in which I have practised for 20 years, there are at present approximately 60 male medical practitioners. Of these I have definite knowledge of the homosexuality...of five. There are three others of whose homosexuality I have suspicions but not definite knowledge.

...

I would classify male homosexuals as follows:

1. Completely homosexual men i.e. inverts

Such men are incapable of any form of sexual or emotional relationship with women—about 5% of adult men.

2. Bisexual men with a predominance of homosexuality

This group includes many men who only discover their true inclinations after marriage and perhaps parenthood.

Others marry in the misguided hope that this will cure their anomaly or as a 'camouflage' protection against the gossip of their friends.

Such men are usually potent for a short time (i.e. whilst the 'novelty' lasts) but are unlikely to form any satisfactory emotional and sexual relationship with a woman on a permanent basis—at least 5% of adult men.

3. Bisexual men with a predominance of heterosexuality but with definite homosexual tendencies of which they may, or may not, be aware:-

Such men may indulge in homosexual activity occasionally, incidentally, or for limited periods depending upon temptation, opportunity and the existence or not of a heterosexual partner.

This type of activity is a question of "Faute de mieux" and is encouraged by conditions in P.o.W. camps; National Servicemen barracks especially abroad; in all boarding Institutions where there is segregation of the sexes—probably about 20% of adult men.

For practical purposes groups 1 and 2 above constitute the homosexual proportion of the male population and amount to about 10% of the male population. That few people are prepared to accept such an estimate does not prove—through lack of information available to them—that the estimate is incorrect.

Note: (a) The opinions of psycho-sexual experts, prison medical officers, social workers, magistrates, lawyers, etc. are distorted by the fact that they only see the homosexuals who are neurotic, mal-adjusted or the object of a police prosecution.

It is my opinion that less (perhaps much less) than 10% of male homosexuals have ever:

(i) consulted medical opinion regarding their anomaly, or,
(ii) been before the police on a sexual charge.

(b) It is completely fallacious to believe that more than a minority of male homosexuals can be detected by physical or other external characteristics. Most homosexuals dislike male effeminacy. Transvestites are by no means always homo-sexual: (vide Kinsey 1953: p. 680[47] and "Roberta Cowell's story"[48]) I have known several effeminate men who were not homosexual.

4. Ethical considerations:

No Man has yet produced evidence and proof that the homosexual can be held personally responsible for the anomalous direction of his sexual desires

(a) By what right, therefore, do the Law, the Church, and heterosexual society deny to the invert the only form of overt sexual activity of which he is capable providing:–

   (i)   no offence to public decency is involved
  (ii)   the homosexual partner is a freely consenting adult
 (iii)   no harm is being caused to any third party?

(b) By what principles of Justice does the Law of Britain commit to prison for private homosexual acts consenting adult males whilst it permits the following anti-social activities:–

   (i)   fornicating with the procreation of illegitimate children
  (ii)   adultery with the disruption of family life if divorce results therefrom?

If it should be alleged that such homosexual acts are also anti-social activities, I would ask in what way?

   (i)   The world faces over, not under, population in the future
  (ii)   Marriage for the homosexual man is rarely either successful or happy. When separation, divorce or tragedy results the wife and family—if any—are also involved.
 (iii)   Anti-social and immoral are not synonymous terms

(c) By what principles of Justice does the Law of Britain

   (i)   permit female homosexuality whilst punishing with imprisonment male homosexuality?
  (ii)   permit a man to have intercourse with a prostitute with impunity? The woman is liable to be fined 40/- "for being a nuisance!" Yet consenting adult homosexual males are sent to prison for their acts in private.

If it be argued that heterosexual intercourse is a natural activity designed for the procreation of children whilst homosexual activity is unnatural or "against Nature" I would ask the following questions:–

(i) <u>Contraception</u> deliberately renders intercourse sterile. By no process of logically [*sic*] reasoning can it be classified as biologically natural. Why therefore is it legally permitted (and incidentally socially approved)?

(ii) Is masturbation natural biologically if sex was only intended to be used for the propagation of the species? It is legally permitted.

(iii) <u>Are the widely practiced fellatio and/or cunnilinctus as performed between a man and a woman criminal offences in Britain?</u>

(a) <u>If not, WHY?</u>

They are as immoral, as sterile, as unnatural as the majority of homosexual acts between consenting adults for which the latter are sent to prison.

(b) <u>If they are Criminal offences under British "Justice"</u>

How many men and women have been committed to prison for these offences during the last 25 years?

There is only one reasonable explanation for the extraordinarily illogical attitude in Britain, not only of heterosexual society, but of the Law. It can best be summarised in the words of Samuel Butler:–

Heterosexuals

> "COMPOUND FOR THE SINS THEY ARE INCLIN'D TO
>
> BY DAMNING THOSE THEY HAVE NO MIND TO."[49]

...

9. <u>Concerning a homosexual "Age of Consent":</u>

It has been suggested[50] that, should adult male homosexuality become legalised under certain conditions, a higher age of consent should apply, e.g. 21 years instead of 16 years.

<u>I submit that the arguments for so doing would be emotional rather than logical:</u>

(a) There is no evidence that a youth of 16 can be made into a homosexual adult unless he already has homosexual or bisexual tendencies ...

...

(d) Large numbers of National Servicemen indulge in homosexual activities, especially abroad in places where "palatable" and "hygienic" heterosexual partners are scarce, e.g. Korea, Africa, etc...

In my opinion no permanent damage is caused by such activity. Those who have predominantly heterosexual instincts will inevitably revert to heterosexual behaviour...

<u>To make such homosexual activity illegal does not prevent it as is already obvious.</u>
But making it illegal does lead occasionally to Courts martial, police prosecutions,

prison sentences, wrecked lives, even suicide. When such disasters occur—and they do regularly—it is my opinion that the punishment has caused greater harm than the crime.

...

12.   Personal history of the writer:

Bachelor; aged 46; educated in a Roman Catholic (Jesuit) boarding school. Qualified Bachelor of Medicine, Bachelor of Surgery in a British University (Scotland).

...

I have neither attempted—nor felt any desire for—heterosexual intercourse for about 20 years. I have during my life had very extensive (and from necessity promiscuous) homosexual experience in Britain, France, Italy, Scandinavia, Africa and other parts of the world...

I believe I show no traces of effeminacy either physical or in my interests. I am not attracted by effeminate men.

In the Kinsey Heterosexual-Homosexual rating scale (p. 638, 1948)[51] I would place myself in <u>column</u> 5, i.e. predominantly homosexual but incidentally heterosexual.

Looking back upon childhood I can discover no factor to which I can attribute my homosexuality. Perhaps for this reason I have never experienced any feeling of "guilt" in the matter for which I am convinced I was in no way responsible.

...

I believe the British Laws against male homosexuality are barbaric and inhumane. I have person [sic] knowledge of suicides[,] alcoholism, wrecked lives, which—quite unnecessarily and unjustifiably—this ignorant and prejudiced legislation has caused.

I have lost most of the respect I once had for "British Justice".

...

The British Police Force—reputedly incorruptible—has long ago realised the unique possibilities for undetected blackmail which are open to <u>two</u> police officers who decide to blackmail not only the homosexual man, but any man they choose to incriminate...It is my opinion that probably not one in one thousand of such cases ever reaches the police court. The risks are too great: the victim pays up.

...

4

Christians, Moralists and Reformers

Introduction

This section includes a miscellany of concerned voices in the public arena. The Public Morality Council's report [(a)] relied largely upon medical expertise and therefore produced a variation on a familiar theme: homosexuals should be divided into inverts (untreatable; they should not be prosecuted) and a variety of others (many of whom should be treated—rather than imprisoned—before they became habituated to their homosexual practices). The Ethical Union also advocated reform [(b)], making a clear distinction between the law and morality. What others considered to be morally wrong or distasteful, it argued, was no grounds for the law to interfere in private, adult, consenting conduct. This was a straightforward echo of John Stuart Mill's *On Liberty*, which was to be reflected in the logic of the Wolfenden Report and the Hart–Devlin debate (p. 262) as well.

Of the churches, only the Anglicans and the Catholics submitted evidence. The Church of England Moral Welfare Council's memorandum [(c)] (published as *Sexual Offenders and Social Punishment* in 1956) was drafted by the Rev. Dr Derrick Sherwin Bailey and—although it did not reflect the opinion of the Church as a whole[1]—it spoke for an influential strand of thought. Bailey had also drafted the Moral Welfare Council's *The Problem of Homosexuality* in 1954, which was one of the prompts leading to the setting up of the Wolfenden Committee, and he submitted his larger historical study *Homosexuality and the Western Christian Tradition* (1955) as evidence to the committee.[2] He, like most other commentators, divided his homosexuals and mingled his Ellis and Freud: genuine inverts were probably beyond cure, but skilled treatment might be able to coax along to emotional maturity those suffering from arrested development. Since the immoral practices of fornicators and adulterers were more harmful to society than those of homosexuals, but remained legal, the law needed to change. It was the province of the Church, and not the state, to deal with sin and be the guardians of private morality—a notion with which the Roman Catholic Advisory Committee concurred [(d)]. So this was not a call for an acceptance of the validity of homosexual acts—they still remained sinful. But, in his suggestion that two inverts might find

233

salvation in chaste 'home-making', Sherwin was an early advocate of a form of gay civil partnership.

The Progressive League [(e)] and the Howard League [(f)] both unsurprisingly advocated reform. The main novelty in the former's memorandum, again emphasizing the frantic drawing of lines between homosexuals and paedophiles, was the suggestion that criminalization, by outlawing both groups, had caused them to bind together in a 'closely knit freemasonry'; reform would break them apart by co-opting decent homosexuals. Protection of youths naturally preoccupied the National Voluntary Youth Organisations as well [(g)], but their memorandum is symptomatic of the kind of logical semi-coherence or plain incoherence that characterized many of the witnesses' statements. They could not agree on whether to recommend reform, and certainly did not want the law changed for youths in the age group with which they were concerned: that is, up to the age of 21. They had absorbed enough Freud to believe that many boys and girls went through a homosexual phase during puberty and early adolescence, and favoured education over prosecution for those aged 15 or under who committed homosexual acts, on the grounds that they probably were not genuine inverts. At the same time they stressed the responsibility of parents, clergymen, teachers and youth leaders to inculcate the necessary religious and ethical values to prevent boys from going astray. All of this would appear to leave young men aged between 16 and 20 vulnerable to prosecution if they acted on their 'genuine' inversion or if the forces of clean living that surrounded them failed in their quest to guide them to a secure heterosexuality. This was the necessary if unintended corollary of the stated aim 'that our young members must be protected against sin and perversion'.

The final witness included here was Conservative politician and lawyer Quintin Hogg, 2nd Viscount Hailsham [(h)]. He was the only politician to write a memorandum and to appear before the committee, and it is not clear what drove him to do so.[3] He was widely credited with having a brilliant intellect, but on this occasion it seems to have let him down rather badly: his views on homosexuality were strong on assertion and strikingly deficient in evidence. Since every 'normal' person possessed homosexual tendencies, he claimed, he was 'quite certain' that homosexuals were made—initiated by older homosexuals while the personality was still pliable—and not born. Only that could possibly explain the recent spike in police statistics. Since they used their bodily organs for physically incompatible functions, nearly all the homosexuals he had known were emotionally unstable and profoundly unhappy. As homosexuality was a proselytizing religion, and the consequences of contagion profoundly unsettling for both individuals and society, continued criminalization of even private, consenting, adult homosexual acts was sadly necessary. And he returned to the old chestnut that discreet, respectable queers had little to fear since the law was rarely applied to them. The committee tacitly gave its verdict on the eminent statesman's opinions by adopting none of them in the Wolfenden Report.[4]

*

(a) <u>HO 345/7: Interim Report of a Sub-Committee of the Public Morality Council[5]</u>
<u>Appointed to Study the Problem of Homosexuality, January 1954[6]</u>

...

TREATMENT OF VARIOUS CATEGORIES

<u>Category I</u>	The invert. He should not be prosecuted, provided that his sexual activities are not anti-social, do not lead to delinquency, do not involve pederasty or seduction of youth, and do not cause a breach of the Peace. Punishment cannot be a deterrent to the invert; it can only make him more wary.
<u>Category II</u>	The immature or oscillating type. These need treatment before they become habituated. If there is no breach of the peace, no anti-social act, no pederasty, no bribe and no associated delinquency, they should not be sent to prison.

...

<u>Category III</u>	The Psychopath. He commonly needs detention to protect the public...
<u>Category IV</u>	There are other easily defined psychiatric cases where the mental abnormality needs treatment in a Mental Hospital, and those with a marked degree of mental backwardness need special provision.
<u>Category V</u>	The bisexual. There is a very big group that could be said to be in an oscillating bisexual phase, rather than truly homosexual. Unless these are properly treated, they are likely to become truly homosexual and irreversible.

COMMENTS

...

A great deal of simplification could be arrived at, if a code were drawn up in regard to sexual behaviour which applied equally to both sexes. Anti-social behaviour should be suppressed by punishment:–

a　Where pederasty or corruption of youth occurs.
b　Where there is duress, influence of any kind whereby the victim gives way through force, fear, bribe, deception, etc.
c　Where there is indecency that threatens a breach of the peace, or public order; or perversions of a gross or undisciplined kind.
d　Sexual commercialization.

At present there is no clear distinction made between the invert whose homosexuality is not open to treatment, and the other forms of homosexuality. Where these show clearly nurturally and environmentally acquired causes in their make-up, they are, in the early stages open to treatment because the individual is not habituated. Even where treatment is available in prison the unbiological and custodial regime is prejudicial to its success.

What is surprising is that the law which holds homosexuality to be highly reprehensible in all its forms, does not consider the effect of homosexuals in prisons
on other prisoners. They are an extremely unsettling factor in prison.

*

(b) HO 345/8: Memorandum of Evidence from the Ethical Union, April 1955

The Ethical Union was founded in 1894[7] ...

The object of the Ethical Union is to state ethical principles, advocate a religion
of human fellowship and service based upon the principle that the supreme aim
of religion is the love of goodness and, by human and natural means, to help man
to love, know and to do the right in all relations of life.

Legislation making certain types of sexual conduct criminal can only be justified, in the view of the Union, if the conduct is either socially dangerous or
a nuisance to other people... Law should not be an instrument for penalising
actions which are considered, even by the majority to be morally wrong, or which
are abhorrent to most people.

It is necessary to recognise that a large, and especially well informed, body of
professional opinion now accepts very great divergencies in what may be considered norms of sexual behaviour. It is also established that there is a wide variation
of personal conduct amongst individuals in their sexual activities.

...

[The existing homosexual laws] have not changed fundamentally since the
nineteenth century. Many people now consider that they are cruel and cause
punishment to be inflicted upon individuals who need treatment and not punishment. Moreover, a section of public opinion has come to see nothing wrong in
many of the acts made criminal, e.g., homosexual relationships between adults in
private. These laws are, therefore, ineffective and cannot properly be enforced.

...

It is impossible to see any rational ground for the penalising of homosexual
acts of any nature partaken of by consenting adults in private. The distaste that a
normal person may feel for such conduct should not be a basis for criminal legislation. Admittedly there is a long history of legislation against homosexuality in this
country, but history also shows the inability of the law to suppress homosexuality.
Indeed, homosexuality, so far as evidence goes, seems to be a fairly constant and
consistent thread in the fabric of human society and behaviour.

...

[H]omosexual relations with the young should continue to be criminal. Homosexual conduct in public places, such as lavatories, the street or parks should also
remain subject to the criminal law, but what two adults choose to do in private
should have nothing to do with the criminal law ... In such circumstances they

are the only judges of what is right for them and, if the criminal law interferes, it only reflects the moral prejudice of the majority. We deny any right in majorities to translate their prejudices into law in this way.

The age of consent should perhaps be 21. This may seem a little high, but it is doubtful if public opinion would at present accept a lower age. But we believe there is a provision in Swedish law, which seems to us of importance, dealing with the question of homosexual acts between persons of the same sex, both of whom are between the ages of 16 and 21. In Sweden, we are informed, homosexual acts committed by such persons in private are not illegal provided that the elder of the two has not abused the dependence of the younger person upon him.[8]

It is, moreover, undesirable that homosexual acts between schoolboys should be the subject of the criminal law. Acts between persons of the same sex, both of whom are under 16 and neither of whom is under compulsion, should not be criminal. Such cases should be dealt with either by parents or schoolteachers...

...

*

(c) HO 345/7: *The Homosexual, the Law, and Society*: Evidence Submitted by the Church of England Moral Welfare Council[9]

...

...[C]onfusion has resulted from failure to distinguish between the homosexual condition and homosexual acts...

...

...Although most males and females exhibit that decided propensity towards members of the complementary sex which is rightly regarded as normal and natural in human beings, it is incontestable that a minority displays an equally marked orientation towards members of the same sex. Inversion as a personal and social problem, therefore, must be carefully distinguished from the homosexual practices which may (or may not) be 'offences' of which the law takes cognizance, or sins upon which moral and theological judgements must be pronounced. The fact that certain homosexual acts committed in certain circumstances may be penalized by statute or condemned by religion and morality does not imply that the homosexual condition, per se, is immoral or culpable.

...We may distinguish two kinds of personal sexual condition, both of which are morally neutral, and may vary in intensity from one individual to another:

(a) the normal and natural, or heterosexual condition;
(b) the abnormal and (by comparison with heterosexuality) relatively uncommon condition of homosexuality, or inversion.

But no such precise differentiation can be made between types of sexual conduct. At the two extremes there are, as we have noted, inverts who have had no heterosexual experience, and normal persons who have had no homosexual experience. Between these extremes a great range and variety of sexual behaviour will be found...

This has led to the postulation of a 'bisexual' condition which finds expression indiscriminately in heterosexual and homosexual acts; but it is doubtful whether the so-called 'bisexual' is, in fact, more than a convenient fiction. In practically all alleged cases of 'bisexuality' there is a definite basic personal sexual orientation, but the individual's adaptability and inclination to experiment produce deviations in behaviour which may mislead those who try to classify him or her according to 'type' on the evidence of conduct alone...

...

The evidence which we have received confirms the view that one of the most frequent predisposing or precipitating causes of inversion is an unsatisfactory emotional adjustment in childhood. This may be due to an unhappy marriage, to a faulty relation between a child and one or both of its parents, or to circumstances such as death, divorce, or prolonged war service by which a child has been deprived of father or mother...

Another cause to which inverts attribute their condition is sexual segregation during childhood and adolescence...

...

"Society gets the homosexuals it deserves", declares one invert—and his charge is well-founded. By unhappy marriages and homes, by inept handling of youthful problems, by prolonged segregation of the sexes, and by war and its consequences (to mention only a few factors), society itself creates just those situations which cause inversion, or lead to the adoption of a homosexual attitude to life. Yet should the male invert give physico-sexual expression to his impulses, even by way of demonstrating his affection for another man, society treats his conduct as criminal and imposes penalties which many regard as excessive for the 'offence' committed, and unworthy of a civilized people. Since the whole discussion of homosexuality tends to revolve around this question of homosexual practices, especially between men, there are one or two points to which it may be helpful to draw attention here.

It is still commonly supposed that the practising male invert favours only one form of physical expression—namely, sodomy. It is worth placing on record, therefore, the protestation which we have received from many male homosexuals that sodomy is often quite uncongenial, if not repulsive, to the invert—though he may desire to show affection in other physico-sexual acts...

There is no evidence that homosexual practices themselves are either more or less harmful (according to circumstances) than heterosexual practices. Many undoubtedly connect homosexual practices with the danger of corrupting the young, and appear to take it for granted that every invert is actually or potentially a paederast, and liable to attempt indecent behaviour with any boy or youth who associates with him. While we have every sympathy with those who are anxious to protect the young, it must be said that there is no ground for such an assumption as this; in itself it is productive of harmful consequences, and can cause embarrassment in the most honourable associations between man and boy. There even seems to be abroad the notion that inversion is actually a contagious condition—which is nonsense; at most, a boy's or a girl's homosexual experience can only precipitate a pre-existent but latent condition of inversion. Inverts have repeatedly stressed the fact that the genuine male homosexual is drawn towards the older youth and the mature man, and not towards the boy—though some may associate with and seek physical satisfaction from boys because such contacts are easier and less fraught with the danger of exposure or blackmail. The paederast proper, who seeks none but the young, and often pre-pubertial boy, constitutes a distinct problem, which should not be confused with that of ordinary homosexuality.

...Some experts claim that 'cures' have been effected, and are possible in many cases; others hold that little, if anything, can be done to reorientate the homosexual. This is clearly a question for the experts themselves, and we would only stress the paucity of the data upon which any opinion must at present be based. It is certain that marriage is no 'cure', and may have disastrous consequences. We believe, too, that in most cases of genuine inversion it is not kind to raise what will generally be false hopes of a 'cure'; the best help that the real invert can be given is the promotion of mutual sympathetic adjustment between himself or herself and society.

There are, however, cases which appear to be homosexual, but actually do not come within that category. Often the subjects are simply the victims of an arrested emotional development, and skilled treatment can generally assist them to attain emotional maturity, provided they are willing to co-operate. Such persons may experience both homosexual impulses and a real desire to marry, and here, adjustment is possible; but successful treatment will not constitute a 'cure' of a homosexual condition, for strictly speaking that condition was never present. It is important that arrested emotional development (which may go with considerable intellectual maturity) should not be confused with inversion.

Adult inverts are not only doubtful of the efficacy of curative treatment, but often unwilling to submit to it. They recognize that their condition, for better or worse, is part of their personality, and they relish the idea of being 'tampered with' no more than would the heterosexual, if offered the chance of becoming an invert. They appreciate that they are 'different' through no fault of their own, but are not convinced that this 'difference' constitutes any failing or deprivation. On the contrary, many inverts feel that their condition gives them the opportunity to make a definite and unique contribution to the good of the community, if only society would accept them and their gift with understanding and sympathy ... They only ask that society will remember its debt to many homosexuals of genius and distinction—and that it will also give them their chance.

...The homosexual's hope is not that he may become other than he is, but that he may be accepted by his fellows <u>as he is</u>. Society and the law take no cognizance of the invert until he commits an 'offence', with the result that all inverts are assumed to be sodomists or the like. It is necessary, therefore, to recognize that there is a large body of inverts who never obtrude themselves upon the notice of the public, and who are concerned far less with freedom to indulge in physical acts than with removal of the stigma which brands them as outcasts and 'queer'. For them, even cessation of the revolting abuse heaped upon the homosexual is not enough; they wish to be received by society, not with smug condescension and charitable virtue, but with kindness and understanding. The invert sees himself (to quote one of them) "a proud and honourable man... [sic] not ready to be patronized as an unfortunate"—yet for all that, he is an unfortunate, and we have learnt to deal kindly with other unfortunates in our midst, without causing them embarrassment. His plea is for justice and friendship. He sees that even the worst of criminals gets a fair hearing, and that for some offenders there is even a ready tolerance and a search for extenuations. But for him, saddled with a burden the pressure of which has moulded, but not necessarily deformed his personality, there is nothing but the loneliness of a restricted and sometimes unnatural circle of companions, if not the curse of an almost solitary existence. And should he slip ... there is the threat of prosecution, heavy penalty, and social ruin ...

. . .

... Why should the community look with suspicion upon two men who do what women can do without exciting any comment—that is, set up house together? Two inverts who are congenial may find their salvation in the enterprize [sic] of 'home-making'—and it is arguable that society should encourage them, and should not impute to them the basest of motives.

When we consider the law as it stands, we find that it is not even equal in its injustice. While the male homosexual is heavily penalized for his offences, the female homosexual is ignored, and can do what she wishes with impunity. Yet socially she is often dangerous. An older woman can dominate a younger, and can compel her to acquiesce in a lesbian liaison which may ruin her life. Even more serious, a persistent lesbian can break up a marriage by seducing the wife, or by insinuating herself into the home—and not a few such cases come to the notice of the pastor and the social worker. Offences against the young are grave, no matter by whom they are committed; yet it is noticeable that sexual crimes of violence (and even murder) are almost always committed by men against young girls, and very rarely (if ever?) by paederasts against young boys. Fornication between men and women in 1952 resulted in over 32,000 illegitimate births; and adultery was no doubt one of the principal factors in a majority of the 33,922 divorces made absolute in that year. Yet the adulterer, a serious social menace, goes unpenalized, while the fornicator is accepted with something approaching complacency by society at large. It is hard to deny the logic of the practising invert that he should at least receive treatment commensurate with the proportionate gravity of his offence. He contends, with some reason, that his private practices with consenting adults are harmless compared with the activities of the adulterer and the fornicator... If the law is to penalize sexual 'offences' as crimes, this should be done on a logical, equitable, and rational basis, regardless of historical precedents whose validity is doubtful, and of the emotional prejudices which tend to cloud discussion of the whole question.

Some homosexuals are undeniably vicious, and some are dangerous to society; and this is no less true of those heterosexuals who indulge in homosexual practices. The community must be protected against the paederast, the seducer, the exhibitionist, the ruthless lesbian, and the blackmailer; yet it must be recognized that in certain cases the offence may not be due to vice, but simply to psychological causes. To distinguish, as we have done, between condition and conduct, and to plead for sympathy and help for the invert, does not mean that his anti-social conduct must be condoned or excused. It is, we believe, the duty of the state to protect by its laws and its police the young, and the institution of marriage; it is its duty to maintain public order and decency; but it has a responsibility to see that its provisions are framed and executed not only effectively, but also equitably. On the other hand, it is the responsibility of society at large to see that those of its members who are handicapped by inversion are assisted to a constructive acceptance of their condition, and are helped to lead useful and creative lives—thus benefitting both themselves and the community, to the service of which their special gifts can often make an important contribution.

...

... [I]t is not the function of the state and the law to constitute themselves the guardians of <u>private</u> morality, and that to deal with <u>sin as such</u> belongs to the province of the Church...

...

We beg to submit the following specific recommendations for the Committee's consideration:

1. That sections 61 and 62 of the Offences against the Person Act, 1861; section 11 of the Criminal Law Amendment Act, 1885; section 1 (1) b of the Vagrancy Act, 1898; and section 1 (1) b of the Immoral Traffic (Scotland) Act, 1902, be repealed.
2. That legislation be introduced to penalize any male or female person who commits, or attempts to commit, or is a party to the commission of, or procures or attempts to procure the commission by any person of, any homosexual act,

   (i) with any person under the legal age of consent; or
   (ii) in circumstances constituting a public nuisance, or an infringement of public decency; or
   (iii) involving assault, violence, fraud, or duress.

...

*

(d) Report of the Roman Catholic Advisory Committee on Prostitution and Homosexual Offences and the Existing Law[10]

Section I

CATHOLIC TEACHING ON HOMOSEXUAL OFFENCES

I. Homosexual activities and desires to which the informed will gives full consent involve grave sin.

II. . . .

It is desirable to emphasize that sins of a sexual nature like any others derive from original sin and man who is endowed with free will is not the sport of fate in these matters.

III. The fundamental governing force in man is the soul with its free will and not the instinctive drives of the unconscious. However strong the instinctive drives in man they can be ordinarily governed and controlled by the will informed by grace.

. . .

VI. Whilst every sympathy must be shown towards homosexual persons, such persons must not be led to believe that they are doing no wrong when they commit homosexual acts . . .

VII. It is not the business of the State to intervene in the purely private sphere but to act solely as the defender of the common good. Morally evil things so far as they do not affect the common good are not the concern of the human legislator.

. . .

X. All directly voluntary sexual pleasure outside marriage is sinful.

. . .

XII. . . .

. . . Attempts by the State to enlarge its authority and invade the individual conscience, however high-minded, always fail and frequently do positive harm. The Volstead Act in the U.S.A.[11] affords the best recent illustration of this principle. It should accordingly be clearly stated that penal sanctions are not justified for the purpose of attempting to restrain sins against sexual morality committed in private by responsible adults. They . . . should be discontinued because:

(a) they are ineffectual;
(b) they are inequitable in their incidence;
(c) they involve severities disproportionate to the offence committed;
(d) they undoubtedly give scope for blackmail and other forms of corruption.

XIII. It is accordingly recommended that the Criminal Law should be amended in order to restrict penal sanctions for homosexual offences as follows, namely to prevent:

(a) the corruption of youth;
(b) offences against public decency;
(c) the exploitation of vice for the purpose of gain.

Section II

THE NATURE OF SEX INVERSION

. . .

II. Various views exist on the nature of sex inversion.

(*a*) Some authorities regard it as a congenital anomaly. According to this view a certain proportion of the population—a fairly conservative estimate would be 4 per cent—are so constituted that quite regardless of training, environment and in fact all external influences or individual experiences they are quite unable to develop normal heterosexual desires. Such persons, termed 'true inverts', may or may not exhibit physical or temperamental characteristics proper to the opposite sex. All grades are found from physical pseudo-hermaphroditism down to apparent physical and temperamental normality. The origin of this condition is not always clear, although there is much to suggest that it depends primarily on a lack of balance between the various glands of internal secretion. Apart from the misdirection of sexual impulse, inverts may be perfectly normal in every other respect although it is true that moral and social conflicts arising equally from the frustration of their sexual lives and the gratification of their—to them perfectly normal—impulses frequently lead to neurosis, but such neurosis is adventitious rather than part and parcel of their condition.

(*b*) Others who regard sex inversion as a psycho-genetically acquired misdirection of the sexual impulse form three main groups:

(1) Various members of the Freudian School maintain that we are all potentially bisexual, and that we all pass consciously or unconsciously through a homosexual phase. Sexual inverts are peculiar in that they remain stuck at or 'fixated' as the technical term is, or suffer a regression to this level of development. The cause of this developmental anomaly is to be found in the failure to resolve the well-known 'Edipus [*sic*] complex situation'.
(2) The Adlerians[12] interpret sexual inversion in terms of the inferiority complex according to which view homosexual males have a profound distrust in their own essential virility and ability to dominate the opposite sex. They attempt the treatment of inverts by their re-education technique.
(3) A third group maintains that sexual inversion results from the psychical shock associated with homosexual experience occurring in early or adolescent years. However, all the evidence tends to show that such experience in no way determines the direction that the sexual impulse will take in later years. The number of perfectly normal heterosexual adults who have had homosexual experience at various times during childhood and adolescence proves that that factor alone cannot account for inversion. Further we come across a large number of inverts who have either never indulged their homosexual tendencies or who have remained perfectly chaste till adulthood.

III. It is desirable to establish a distinction between sex inversion and homosexuality. The evidence is almost overwhelmingly in favour of the view that a conscious or unconscious phase of homosexuality is common to the race during the years of late childhood or adolescence. Further, a very large number of people experience sexual attraction towards members of both sexes. The term homosexuality should be restricted to homosexual desires or activities occurring in potentially heterosexual or in bisexual individuals. 'Sex inversion' should be applied to a sexual impulse which appears to be congenitally and ineradicably homosexual.

IV. Homosexuality and sex inversion have been known to occur from the dawn of human history. In Athens, for example, and more or less throughout Hellas for two or three centuries prior to her decadence, pederasty was encouraged until a man reached the age of about thirty, when he married as a civic duty. The same practice has been tolerated and is still exceedingly common throughout Islam.

Romantic friendships of the same kind were also the rule between adult Samurai and their pages in feudal Japan. Every culture develops its own code and standards. Our own Christian civilization is based entirely upon a heterosexual attitude and there is no danger of homosexuality being absorbed into its structure...

. . .

Section III

SUMMARY OF CONCLUSIONS AND RECOMMENDATIONS

. . .

IV. ... [T]he Committee, having taken note of the fact that a valid marriage may be contracted in this country at the age of 16, that the age of consent in France under the Napoleonic Code is 18, and that children are only amenable to the 'care and protection' provisions of the Children and Young Persons Act, 1933, up to the age of 17, nevertheless recommend that for the present purpose male persons should be deemed to be adult at the age of 21.

V. The Committee has reached the conclusions *(a)* that imprisonment is largely ineffectual to reorientate persons with homosexual tendencies and usually has deleterious effects upon them, and *(b)* that a satisfactory solution of the problem is unlikely to be found in places of confinement exclusively reserved for homosexuals. Accordingly, no positive recommendation is made with regard to methods of detention.

VI. The Committee regards with abhorrence arrangements understood to obtain in Denmark whereby homosexuals condemned to imprisonment may obtain release by voluntarily submitting to castration.[13]

VII. The Committee accept the propriety of the use for good cause under medical supervision of drugs to suppress sexual desire and activity, with the consent of the patient. Such treatment is permissible where serious pathological conditions obtain and when other remedies have proved ineffectual.

. . .

*

(e) HO 345/8: Evidence submitted by the Progressive League, March 1955

. . .

SECTION ONE

INTRODUCTION

1   The Progressive League was founded in 1932 for the study of political, economic and social problems and for the application of rational principles to their solution.[14] Its membership is largely drawn from the professional classes and includes teachers, lawyers, social workers, doctors, civil servants, psychiatrists and psychologists.

. . .

3   ... [T]he subjects which are under consideration by the Home Office Committee were dealt with by a conference held in London on March 20th and 21st, 1954. The Conference passed a resolution which was sent to the Home Secretary and other Members of Parliament and to the Press which included the following passages:

"Homosexual acts between consenting adults carried out in private should cease to be criminal".

. . .

SECTION TWO

HOMOSEXUAL OFFENCES

(A) GENERAL

5   When the League came to consider their evidence on homosexuality they were immediately struck by the fact that although the criminal law on the subject could only be described as draconic, the subject had never been discussed at length in the legislature or by an official commission or committee and that the paucity of factual information as distinct from expressions of unsupported opinion, either in literature or in the press, was alarming.

They therefore decided to carry out investigations in two ways:–

(i) By enquiries addressed to lawyers and suitable bodies abroad with regard to law, practice and public opinion in this subject in foreign countries:
(ii) By means of a questionnaire addressed to ordinary men and women of socially responsible categories in this country.

. . .

(B) ENQUIRIES IN FOREIGN COUNTRIES

7   We directed our enquiries abroad to France, where the law is essentially Roman in character, and to Sweden whose legal position is more similar to our own.

. . .

8   It will be seen that in both these countries, homosexual acts between consenting adults are not criminal, but that the protection of minors from homosexual seduction is a matter of grave concern.

These examples seem to establish that it is possible to combine a genuine and effective concern for the protection of minors with a legal attitude of neutrality to homosexual acts between consenting adults. Furthermore, it appears from the example of France . . . that where this is so the state of public opinion towards homosexuality is more healthy than in this country.

9   This result receives some confirmation from the following general considerations:–

(a) The English law, by treating as criminals all homosexuals who yield to their impulses, allows them no possibility of finding a non-criminal outlet by confining their attentions to adults:

(b) The operation of the English law makes many homosexuals martyrs in the eyes of adolescents and surrounds the whole subject of homosexuality with an aura of fascinating horror. This creates an unbalanced and unhealthy atmosphere which makes it difficult to deal with manifestations of homosexuality between juveniles in a sane and educative manner.

As an example of this tendency, it is well to consider the effect of the Oscar Wilde trials on successive generations of youth. Every educated boy sooner or later comes across Oscar Wilde as a minor figure in English literature, and learns at the same time that he suffered a harsh term of imprisonment and went to an early grave because of "serious offences". The student soon learns that these "serious offences", however repellent, did not [sic] harm to anybody and, if his reading extends to foreign languages, that Wilde had the sympathy of the whole civilised world. This situation handicaps those trying to help adolescents in their difficulties by investing them with the odium attached to persecutors and fanatics.

(c) The prosecution of homosexuals who are innocent of any relations with youths also leads to confusion in the popular mind, by making them objects of sympathy which often flows over to the whole gamut of homosexual conduct and weakens moral indignation against those who are a menace to youth.

The failure of the administration of the present law to enlist the moral support of the public at large was illustrated in a recent case[15] where convictions were secured in respect of homosexual acts with other adults on the word of accomplices. Many people considered that to incite young men to treachery against former friends was more damaging to them morally than repetition of homosexual acts to which they were already habituated.

(d) The present draconic state of the law tends to drive all homosexuals into a closely knit freemasonry irrespective of distinction between those who menace youth and those who do not. Even homosexuals who confine their practices to adults feel that they are outlaws and are driven to combat what they consider to be social injustice with any allies they can find. We believe this freemasonry to be extensive and cohesive.

(C) THE PROGRESSIVE LEAGUE'S ENQUIRY

10   The Progressive League's questionnaire was issued to all readers of the league's organ "Plan" and copies were also made available to members of the Ethical Union,[16] the Eugenics Society,[17] the Society of Labour Lawyers,[18] the Rationalist Press Association,[19] the Personalist Group,[20] and to the Lecturers and senior students of a provincial university ...

...

IV Comments on Replies to Questionnaire

(A) General

...

Opinion as to the Present Law.

The great majority support the Progressive League view as set out earlier in this evidence. Of these, however, about a quarter have some doubts or qualifications in their minds. Only three [out of 106] support the law as it is today ...

...

Returns of Special Interest

No. 3 This is a return indicating that the writer has practised homosexual acts extensively. The record begins in boyhood and continues up to the time of his marriage. Since his marriage homosexual acts have apparently ceased, and he enjoys normal sex life. He is, however, still drawn strongly to homosexual acts and anticipates that he will one day repeat them. He blames his parents for excessive emotional reaction when they learnt of a minor incident when he was a boy.

No. 10 The writer describes homosexual experience as a boy which he accepted unresistingly and later approaches which he repelled. He does not think he suffered any harm. He thinks that such acts among boys are a normal phase of development and call for no public attention.

...

No. 13 This return is of special interest. The writer, a lady, has furnished the following particulars of herself:– She started her career as a journalist and dramatic critic and later went into the theatre where she was a producer and playwright for the "little" theatres and in repertory and touring companies. She has been on the L.C.C. panel for years as a lecturer on dramatic art and has also lectured in the British Drama League, various Co-operative Societies and many other organisations connected with drama. She also coaches professional actors for stage and radio parts. She is the author of two text-books on acting and make-up and of two other books written under a pen-name.

On the basis of this considerable experience, she asserts that parts of the theatrical world are dominated by homosexuals. Homosexuals often make good actors. They tend to be emotionally sensitive perceptive [*sic*]. Let one such get the control of a company and he does his best to bring other homosexuals into all the more important positions. Newcomers of unknown type will receive homosexual advances and if they are rejected the chance of advancement, possibly even the post itself, is lost. Because the outside world rejects them, they tend to cling strongly together and support each other as and when they can.[21]

The writer describes one particular case of attempted homosexual assault by a man in the theatrical world aged about 30 on a youth of 17 whom he was interviewing for employment. The youth resisted and escaped at the time but later became a homosexual.

The writer intends to elaborate this theme in a forthcoming book.

No. 16 Describes an incident of mutual masturbation between two soldiers leading to one year's imprisonment for one and two year's [*sic*] imprisonment for the other.

. . .

No. 44 The writer appears to be mentally abnormal. He has a persecution complex. He is strongly opposed to any relaxation of the laws against homosexual acts.

No. 46 Regards homosexuality as harmful because it tends to lessen normal heterosexual affection and action which is essential for social well-being. The writer is a woman who describes a loss of sense of responsibility in her husband following a homosexual act. She suggests homosexuality is encouraged by our unsatisfactory standards in normal sex relations. If convention prevents a man from having sex relations except with a prostitute, then a homosexual affair may appear more romantic and expressive of genuine feeling. The remedy lies in facilitating normal sex relations on a non-commercial basis.

. . .

No. 60 As a practising physiotherapist [*sic*] has come to know of many cases of all types. Says that the law is extremely harmful because punishment or social condemnation brings "an increased sense of guilt which is itself the unconscious origin of homosexuality in its compulsive form. Guilt and resentment aroused in more superficial cases where it did not exist to really damaging extent before. Paranoid tendencies created or increased".

No. 66 The writer is well-known as an authority on sex questions. His comment on the present law is "that it is illogical, unjust and obsolete insofar as it is related to private acts between consenting adults but that it should be strengthened as regards seduction of minors." He recalls one case of suicide and one of imprisonment.

No. 68 Says that the law should protect girls under 21 from Lesbian women.

<u>No. 76</u> The writer is an avowed homosexual and gives a lengthy and well-reasoned defence of his views. In summary he asserts that his conduct harms nobody and by fulfilling his nature enables him to be a better citizen than he otherwise would be.

. . .

CONCLUSION

. . .

18   We submit that homosexuality is present in all human societies and that the extent of its manifestation varies from time to time and place to place owing to causes about which there is little or no knowledge. To talk about "stamping it out" altogether is idle and attempts on the part of the criminal law to attempt this end do more harm than good.

19   We recognise that the existence of homosexuality presents a problem to society and in certain circumstances may constitute a menace but not a serious one. We consider that the principal means of dealing with the problem should be through rational sex education, guidance of youth by parents, schoolteachers and others who have charge of them; and the inculcation of a sane and not a vindictive attitude to homosexuals by the leaders of public opinion.

20   While insisting that very little is known about the effects of homosexual interference with boys and youths by adult homosexuals, we do recognise that the present state of public opinion demands the penalisation of such interference by law.

21   We are of the opinion that legal penalties on homosexual acts should be confined to the punishment of adults who interfere with juveniles and that homosexual acts between consenting adults and between juveniles should not be legally penalised.

22   We cannot accept any differentiation in principle between street offences committed by men or by women. In our view the offences and penalties should be the same. Should, however, the Committee not agree with us and consider that public opinion regrettably still demands some distinction then we propose certain changes in the present law relating to offences by homosexuals in streets. For instance, to constitute an offence, there should be an element of definite nuisance to the public. We suggest that imprisonment of first offenders should not be allowed and that the accused should have a right to trial by jury whatever the charge.

23   The fact that the present state of the law provides opportunities for, and is even an incentive to, blackmail, has been so often emphasised by judges and other authoritative persons that it is only necessary for us to allude to it in order to complete our picture of our subject. We have not thought it necessary to collect any evidence in this respect nor in the related one of police corruption.[22]

24   We find that there is some evidence that homosexuals exercise a deleterious influence in certain professions, and urge that the attention of professional

organisations and of those prominent in the several professions should be drawn to this matter particularly as regards the theatrical profession.

...

APPENDIX B

Extract from a memorandum sent by the
  Institut de Droit Comparé
  University of Paris
  Faculty of Law...

In France, the question does not play a great part in public opinion, in the sense that while this vice is known to exist in certain artistic and literary circles—the example of Gide[23] is in everyone's mind—it has no repercussions in the life of the nation. It is considered to be a sort of snobbish cult restricted to very small area, and in France people do not get very worked up about this kind of question. It is generally considered that it is bound up with special physiological conditions, and that for the most part it is a medical rather than a social question. The few spectacular operations which have transformed men into women have had the effect of confirming this opinion in people's minds.

APPENDIX C

Extract from a Letter dated 29.9.54 from M. Robert Vouin, Faculty of Law, University of Bordeaux

---

In France, homosexual relationships are considered to be socially dangerous because of the ties of narrow solidarity which they create between those who indulge in them. This point has been evoked when opportunity offered at the time of one or two criminal cases. But convictions under Article 331, paragraph 3, are relatively few.[24] It is not proposed to extend the application of the penalty to acts of homosexuality committed with a complete absence of publicity by adult persons. And it does not seem that the impunity accorded by the law has had the effect of perceptibly increasing the number of those culpable acts which public opinion itself represses by mockery or contempt.

...

*

(f) HO 345/8: Memorandum submitted by the Howard League for Penal Reform[25]

...

A. HOMOSEXUAL OFFENCES

I. General Remarks.

1. Broadly, our position regarding the law is that we feel there is a need to distinguish between acts which are injurious to individuals or which offend public

decency, and acts which, whatever the morality of the matter, are performed by consenting adults in private. This is in line with the law as it stands at present, except for homosexual conduct between consenting male adults in private. Neither adultery nor fornication, for instance, is subject to criminal proceedings. Even though adultery may cause suffering, and fornication result in the birth of illegitimate children, the wisdom of allowing the State to regulate private sexual conduct is, quite properly, doubted. Where such laws do exist, as, for instance, in some of the United States, their effective implementation has always presented difficulties. It has also been put to us that, as far as protection of the young is concerned, the law on occasion achieves the opposite of what it sets out to do, in that some homosexuals who are attracted by minors yield to temptation more easily, since they render themselves punishable even if they find their partners exclusively among consenting adults...

2.... [I]t is certainly an anomaly that the law as it now stands discriminates between male and female homosexuality, a distinction which certainly does nothing to help the social readjustment of male homosexuals. Such a discrimination only underlines the sense of alienation from the community which such men experience already, and may reinforce anti-social tendencies. And in so far as certain provisions in the present law result in a number of homosexuals being needlessly sent to jail, it emphasizes one of the ironies of the penal system which causes homosexuals to be confined in the all-male atmosphere of prison, and prisoners without sexual outlet to be thrown into the company of homosexuals. Finally, the opportunities for blackmail which the present law affords would at least be diminished by the amendment we suggest.

...

III. Proposals for Changing the Law

9.   We would propose that it should no longer be an offence for male persons above the age of 21 to have sexual relations with each other, voluntarily and in private; and we recommend that in every other respect homosexual offences should be dealt with as if they were heterosexual offences, save that sexual relations between any male person above the age of 21 with any male person under that age should remain an offence. In view of the difficulties which arise out of the imprisonment of homosexuals, we further recommend that prison sentences should be imposed only as a last resort and should not exceed two years.

...

*

(g) HO 345/8: Memorandum of the Standing Conference of National Voluntary Youth Organisations (in association with the National Council of Social Service), 15 June 1955[26]

...

1. Introduction:

We are, as a group, concerned with the welfare of boys between the ages of 8 and 21 who are members of youth organisations.[27] We have a responsibility for their moral upbringing and are, consequently, concerned to see that the adults who act as leaders are men of good character ...

...

2. Extent of the Problem:

...

(a) [A]lthough more cases of homosexual practices between men and boys (including leaders and members of youth organisations) are now reported, we wonder whether these practices are in fact any more prevalent now than they were twenty, thirty or forty years ago. It is our impression that such cases are now more readily talked about and brought to light so that legal or other action results. The problem itself, as far as youth organisations are concerned, has not in our opinion suddenly become acute or shown any marked variation in recent years and, while the strictest vigilance remains essential, it would be quite wrong to imagine that the evil was widespread.

...

4. The Legal Position: (1) offences involving an adult and a young person:

The care and protection of the immature is [*sic*] a plain social obligation and we wholly share the common view that the corruption of the young should remain a criminal offence.

5. The Legal Position: (2) offences involving consenting adults:

It has been publicly suggested that private homosexual activities between consenting male adults should cease to be criminal offences. We have no expert knowledge about this aspect of the problem. Our views on it diverge widely and we are, therefore, unable as a group to express an opinion.

We are, however, agreed:

(a) that if the law were changed, youth organisations would be presented with a greater problem in detecting and excluding homosexuals.
(b) That it would be very important to prevent any widespread misunderstanding to the effect that a change in the law implied that the behaviour previously considered criminal had suddenly become socially and morally acceptable.
(c) That, should any change in the law be contemplated, we should view with great misgiving any legalisation of homosexual practices involving young people in the age-group with which we are concerned, i.e. up to 21.

6. The Legal Position: (3) offences between Young Persons.

It is generally recognised that during puberty and early adolescence, many boys and girls pass through a "homosexual" phase; their interests and emotions are temporarily engaged with others of the same sex, but normally they grow quickly out of this stage. Sometimes, however, undesirable practices occur; these are only rarely the symptoms of a genuine inversion, yet at present they remain, at any rate technically, criminal offences and may result in police court proceedings. We are of the opinion that up to the age of about 15 and where both persons are of about the same age, prosecution is not generally the right course. Education is needed rather than prosecution and the offenders are best left to be dealt with by those who have responsibility for the boys' moral and spiritual welfare.

Where an older adolescent, of perhaps 17 or 18, is concerned in active homosexual practices with a younger boy, of perhaps 13 or 14, the psychological background is different. Our view is that such cases are best left, as at present, to the discretion of the police. It may still often be true especially where the boy comes from a good home or where good educational resources are available, that prosecution is better avoided. A careful study of each case is necessary before the most appropriate action can be determined.

. . .

8. Home Life and Education.

(a) The aspects of the problem which seem to us by far the most important are those of upbringing and education. Whatever the legal position, the surest defence against wrongdoing is that built up by each individual within himself, and it is here that the parents have the major part to play. If a boy is brought up in a home motivated by religious faith and values, where affection and stability are assured, where the parents are emotionally in harmony and play their proper part in the upbringing of their children, the chances of the boy's going astray are lessened. Nothing can wholly take the place of a good home life. It is then for the community to support the parents' efforts by providing a wholesome environment and an understanding education.

(b) Homosexuality has to be seen, as one possible form of deviation, against the whole framework of normal sex and personal relationships. Education designed to safeguard or fortify the young against homosexual and other undesirable practices cannot be something narrowly conceived, a detached fragment of instruction. It is based on general education of a religious, ethical and aesthetic kind which nurtures the young in an appreciation of values. This is a continuous process and all who at any stage are in charge of young people—as parents, clergy, teachers, youth leaders, those in authority in industry—share the responsibility for it.

. . .

9. Conclusion:

We are bound to uphold the position that our young members must be protected against sin and perversion and must also be fortified to overcome these if they meet them. While we recognise that the genuine invert may be so congenitally, we are sure that others become homosexual through circumstance—perhaps emotional entanglement or immaturity, perhaps corruption by others. Therefore, in the formative years up to 21 (and leaving on one side the inverts who require special treatment), the principal concern should be to strengthen and improve home life to prevent corrupting circumstances and to provide the right educational and emotional support. The legal question should, in our view, be seen against this background.

*

(h) HO 345/9: A Statement by Viscount Hailsham, Q.C.[28]

Out of the welter of conflicting opinions and prejudices, one fact emerges beyond dispute. Male homosexual practices known to the police are running at a rate between four and five times that of 1938...

...Since there is no evidence of any change in the detection rate for homosexual offences, and, since it is quite impossible to postulate in so short a period a change in the congenital inheritance of human beings, it follows quite certainly that active male homosexuals are made and not born...In so far as active homosexuality is a problem at all, it is a problem of social environment and not of congenital make up.

...Although both homosexuals and their critics tend from time to time to advance the view that homosexual impulses are of a nature to separate active homosexuals from the common run of men, in truth the opposite appears to be the case. Homosexual tendencies are, at some time or another, present in almost every normal individual, and, during adolescence, they are often the prevalent emotional tendency. What makes an active homosexual out of an otherwise normal individual is the predominance and fixation of this tendency in adult life, coupled with the acquisition of the habit of securing satisfaction of it by physical homosexual practices. Contrary to what is implied by many classifications, such as that into "perverts", "inverts", and "casuals", active homosexuality can exist in some degree either to the exclusion of, or side by side with, normal heterosexual activity.

If, however, homosexuality is something which is acquired from environment by the fixation in a false predominance of a tendency almost always existing in normal individuals, it is unfortunately also true that, once permanently fixed by an established routine of sexual satisfaction, a homosexual can never be "cured" in the sense of making him invulnerable to temptation by members of his own sex...The psychiatrist can, it is true, with the conscious co-operation of the patient, and only in some cases, lead a homosexual to accept and adjust himself to his homosexual impulses in such a way as to sublimate and control their physical expression...The demand for "medical treatment" for homosexuals as a means of curing them of the inclination does not, therefore, come from well informed or professional sources, but is largely a sentimental demand born of an unwillingness to face the hard choices presented by an intractable problem.

A last fact which must be faced is that, at any rate as regards the great majority of active homosexuals, the precipitating factor in their abnormality has been initiation by older homosexuals whilst the personality is still pliable... [T]here is no single factor except direct initiation which can account for the phenomenal increase since 1938. Homosexuality is a proselytising religion, and initiation by an adept is at once the cause and the occasion of the type of fixation which has led to the increase in homosexual practices... Unless the deliberate communication of homosexuality is discouraged by some means or another, it may be assumed that the recent increase in homosexuality will continue, and although, no doubt, there comes a point of saturation, an acquaintance with classical literature would seem at least to suggest that such a point would involve a degree of corruption quite beyond the experience of any contemporary civilised society of Christian origins.

...

...If homosexuality were of its nature congenital, and the impulse irresistible, the problem would not concern the criminologist or even the moralist at all. If it could be cured by medical treatment, it would primarily be an affair for the physician. If it were induced by circumstances less within the control of individuals than deliberate initiation, it might safely be left to the individual conscience.

But none of these conditions obtain. Homosexuality is the result of environment, and therefore is within the field of social science. Homosexual practices are both contagious, incurable, and self perpetuating, and therefore not without their social consequences... Homosexuality is, and for fundamentally the same reasons, as much a moral and social issue as heroin addiction.

...

... [T]here is nothing necessarily sinful or immoral in being the subject of homosexual inclinations. All healthy people, homosexual or otherwise, have a large number of sexual inclinations which they are compelled to repress or sublimate. An emotional affection for the members of one's own sex may be the occasion of moral danger, but in itself it is no more sinful, still less criminal, than the love of a woman that cannot be satisfied. Nor, in fact, since the greater number of human sex impulses remain unavowed does it place the homosexual in a dramatic situation differing fundamentally from many in which all of us find ourselves from time to time.[29]

Although this is a matter quite impossible of demonstration, I feel myself wholly convinced that the lives of many of the most respected, and even saintly, educationalists, social workers, and others have been inspired by a dedicated and ascetic response by good men and women to sexual impulses, many of them caused by attractions to members of their own sex. Morality is concerned with the response to inclination, and not with the inclination itself.

...Whatever meaning is to be attached to the much abused word "natural", the instinct of mankind to describe homosexual acts as "unnatural" is not based on mere prejudice...Adultery and fornication may be immoral, but, on the lowest physical plane, they both involve the use of the complementary physical organs of male and female in the sense in which they are complementary...Homosexual practices necessarily involve the use of non-complementary physical organs in a manner which no less necessarily accentuates their non-complementary character. The psychological consequences of this physical misuse of the bodily organs cannot in the long run be ignored. It is certainly my experience, and I do not believe it to be a coincidence, that nearly all the homosexuals I have known have been emotionally unbalanced and profoundly unhappy.[30] I do not believe that this is solely or exclusively due to the fear of detection, or of [sic] the sense of guilt attaching to practices in fact disapproved of by society. It is inherent in an activity which seeks a satisfaction for which the bodily organs employed are physically unsuited.

...The unsatisfactory physical basis for a homosexual relationship, to which I have alluded, cannot form the basis of a lasting relationship physical or spiritual, and this is the end of true love.[31] Its necessarily sterile outcome from the point of view of the procreation of children also deprives it of the basis of lasting comradeship which in natural parenthood often succeeds the passionate romance of earlier days. Nor, I think, does a homosexual relationship ever flower in this way...

...

...[S]ociety must necessarily consider how far it is desirable to tolerate practices which develop within the body of society a self-perpetuating and potentially widely expansible secret society of addicts to a practice intimately harmful to the adjustment of the individual to his surroundings and effecting a permanent and detrimental change in his personality.[32]

...

It must...be seriously considered how far it is reasonable to tolerate homosexual practices in view of the danger to society of the corruption of youth...

...

...[T]he more I think about this difficult matter, the more convinced I become that homosexuality is to be treated as a socially undesirable activity, and that, on balance, we have been right in attaching to male homosexuality both criminal and social sanctions.

In the long run, I do not think that the admitted disadvantages of taking this course outweigh the solid advantages of shaping the public conscience which is obtained by stigmatising as inherently unlawful an activity which it is of serious consequence to society to discourage and prevent where possible.

...

To say [to] a confirmed homosexual not merely that the one satisfactory way out for him is to suppress all physical satisfaction of his sexual nature, (which, though hard doctrine, is no more than the truth, whatever the law may be), but that if he does not take this stern advice he renders himself liable to serious criminal penalties, is, undoubtedly, at first sight a Draconian precept. In practice, it is not so bad as it sounds, since ... it is relatively rare for homosexual acts between consenting adults to be the subject of criminal proceedings where no element of corruption or public indecency is concerned, and, where it happens or appears to happen to the contrary, there are usually special circumstances existing which make the case rather the exception which proves the rule. This, I am aware, is not an altogether satisfactory answer. It is dangerous teaching that a law may be justified because its true vigour is seldom, if ever, invoked. Indeed, the purist is entitled to regard this as a good theoretical argument for mitigating the rigour of the law so as to conform with existing practice. But law is not an exact science. It is necessarily a compromise between morality and expediency, and, like most other points at which the organising activities of man come into conflict with human folly and human weakness, the most advantageous and practical course is seldom that which gives the neatest and most logical theoretical solution.

...

Part III
The Wolfenden Report

Conclusion

Introduction

The Wolfenden Report, largely drafted by Wolfenden and Conwy Roberts, with input, reservations and stylistic suggestions from the other committee members, was published on 4 September 1957.[1] This introduction and the excerpts that follow will focus on the most significant parts of the report: the committee's understanding of homosexuality; its recommendations for a limited legalization of private homosexual acts; and its discussion of an age of consent.

After some prefatory comments in Part I, the report delved into an extensive discussion about the nature and extent of homosexuality in Part II. Jack Wolfenden had telegraphed his evolving thinking early the previous year in some remarks to witnesses from the Magistrates' Association about the use of the term 'true invert':

> CHAIRMAN:...I have got so cautious about the terminology in the whole of this matter by now that I would much prefer not to use these words at all, because I think as soon as one begins to use 'inverts', especially if one uses it in this junction from pervert one is already prejudging several questions, and if one then adds the adjective 'true' to the noun 'invert' one suggests that there is some other kind of invert which is not a true invert, and I do not know who that is at all. I think that if one is coming to a conclusion at all about this descriptive part it is that there are a great many shades and varieties of this business, either on a Kinsey scale, or on something like it, and that whether one can say of any one particular individual that he is an invert, or that he is a pervert, or that his state was caused by this, that or the other, the only advice I could give on all that is extreme caution in the use of words.[2]

The report reflected this cautious, equivocal and sceptical approach. Members of the committee, having filtered all the memoranda and witness statements, concluded that there was no single cause of homosexuality. They were not persuaded that it was a disease, illness or pathology. They resisted attempts to categorize homosexuals as a class apart or as divided into readily identifiable subsets. As for the incidence of homosexuality, they discovered no firm basis for judging whether

261

the Kinsey figures or the Swedish figures were or were not applicable to Britain. And while they conceded that the rise of homosexual offences known to the law owed much to more vigorous policing, they doubted whether this alone could explain the entire increase.

The report's principal recommendation was much more categorical: that (to paraphrase Pierre Trudeau) the law has no place in the bedrooms of the nation.[3] One committee member, James Adair, could not be reconciled to this and signalled his dissent in a lengthy 'Reservation'. 'The fact that activities inherently hurtful to community life are carried out clandestinely and in privacy does not adequately justify the removal of such conduct from the criminal code', he wrote. 'It is indisputable that many acts committed in private may be contrary to the public good and as such fall under the criminal law. In my view, homosexual acts are of this class.' One of his major concerns was for the children: 'The presence in a district of, for example, adult male lovers living openly and notoriously under the approval of the law is bound to have a regrettable and pernicious effect on the young people of the community.'[4]

The public/private distinction was to be thrashed out at length in the celebrated Hart–Devlin debate in the wake of Wolfenden. Sir Patrick Devlin, a high court judge, in the Maccabaean Lecture in Jurisprudence at the British Academy in 1958, found substantial fault with Wolfenden's reasoning. His central argument was that no society could exist without shared ideas on politics, morals and ethics. The moral judgements of society were to be determined by the standard of the reasonable man—the man on the Clapham omnibus or in the jury box. If immorality (read homosexuality) posed an existential threat to society, it would not be possible to set theoretical limits to the power of the state to legislate against it.[5] If Devlin drew on James Fitzjames Stephen to suggest that the law might justifiably enforce morality, H. L. A. Hart, Professor of Jurisprudence at Oxford, based his argument on John Stuart Mill, just as Wolfenden had done.[6] His conclusion laid down a robust defence of individual liberty: 'Whatever other arguments there may be for the enforcement of morality, no one should think even when popular morality is supported by an "overwhelming majority" or marked by widespread "intolerance, indignation, and disgust" that loyalty to democratic principles requires him to admit that its imposition on a minority is justified.'[7]

The Mill/Wolfenden/Hart principles—separating public and private, crime and sin, illegality and immorality—won out. A stream of significant legislation on social questions throughout the late 1950s and 1960s bears their imprint.[8] But the Wolfenden Committee was quite clear that to suggest a limited tolerance for the poor buggers by repealing archaic laws was in no way intended 'to condone or encourage private immorality' (s. 61). And there was to be no relaxation in the policing of public misconduct such as homosexual importuning either: 'It is important that the limited modification of the law which we propose should not be interpreted as an indication that the law can be indifferent to other forms of homosexual behaviour, or as a general licence to adult homosexuals to behave as they please' (s. 124).

In the wake of publication, members of the committee hammered home this point, making it 'quite clear that removing adult homosexual behaviour from the law was not meant to condone it in any way' or to give it 'moral approval'.[9] In the first Commons debate on the report, on 26 November 1958, the Home Secretary, R. A. Butler, also sought to exonerate the committee from the widespread misunderstanding that it wished to make homosexuality easier: 'In fact, what the members of the Committee wished to do was to alter the law, not expressly to encourage or legalise such practices, but to remove them, like adultery and other sins, from the realm of the law.' In declaring the government's intention not to act on the recommendations, Butler's stated reason was that so many people considered homosexuality to be 'a great social evil' that 'education and time' would be needed to persuade them of the committee's logic.[10]

In a debate in the Lords on 12 May 1965, by which time parliament was more receptive, one of the committee members, the Marquess of Lothian, reminisced about his and his colleagues' intentions:

> I am certain that, without exception, we took the view that homosexual acts are wrong and harmful—some very gravely so—and, therefore, to be deplored. I think we also agreed that, in general, homosexuals are more often than not unhappy people, maladjusted and sometimes degraded. This, I suggest, is in line, not only with ordinary, decent, Christian opinion, but also with a good deal of medical evidence as well. From this, two things follow: first, that the young and the weak must be protected; and, secondly, that homosexuals must, so far as is practically and medically possible, be assisted towards cure.[11]

He at least was convinced that this 'constructive, curative approach' formed the basis of their recommendations; and, once more, the notion that they were 'in any way condoning homosexual acts' was 'very far from the case'.[12]

This was all well and good, whether one agreed or not: conclusions following premises in a clear, logical, easy-to-comprehend fashion. But the careful reasoning appeared to fall apart with the recommendation of a minimum age of 21 for private consensual sex between males. The passages devoted to the age of consent (ss. 65–71) are among the weakest in the report, and, given the lack of consensus in the committee, this is not surprising. Linstead advocated 17, Curran, Demant, Diplock, Lothian, Lovibond, Rees and Whitby 18, and only Cohen, Stopford, Wells and the chairman himself wanted 21.[13] Conwy Roberts gave a succinct summary of the arguments:

> It seems to be evident that any age limit would have to be arbitrary:
> Should it be
> 16—to bring it into line with girls;
> 18—because boys develop a little later;
> 21—to protect the national service man;
> 25—for late developers;
> 30—for safety!

To put it too low might arouse public indignation.
To put it too high might militate against the possible protection of youth.[14]

Wolfenden recognized well enough the problems with drawing an age line: '[I]f you have got two twenty year olds, you are in a position to say to them "Now steady chaps, hold it. You must not do it now but it will be all right in a fortnight." '[15] And he was not persuaded by the oft-touted notion of protecting young men in National Service, since 'the age at which National Service becomes an obligation may well change' and 'I am not clear why a boy of nineteen needs more legal protection than a girl of the same age'.[16] Nevertheless, through a combination of caution and calculation, he persuaded the committee to back the expedient age of 21.

But it was all still too radical in the short run. In parliament, neither the Conservative government nor the Labour opposition was prepared to touch the report's recommendations regarding homosexuality for fear of alienating their supporters.[17] The higher-brow national press tended to support reform, the lower-brow (and most of the provincial and Scottish press) to deprecate it.[18] For the *Daily Mail*, 'If the law were to tolerate homosexual acts a great barrier against depravity would be swept aside' and the potential consequences would be incalculable: 'great nations have fallen and empires decayed because corruption became socially acceptable'.[19] The Church of England's Church Assembly voted narrowly (155 to 138) to support the report, but with the inevitable caveat from the mover of the motion, the Bishop of Exeter, that homosexual behaviour 'is gravely sinful' and that the assembled delegates 'were all united in their disgust and condemnation'.[20] After the government refused to legislate, the *Economist* called it correctly: 'The matter now goes into that extensive limbo of reforms that are supported by almost everybody who has seriously studied the subject, by the Archbishop of Canterbury (except that he would still ban sodomy), and by penological common sense—but are rejected by common emotion. It will be enacted in the end, but past experience suggests that the end may be a decade or so away.'[21]

The end, when it came, was thoroughly Wolfendenian—that is, no more radical than politicians and public could stomach.[22] The legacy of the Wolfenden Report as embodied in the 1967 Sexual Offences Act was a carefully demarcated private space in which the poor homosexual, who couldn't help himself, could behave deplorably with another consenting adult male, without fear of a policeman feeling his collar. And yet the attempt at containment failed. The Babel of competing discourses around the origins and nature of homosexualities that the Wolfenden Committee had helped conjure up and broadcast could not be restrained by the sorcerer's apprentice or stuffed back into Pandora's box. No single hegemonic interpretation could cement over the discursive fissures. Far from eradicating homosexuality to the fullest extent possible, through the prevention of initial slippage and the treatment of those who had slipped, Wolfenden began to open up the possibility of sexual minorities breaking out of the newly legitimated spaces.[23] As James Adair and the moral conservatives had feared, the report and the 1967 Act helped ease open the floodgates to the proliferation of a very visible and vibrant associational and commercial gay scene in the 1970s and to alter-native discourses—in the Gay Liberation Front and elsewhere—that attacked and (in time) demolished the notion that homosexuality was a medical or psychiatric disorder.[24]

Not that Wolfenden failed entirely in his broader mission. The figure of the straight-mimicking, respectable homosexual, given powerful form by the Wolfenden Report, has had a continuous presence in gay rights discourse through-out all the campaigns for recognition and inclusion, culminating in marriage equality in 2013 under a Conservative-led government. When Prime Minister David Cameron backed gay marriage he famously declared, 'I don't support gay marriage despite being a Conservative. I support gay marriage because I'm a Conservative.'[25] Conservatives of the 1950s would have been appalled; but one imagines that a smile of recognition would have played around Sir John Wolfenden's lips.

*

Excerpts from the Wolfenden Report
___

PART ONE—INTRODUCTORY

. . .

CHAPTER II
OUR APPROACH TO THE PROBLEM

12. It will be apparent from our terms of reference that we are concerned throughout with the law and offences against it. We clearly recognize that the laws of any society must be acceptable to the general moral sense of the commu-nity if they are to be respected and enforced. But we are not charged to enter into matters of private moral conduct except in so far as they directly affect the public good...

13. Further, we do not consider it to be within our province or competence to make a full examination of the moral, social, psychological and biological causes of homosexuality or prostitution, or of the many theories advanced about these causes. Our primary duty has been to consider the extent to which homosexual behaviour and female prostitution should come under the condemnation of the criminal law, and this has presented us with the difficulty of deciding what are the essential elements of a criminal offence. There appears to be no unquestioned definition of what constitutes or ought to constitute a crime ... We have therefore worked with our own formulation of the function of the criminal law so far as it concerns the subjects of this enquiry. In this field, its function, as we see it, is to preserve public order and decency, to protect the citizen from what is offensive or injurious, and to provide sufficient safeguards against exploitation and corruption of others, particularly those who are specially vulnerable because they are young, weak in body or mind, inexperienced, or in a state of special physical, official or economic dependence.

14.   It is not, in our view, the function of the law to intervene in the private lives of citizens, or to seek to enforce any particular pattern of behaviour, further than is necessary to carry out the purposes we have outlined...

CHAPTER III

HOMOSEXUALITY

...

18.   It is important to make a clear distinction between 'homosexual offences' and 'homosexuality.'... For the latter, we are content to rely on the dictionary definition that homosexuality is a sexual propensity for persons of one's own sex. Homosexuality, then, is a state or condition, and as such does not, and cannot, come within the purview of the criminal law.

...

22.   ... [H]omosexuality as a propensity is not an 'all or none' condition, and this view has been abundantly confirmed by the evidence submitted to us. All gradations can exist from apparently exclusive homosexuality without any conscious capacity for arousal by heterosexual stimuli to apparently exclusive heterosexuality, though in the latter case there may be transient and minor homosexual inclinations, for instance in adolescence. According to the psycho-analytic school, all individuals pass through a homosexual phase. Be this as it may, we would agree that a transient homosexual phase in development is very common and should usually cause neither surprise nor concern.

It is interesting that the late Dr. Kinsey, in his study entitled 'The Sexual Behaviour of the Human Male,' formulated this homosexual–heterosexual continuum on a 7-point scale, with a rating of 6 for sexual arousal and activity with other males only, 3 for arousals and acts equally with either sex, 0 for exclusive heterosexuality, and intermediate ratings accordingly. The recognition of the existence of this continuum is, in our opinion, important for two reasons. First, it leads to the conclusion that homosexuals cannot reasonably be regarded as quite separate from the rest of mankind. Secondly... it has some relevance in connection with claims made for the success of various forms of treatment.

23.   ...

It must not be thought that the existence of the homosexual propensity necessarily leads to homosexual behaviour of an overtly sexual kind. Even where it does, this behaviour does not necessarily amount to a homosexual offence; for instance, solitary masturbation with homosexual fantasies is probably the most common homosexual act. Many persons, though they are aware of the existence within themselves of the propensity, and though they may be conscious of sexual arousal in the presence of homosexual stimuli, successfully control their urges towards overtly homosexual acts with others, either because of their ethical standards or from fear of social or penal consequences, so that their homosexual condition never manifests itself in overtly sexual behaviour. There are others who, though aware of the existence within themselves of the propensity, are helped by a happy family life, a satisfying vocation, or a well-balanced social life to live happily without any urge to indulge in homosexual acts. Our evidence suggests however that complete continence in the homosexual is relatively uncommon—as, indeed, it is in the heterosexual—and that even where the individual is by disposition continent, self-control may break down temporarily under the influence of factors like alcohol, emotional distress or mental or physical disorder or disease.

24. Moreover, it is clear that homosexuals differ one from another in the extent to which they are aware of the existence within themselves of the propensity. Some are, indeed, quite unaware of it, and where this is so the homosexuality is technically described as latent, its existence being inferred from the individual's behaviour in spheres not obviously sexual. Although there is room for dispute as to the extent and variety of behaviour of this kind which may legitimately be included in the making of this inference, there is general agreement that the existence of a latent homosexuality is an inference validly to be drawn in certain cases. Sometimes, for example, a doctor can infer a homosexual component which accounts for the condition of a patient who has consulted him because of some symptom, discomfort or difficulty, though the patient himself is completely unaware of the existence within himself of any homosexual inclinations. There are other cases in which the existence of a latent homosexuality may be inferred from an individual's outlook or judgment: for instance, a persistent and indignant preoccupation with the subject of homosexuality has been taken to suggest in some cases the existence of repressed homosexuality. Thirdly, among those who work with notable success in occupations which call for service to others, there are some in whom a latent homosexuality provides the motivation for activities of the greatest value to society. Examples of this are to be found among teachers, clergy, nurses and those who are interested in youth movements and the care of the aged.

25. We believe that there would be a wide measure of agreement on the general account of homosexuality and its manifestations that we have given above. On the other hand, the general position which we have tried to summarise permits the drawing of many different inferences, not all of them in our opinion justified. Especially is this so in connection with the concept of 'disease.' There is a tendency, noticeably increasing in strength over recent years, to label homosexuality as a 'disease' or 'illness.' This may be no more than a particular manifestation of a general tendency discernible in modern society by which, as one leading sociologist puts it, 'the concept of illness expands continually at the expense of the concept of moral failure.'[26] There are two important practical consequences which are often thought to follow from regarding homosexuality as an illness. The first is that those in whom the condition exists are sick persons and should therefore be regarded as medical problems and consequently as primarily a medical responsibility. The second is that sickness implies irresponsibility, or at least diminished responsibility. Hence it becomes important in this connection to examine the criteria of 'disease,' and also to examine the claim that these consequences follow.

26. ...

The traditional view seems to be that for a condition to be recognized as a disease, three criteria must be satisfied, namely (i) the presence of abnormal symptoms, which are caused by (ii) a demonstrable pathological condition, in turn caused by (iii) some factor called 'the cause,' each link in this causal chain being understood as something necessarily antecedent to the next...

27. ...On the criterion of symptoms...homosexuality cannot legitimately be regarded as a disease, because in many cases it is the only symptom and is compatible with full mental health in other respects. In some cases, associated psychiatric abnormalities do occur, and it seems to us that if, as has been suggested, they occur with greater frequency in the homosexual, this may be because they are products of the strain and conflict brought about by the homosexual condition and not because they are causal factors. It has been suggested to us that associated psychiatric abnormalities are less prominent, or even absent, in countries where the homosexual is regarded with more tolerance.

28. As regards the second criterion, namely, the presence of a demonstrable pathological condition, some, though not all, cases of mental illness are accompanied by a demonstrable physical pathology. We have heard no convincing evidence that this has yet been demonstrated in relation to homosexuality. Biochemical and endocrine studies so far carried out in this field have, it appears, proved negative, and investigations of body-build and the like have also so far proved inconclusive. We are aware that studies carried out on sets of twins suggest that certain genes lay down a potentiality which will lead to homosexuality in the person who possesses them, but even if this were established (and the results of these studies have not commanded universal acceptance), a genetic predisposition would not necessarily amount to a pathological condition, since it may be no more than a natural biological variation comparable with variations in stature, hair pigmentation, handedness and so on.

In the absence of a physical pathology, psychopathological theories have been constructed to explain the symptoms of various forms of abnormal behaviour or mental illness. These theories range from rather primitive formulations like a repressed complex or a mental 'abscess' to elaborate systems. They are theoretical constructions to explain observed facts, not the facts themselves, and similar theories have been constructed to explain 'normal' behaviour. These theoretical constructions differ from school to school. The alleged psychopathological causes adduced for homosexuality have, however, also been found to occur in others besides the homosexual.

29. As regards the third criterion, that is, the 'cause,' there is never a single cause for normal behaviour, abnormal behaviour or mental illness. The causes are always multiple... To speak, as some do, of some single factor such as seduction in youth as the 'cause' of homosexuality is unrealistic unless other factors are taken into account. Besides genetic predisposition, a number of such factors have been suggested, for instance, unbalanced family relationships, faulty sex education, or lack

of opportunity for heterosexual contacts in youth. In the present state of our knowledge, none of these can be held to bear a specific causal relationship to any recognized psychopathology or physical pathology; and to assert a direct and specific causal relationship between these factors and the homosexual condition is to ignore the fact that they have all, including seduction, been observed to occur in persons who became entirely heterosexual in their disposition.

30.   Besides the notion of homosexuality as a disease, there have been alternative hypotheses offered by others of our expert witnesses. Some have preferred to regard it as a state of arrested development. Some, particularly among the biologists, regard it as simply a natural deviation. Others, again, regard it as a universal potentiality which can develop in response to a variety of factors.

We do not consider ourselves qualified to pronounce on controversial and scientific problems of this kind, but we feel bound to say that the evidence put before us has not established to our satisfaction the proposition that homosexuality is a disease...

...

32.   The claim that homosexuality is an illness carries the further implication that the sufferer cannot help it and therefore carries a diminished responsibility for his actions. Even if it were accepted that homosexuality could properly be described as a 'disease,' we should not accept this corollary. There are no *prima facie* grounds for supposing that because a particular person's sexual propensity happens to lie in the direction of persons of his or her own sex it is any less controllable than that of those whose propensity is for persons of the opposite sex...

35.   Some writers on the subject, and some of our witnesses, have drawn a distinction between the 'invert' and the 'pervert.' We have not found this distinction very useful. It suggests that it is possible to distinguish between two men who commit the same offence, the one as the result of his constitution, the other from a perverse and deliberate choice, with the further suggestions that the former is in some sense less culpable than the latter. To make this distinction as a matter of definition seems to prejudge a very difficult question.

Similarly, we have avoided the use of the terms 'natural' and 'unnatural' in relation to sexual behaviour, for they depend for their force upon certain explicit theological or philosophical interpretations, and without these interpretations their use imports an approving or a condemnatory note into a discussion where dispassionate thought and statement should not be hindered by adherence to particular preconceptions.

36.   Homosexuality is not, in spite of widely held belief to the contrary, peculiar to members of particular professions or social classes; nor, as is sometimes supposed, is it peculiar to the *intelligentsia*. Our evidence shows that it exists among all callings and at all levels of society; and that among homosexuals will be found not only those possessing a high degree of intelligence, but also the dullest oafs.

Some homosexuals, it is true, choose to follow occupations which afford opportunities for contact with those of their own sex, and it is not unnatural that those who feel themselves to be 'misfits' in society should gravitate towards occupations offering an atmosphere of tolerance or understanding, with the result that some occupations may appear to attract more homosexuals than do others. Again, the arrest of a prominent national or local figure has greater news value than the arrest of (say) a labourer for a similar offence, and in consequence the press naturally finds room for a report of the one where it might not find room for a report of the other. Factors such as these may well account to some extent for the prevalent misconceptions.

CHAPTER IV
THE EXTENT OF THE PROBLEM

37.   Our consideration of the problems we have had to face would have been made much easier if it had been possible to arrive at some reasonably firm estimate of the prevalence either of the condition of homosexuality or of the commission of homosexual acts. So far as we have been able to discover, there is no precise information about the number of men in Great Britain who either have a homosexual disposition or engage in homosexual behaviour.

38.   No enquiries have been made in this country comparable to those which the late Dr. Kinsey conducted in the United States of America … Dr. Kinsey's findings have aroused opposition and scepticism. But it was noteworthy that some of our medical witnesses expressed the view that something very like these figures would be established in this country if similar enquiries were made. The majority, while stating quite frankly that they did not really know, indicated that their impression was that his figures would be on the high side for Great Britain.

39.   A recent enquiry in Sweden suggested that 1 per cent. of all men were exclusively homosexual and 4 per cent. had both homosexual and heterosexual impulses, and we were interested to learn from official sources in Sweden that other information available seemed to indicate that these figures were too low.[27] But here again, there is no evidence that similar enquiries in this country would yield similar results.

. . .

42.   It is widely believed that the prevalence of homosexuality in this country has greatly increased during the past fifty years and that homosexual behaviour is much more frequent than used to be the case. It is certainly true that the whole subject of homosexuality is much more freely discussed to-day than it was formerly; but this is not in itself evidence that homosexuality is to-day more prevalent, or homosexual behaviour more widespread, than it was when mention of it was less common. Sexual matters in general are more openly talked about to-day than they were in the days of our parents and grandparents; and it is not surprising that homosexuality should take its place, among other sexual topics, in this wider range of permissible subjects of conversation. Public interest in the subject

has undoubtedly increased, with the consequences that court cases are more frequently reported and that responsible papers and magazines give considerable space to its discussion. In general literature, too, there is a growing number of works dealing incidentally or entirely with the subject. All this has no doubt led to a much greater public awareness of the phenomenon and its manifestations. But it does not necessarily follow that the behaviour which is so discussed is more widespread than it was before.

43. It is certainly true also... that the number of homosexual offences known to the police has increased considerably. It does not, however, necessarily follow from these figures that there has been an increase either in homosexuality or in homosexual behaviour; still less can these figures be regarded as an infallible measure of any increase which may have occurred during that period. Unlike some offences (e.g., housebreaking) which, by their nature, tend to be reported to the police as they occur, many sexual offences, particularly those taking place between consenting parties, become 'known to the police' only when they are detected by the police or happen to be reported to them. Any figures relating to homosexual offences known to the police will therefore be conditioned to a large extent both by the efficiency of the police methods of detecting and recording, and by the intensity of police activity. These factors vary from time to time and from place to place...

45. Those who have the impression of a growth in homosexual practices find it supported by at least three wider considerations. First, in the general loosening of former moral standards, it would not be surprising to find that leniency towards sexual irregularities in general included also an increased tolerance of homosexual behaviour and that greater tolerance had encouraged the practice. Secondly, the conditions of war time, with broken families and prolonged separation of the sexes, may well have occasioned homosexual behaviour which in some cases has been carried over into peace time. Thirdly, it is likely that the emotional insecurity, community instability and weakening of the family, inherent in the social changes of our civilisation, have been factors contributing to an increase in homosexual behaviour.

Most of us think it improbable that the increase in the number of offences recorded as known to the police can be explained entirely by greater police activity, though we all think it very unlikely that homosexual behaviour has increased proportionately to the dramatic rise in the number of offences recorded as known to the police.

46. Our medical evidence seems to show three things: first, that in general practice male homosexuals form a very small fraction of the doctor's patients; secondly, that in psychiatric practice male homosexuality is a primary problem in a very small proportion of the cases seen; and thirdly, that only a very small percentage of homosexuals consult doctors about their condition...

47. Our conclusion is that homosexual behaviour is practiced by a small minority of the population, and should be seen in proper perspective, neither ignored

nor given a disproportionate amount of public attention. Especially are we concerned that the principles we have enunciated above on the function of the law should apply to those involved in homosexual behaviour no more and no less than to other persons.

CHAPTER V
THE PRESENT LAW AND PRACTICE

(i) General Review

48.   It is against the foregoing background that we have reviewed the existing provisions of the law in relation to homosexual behaviour between male persons. We have found that with the great majority of these provisions we are in complete agreement. We believe that it is part of the function of the law to safeguard those who need protection by reason of their youth or some mental defect, and we do not wish to see any change in the law that would weaken this protection. Men who commit offences against such persons should be treated as criminal offenders. Whatever may be the causes of their disposition or the proper treatment for it, the law must assume that the responsibility for the overt acts remains theirs, except where there are circumstances which it accepts as exempting them from accountability. Offences of this kind are particularly reprehensible when the men who commit them are in positions of special responsibility or trust. We have been made aware that where a man is involved in an offence with a boy or youth the invitation to the commission of the act sometimes comes from him rather than from the man. But we believe that even when this is so that fact does not serve to exculpate the man.

49.   It is also part of the function of the law to preserve public order and decency. We therefore hold that when homosexual behaviour between males takes place in public it should continue to be dealt with by the criminal law...

50.   Besides the two categories of offence we have just mentioned, namely, offences committed by adults with juveniles and offences committed in public places, there is a third class of offence to which we have to give long and careful consideration. It is that of homosexual acts committed between adults in private.

51.   In England and Wales, during the three years ended March 1956, 480 men aged twenty-one or over were convicted of offences committed in private with consenting partners also aged twenty-one or over. Of these, however, 121 were also convicted of, or admitted, offences in public places (parks, open spaces, lavatories, etc.), and 59 were also convicted of, or admitted, offences with partners under twenty-one. In Scotland, during the same period, 9 men over twenty-one were convicted of offences committed in private with consenting adult partners. Of these, one also admitted offences in public places and one admitted offences with a partner under twenty-one. Thus 307 men (300 in England and Wales and 7 in Scotland), guilty as far as is known only of offences committed in private with consenting adult partners, were convicted by the courts during this period...

52.   We have indicated (in Chapter II above) our opinion as to the province of the law and its sanctions, and how far it properly applies to the sexual behaviour of the individual citizen. On the basis of the considerations there advanced we have reached the conclusion that legislation which covers acts in the third category we have mentioned goes beyond the proper sphere of the law's concern. We do not think that it is proper for the law to concern itself with what a man does in private unless it can be shown to be so contrary to the public good that the law ought to intervene in its function as the guardian of that public good.

53.   In considering whether homosexual acts between consenting adults in private should cease to be criminal offences we have examined the more serious arguments in favour of retaining them as such. We now set out these arguments and our reasons for disagreement with them. In favour of retaining the present law, it has been contended that homosexual behaviour between adult males, in private no less than in public, is contrary to the public good on the grounds that—

  (i) it menaces the health of society;
 (ii) it has damaging effects on adult life;
(iii) a man who indulges in these practices with another man may turn his attention to boys.

54.   As regards the first of these arguments, it is held that conduct of this kind is a cause of the demoralisation and decay of civilisations, and that therefore, unless we wish to see our nation degenerate and decay, such conduct must be stopped, by every possible means. We have found no evidence to support this view, and we cannot feel it right to frame the laws which should govern this country in the present age by reference to hypothetical explanations of the history of other peoples in ages distant in time and different in circumstances from our own. In so far as the basis of this argument can be precisely formulated, it is often no more than the expression of revulsion against what is regarded as unnatural, sinful or disgusting. Many people feel this revulsion, for one or more of these reasons. But moral conviction or instinctive feeling, however strong, is not a valid basis for overriding the individual's privacy and for bringing within the ambit of the criminal law private sexual behaviour of this kind. It is held also that if such men are employed in certain professions or certain branches of the public service their private habits may render them liable to threats of blackmail or to other pressures which may make them 'bad security risks.' If this is true, it is true also of some other categories of person: for example, drunkards, gamblers and those who become involved in compromising situations of a heterosexual kind; and while it may be a valid ground from excluding from certain forms of employment men who indulge in homosexual behaviour, it does not, in our view, constitute a sufficient reason for making their private sexual behaviour an offence in itself.

55.   The second contention, that homosexual behaviour between males has a damaging effect on family life, may well be true. Indeed, we have had evidence that it often is; cases in which homosexual behaviour on the part of the husband

has broken up a marriage are by no means rare, and there are also cases in which a man in whom the homosexual component is relatively weak nevertheless derives such satisfaction from homosexual outlets that he does not enter upon a marriage which might have been successfully and happily consummated. We deplore this damage to what we regard as the basic unit of society; but cases are also frequently encountered in which a marriage has been broken up by homosexual behaviour on the part of the wife, and no doubt some women, too, derive sufficient satisfaction from homosexual outlets to prevent their marrying. We have had no reasons shown to us which would lead us to believe that homosexual behaviour between males inflicts any greater damage on family life than adultery, fornication or lesbian behaviour. These practices are all reprehensible from the point of view of harm to the family, but it is difficult to see why on this ground male homosexual behaviour alone among them should be a criminal offence. This argument is not to be taken as saying that society should condone or approve male homosexual behaviour. But where adultery, fornication and lesbian behaviour are not criminal offences there seems to us to be no valid ground, on the basis of damage to the family, for so regarding homosexual behaviour between men. Moreover, it has to be recognized that the mere existence of the condition of homosexuality in one of the partners can result in an unsatisfactory marriage, so that for a homosexual to marry simply for the sake of conformity with the accepted structure of society or in the hope of curing his condition may result in disaster.

56.   We have given anxious consideration to the third argument, that an adult male who has sought as his partner another adult male may turn from such a relationship and seek as his partner a boy or succession of boys. We should certainly not wish to countenance any proposal which might tend to increase offences against minors. Indeed, if we thought that any recommendation for a change in the law would increase the danger to minors we should not make it. But in this matter we have been much influenced by our expert witnesses. They are in no doubt that whatever may be the origins of the homosexual condition, there are two recognizably different categories among adult male homosexuals. There are those who seek as partners other adult males, and there are paedophiliacs, that is to say men who seek as partners boys who have not reached puberty.

57.   We are authoritatively informed that a man who has homosexual relations with an adult partner seldom turns to boys, and *vice-versa*, though it is apparent from the police reports we have seen and from other evidence submitted to us that such cases do happen … It would be paradoxical if the making legal of an act at present illegal were to turn men towards another kind of act which is, and would remain, contrary to the law …

58.   In addition, an argument of a more general character in favour of retaining the present law has been put to us by some of our witnesses. It is that to change the law in such a way that homosexual acts between consenting adults in private ceased to be criminal offences must suggest to the average citizen a degree of toleration by the Legislature of homosexual behaviour, and that such a change would

'open the floodgates' and result in unbridled licence. It is true that a change of this sort would amount to a limited degree of such toleration, but we do not share the fears of our witnesses that the change would have the effect they expect. This expectation seems to us to exaggerate the effect of the law on human behaviour. It may well be true that the present law deters from homosexual acts some who would otherwise commit them, and that to that extent an increase in homosexual behaviour can be expected. But it is no less true that if the amount of homosexual behaviour has, in fact, increased in recent years, then the law has failed to act as an effective deterrent. It seems to us that the law itself probably makes little difference to the amount of homosexual behaviour which actually occurs; whatever the law may be there will always be strong social forces opposed to homosexual behaviour. It is highly improbable that the man to whom homosexual behaviour is repugnant would find it any less repugnant because the law permitted it in certain circumstances; so that even if, as has been suggested to us, homosexuals tend to proselytise, there is no valid reason for supposing that any considerable number of conversions would follow the change in the law...

61.   ...We have outlined the arguments against a change in the law, and we recognise their weight. We believe, however, that they have been met by the counter-arguments we have already advanced. There remains one additional counter-argument which we believe to be decisive, namely, the importance which society and the law ought to give to individual freedom of choice and action in matters of private morality. Unless a deliberate attempt is to be made by society, acting through the agency of the law, to equate the sphere of crime with that of sin, there must remain a realm of private morality and immorality which is, in brief and crude terms, not the law's business. To say this is not to condone or encourage private immorality. On the contrary, to emphasise the personal and private nature of moral or immoral conduct is to emphasise the personal and private responsibility of the individual for his own actions, and that is a responsibility which a mature agent can properly be expected to carry for himself without the threat of punishment from the law.

62.   We accordingly recommend that homosexual behaviour between consenting adults in private should no longer be a criminal offence.

...

71. There must obviously be an element of arbitrariness in any decision on [the age of consent]; but all things considered the legal age of contractual responsibility seems to us to afford the best criterion for the definition of adulthood in this respect. While there are some grounds for fixing the age as low as sixteen, it is obvious that however 'mature' a boy of that age may be as regards physical development or psycho-sexual make-up, and whatever analogies may be drawn from the law relating to offences against young girls, a boy is incapable, at the age of sixteen, of forming a mature judgment about actions of a kind which might have the effect of setting him apart from the rest of society. The young man between eighteen and twenty-one may be expected to be rather more mature in

this respect. We have, however, encountered several cases in which young men have been induced by means of gifts of money or hospitality to indulge in homosexual behaviour with older men, and we have felt obliged to have regard to the large numbers of young men who leave their homes at or about the age of eighteen and, either for their employment or their education or to fulfil their national service obligations, are then for the first time launched into the world in circumstances which render them particularly vulnerable to advances of this sort. It is arguable that such men should be expected, as one of the conditions of their being considered sufficiently grown-up to leave home, to be able to look after themselves in this respect also, the more so if they are being trained for responsibility in the services or in civil life. Some of us feel, on various grounds, that the age of adulthood should be fixed at eighteen. Nevertheless, most of us would prefer to see the age fixed at twenty-one, not because we think that to fix the age at eighteen would result in any greater readiness on the part of young men between eighteen and twenty-one to lend themselves to homosexual practices than exists at present, but because to fix it at eighteen would lay them open to attentions and pressures of an undesirable kind from which the adoption of the later age would help to protect them, and from which they ought, in view of their special vulnerability, to be protected. We therefore recommend that for the purpose of the amendment of the law which we have proposed, the age at which a man is deemed to be an adult should be twenty-one.

. . .

Notes

Introduction

1. *Times*, 27 May 1954, p. 3; Rupert Croft-Cooke, *The Verdict of You All* (London: Secker and Warburg, 1955), pp. 150–1; David Kynaston, *Family Britain 1951–57* (London: Bloomsbury, 2009), p. 391; Matt Cook, 'Queer Conflicts: Love, Sex and War, 1914–1967', in Matt Cook (ed.), *A Gay History of Britain: Love and Sex between Men since the Middle Ages* (Oxford: Greenwood, 2007), p. 171; *Manchester Guardian*, 27 May 1954, p. 4; 25 June 1954, p. 2; *Daily Mail*, 27 May 1954, p. 5.
2. Home Office and Scottish Home Department, *Report of the Committee on Homosexual Offences and Prostitution* (London: HMSO, 1957) [hereafter *Wolfenden Report*], subsections (ss.) 1–2.
3. Brian Harrison, *Seeking a Role: The United Kingdom, 1951–1970* (Oxford: Clarendon Press, 2009), p. 239; Chris Waters, 'Disorders of the Mind, Disorders of the Body Social: Peter Wildeblood and the Making of the Modern Homosexual', in Becky Conekin, Frank Mort and Chris Waters (eds), *Moments of Modernity: Reconstructing Britain 1945–1964* (London: Rivers Oram Press, 1999), pp. 137–8; Chris Waters, 'The Homosexual as a Social Being in Britain, 1945–1968', in Brian Lewis (ed.), *British Queer History* (Manchester: Manchester University Press, 2013), chap. 9; Michal Shapira, *The War Inside: Psychoanalysis, Total War, and the Making of the Democratic Self in Postwar Britain* (Cambridge: Cambridge University Press, 2013), chap. 6.
4. John Wolfenden, *Turning Points: The Memoirs of Lord Wolfenden* (London: Bodley Head, 1976), p. 130. See also Frank Mort, *Capital Affairs: London and the Making of the Permissive Society* (New Haven, CT: Yale University Press, 2010), pp. 1–3; Peter Wildeblood, *A Way of Life* (London: Weidenfeld and Nicolson, 1956), p. 65; 'Homosexuality, Prostitution and the Law', *Dublin Review*, 230, 471 (Summer 1956), 59.
5. For example, Stuart Hall, 'Reformism and the Legislation of Consent', in National Deviancy Conference (ed.), *Permissiveness and Control: The Fate of the Sixties Legislation* (New York: Barnes and Noble, 1980), p. 8; Tim Newburn, *Permission and Regulation: Law and Morals in Post-War Britain* (London: Routledge, 1992), pp. 49–50; Jeffrey Weeks, *The World We Have Won: The Remaking of Erotic and Intimate Life* (New York: Routledge, 2007), pp. 45–8; Richard Hornsey, *The Spiv and the Architect: Unruly Life in Postwar London* (Minneapolis, MN: University of Minnesota Press, 2010), pp. 83–4.
6. *Wolfenden Report*, appendix I, tables I and II, pp. 130–1.
7. *Times*, 22 Oct. 1953, p. 5; Sheridan Morley, *John Gielgud: The Authorized Biography* (New York: Simon and Schuster, 2010), pp. 266–72. Gielgud's conviction figured prominently in a Lords' debate on homosexual crime instigated by Earl Winterton on 19 May 1954 (*Hansard's Parliamentary Debates*, Lords, 5th ser., vol. 187 (1954), cols. 744, 756–7, 759, 766).
8. *Times*, 26 Jan. 1953, p. 3; 21 Feb. 1953, p. 3.
9. Patrick Higgins, *Heterosexual Dictatorship: Male Homosexuality in Postwar Britain* (London: Fourth Estate, 1996), pp. 65–7.
10. Peter Wildeblood, *Against the Law* (London: Phoenix, 2000; 1st edn, 1955), parts 1–2; Lord Montagu of Beaulieu, *Wheels within Wheels: An Unconventional Life* (London: Weidenfeld and Nicolson, 2000), chap. 8; H. Montgomery Hyde, *The Love That Dared Not Speak Its Name* (Boston, MA: Little, Brown and Company, 1970), pp. 216–24.
11. Croft-Cooke, *The Verdict of You All*, pp. 26–7, 134, 151; Wildeblood, *Against the Law*, pp. 69–70; Morley, *John Gielgud*, pp. 275–6.

12. Higgins, *Heterosexual Dictatorship*, pp. 249–56; Matt Houlbrook, *Queer London: Perils and Pleasures in the Sexual Metropolis, 1918–1957* (Chicago, IL: University of Chicago Press, 2005), pp. 34–6; *Wolfenden Report*, ss. 130–2.
13. Alan Sinfield, *Out on Stage: Lesbian and Gay Theatre in the Twentieth Century* (New Haven, CT: Yale University Press, 1999), pp. 235–8; Cook, *Gay History*, pp. 153, 169–70.
14. Adrian Bingham, *Family Newspapers? Sex, Private Life, and the British Popular Press 1918–1978* (Oxford: Oxford University Press, 2009), chap. 5; Adrian Bingham, 'The "K-Bomb": Social Surveys, the Popular Press, and British Sexual Culture in the 1940s and 1950s', *Journal of British Studies*, 50, 1 (Jan. 2011), 156–79; Lesley Hall, *Sex, Gender and Social Change in Britain Since 1880* (London: Palgrave Macmillan; 2nd edn, 2013), pp. 142–3; Cook, *Gay History*, pp. 169–70.
15. 'A Social Problem', *Sunday Times*, 1 Nov. 1953, p. 6. See also *Sunday Times*, 28 Mar. 1954, p. 6. *The Sunday Times*'s call for action was echoed by Dr Donald Soper, President of the Methodist Conference, who urged the setting up of a Royal Commission on homosexuality (*Times*, 6 Nov. 1953, p. 5).
16. Derrick Sherwin Bailey, 'The Problem of Sexual Inversion', *Theology*, 55, 380 (1952), 47–52; Church of England Moral Welfare Council, *The Problem of Homosexuality: An Interim Report* (Oxford: Church Information Board, 1954); Timothy Willem Jones, *Sexual Politics in the Church of England, 1857–1957* (Oxford: Oxford University Press, 2013), chap. 6; Matthew Grimley, 'Law, Morality and Secularisation: The Church of England and the Wolfenden Report, 1954–1967', *Journal of Ecclesiastical History*, 60, 4 (October 2009), 728–32.
17. Gordon Westwood [Michael Schofield], *Society and the Homosexual* (London: Victor Gollancz, 1952); Hornsey, *The Spiv and the Architect*, pp. 118–19.
18. Rodney Garland [Adam de Hegedus], *The Heart in Exile* (London: W. H. Allen, 1953); Matt Houlbrook and Chris Waters, '*The Heart in Exile*: Detachment and Desire in 1950s London', *History Workshop Journal*, 62 (Autumn 2006), 142–63. Mary Renault's novel *The Charioteer* (London: Longman, 1953) also attempted to do the same work.
19. National Archives [TNA], CAB 129/66/10, memo by Home Secretary, 17 Feb. 1954.
20. CAB 129/66/10, memo by Home Secretary, 17 Feb. 1954; CAB 128/27/11, cabinet conclusions, 24 Feb. 1954; CAB 195/11/94, Cabinet Secretary's notebook, 24 Feb. 1954; CAB 128/27/20, cabinet conclusions, 17 Mar. 1954; CAB 129/67/12, memo by Home Secretary, 1 Apr. 1954; CAB 128/27/29, cabinet conclusions, 15 Apr. 1954; Justin Bengry, 'Queer Profits: Homosexual Scandal and the Origins of Legal Reform in Britain', in Heike Bauer and Matt Cook (eds), *Queer 1950s: Rethinking Sexuality in the Postwar Years* (Houndmills: Palgrave Macmillan, 2012), pp. 169–77. Note that, in a memo of 17 Feb. 1954 (CAB 129/66/11), the Scottish Secretary did not think that prostitution or homosexual offences in Scotland were serious enough problems to justify an inquiry.
21. Wolfenden, *Turning Points*, pp. 133–4. The interviews were recorded in shorthand and typed up more or less verbatim, including 'casual remarks' and 'other little imperfections' that would have been edited out of a published document. The typescripts were 'not intended in any way to provide a permanent record of our proceedings. They are provided only for the convenience of the members, and are intended merely to assist them in recalling what has been said by the witnesses appearing before them' (TNA, HO 345/12: W. Conwy Roberts, secretary to the committee, to Wolfenden, 29 Oct. 1954).
22. Robert Rhodes James, *Robert Boothby: A Portrait of Churchill's Ally* (New York: Viking, 1991), pp. 369–70; Liz Stanley, *Sex Surveyed, 1949–1994: From Mass-Observation's 'Little Kinsey' to the National Survey and the Hite Reports* (London: Taylor and Francis, 1995), pp. 199–200; Jeffrey Weeks, *Coming Out: Homosexual Politics in Britain, from the Nineteenth Century to the Present* (London: Quartet Books, 1977), p. 164.
23. Jeffrey Weeks, 'Wolfenden, John Frederick, Baron Wolfenden (1906–1985)', *Oxford Dictionary of National Biography* (Oxford University Press, 2004), http://www.oxforddnb.

com/view/article/31852 [hereafter *ODNB* online], accessed 11 Apr. 2013. In a radio panel discussion on the Midland Home Service the previous November, Wolfenden had already called for a Royal Commission to investigate homosexuality. See BBC Written Archives Centre, Caversham [hereafter BBC], *Behind the News*, TX 11/11/1953, p. 10.

24. Wolfenden, *Turning Points*, p. 132; Paul Ferris, *Sex and the British: A Twentieth-Century History* (London: Michael Joseph, 1993), p. 158.

25. See his remarks in TNA, HO 345/15, 15 Dec. 1955, QQ4127, 4138, 4140, and in HO 345/16, 31 Jan. 1956, Q4526.

26. Sebastian Faulks, *The Fatal Englishman: Three Short Lives* (London: Hutchinson, 1996), p. 221.

27. Faulks, *Fatal Englishman*, pp. 240–1. See also Philip French, 'We Saw the Light, but Too Late for Some', *Observer*, 24 June 2007.

28. Higgins, *Heterosexual Dictatorship*, pp. 9–10.

29. Roger Davidson and Gayle Davis, *The Sexual State: Sexuality and Scottish Governance, 1950–80* (Edinburgh: Edinburgh University Press, 2012), p. 54; *Glasgow Herald*, 8 January 1982, p. 4 (I am grateful to Jeff Meek for this citation); Jeff Meek, 'Scottish Churches, Morality and Homosexual Law Reform 1957–1980', *Journal of Ecclesiastical History*, 66, 3 (July 2015), 599.

30. 'Cohen, Mary Gwendolen (Mrs Arthur M. Cohen)', *Who Was Who* (online edn, Oxford University Press, 2014), http://www.ukwhoswho.com/view/article/oupww/whowaswho/U163351, accessed 1 Aug. 2014.

31. 'Curran, Desmond', *Who Was Who*, accessed 1 Aug. 2014; Desmond Curran, 'Sexual Perversions', *Practitioner*, 172 (Apr. 1954), 440–5.

32. Angela Cunningham, 'Demant, Vigo Auguste (1893–1983)', *ODNB* online, accessed 11 Apr. 2013.

33. Stephen Sedley, Godfray Le Quesne, 'Diplock, (William John) Kenneth, Baron Diplock (1907–1985)', *ODNB* online, accessed 11 Apr. 2013.

34. 'Linstead, Sir Hugh (Nicholas)', *Who Was Who*, accessed 1 Aug. 2014.

35. 'Lothian', *Who Was Who*, accessed 1 Aug. 2014.

36. Eileen M. Bowlt, *Justice in Middlesex: A Brief History of the Uxbridge Magistrates' Court* (Winchester: Waterside Press, 2007), pp. 71–2.

37. Ross Cranston, 'Mishcon, Victor, Baron Mishcon (1915–2006)', *ODNB* online, accessed 11 Apr. 2013.

38. Kenneth O. Morgan, 'Rees, (Morgan) Goronwy (1909–1979)', *ODNB* online, accessed 11 Apr. 2013.

39. 'Scott, Very Rev. Robert Forrester Victor', *Who Was Who*, accessed 1 Aug. 2014.

40. W. Mansfield Cooper, rev. H. Platt, 'Stopford, John Sebastian Bach, Baron Stopford of Fallowfield (1888–1961)', *ODNB* online, accessed 1 Aug. 2014.

41. 'Wells, William Thomas', *Who Was Who*, accessed 1 Aug. 2014.

42. *British Medical Journal*, 1 (5219), 14 Jan. 1961, p. 135.

43. HO 345/6, minutes of 21st meeting, 12–13 Mar. 1956; HO 345/2, Scott to Wolfenden, 8 Mar. 1956; Higgins, *Heterosexual Dictatorship*, pp. 35–6.

44. Morgan, 'Rees', *ODNB* online.

45. Goronwy Rees, *A Chapter of Accidents* (London: Chatto and Windus, 1972), pp. 62–3, 91–7; Goronwy Rees, *A Bundle of Sensations: Sketches in Autobiography* (London: Chatto and Windus, 1960), pp. 35–6.

46. Jenny Rees, *Looking for Mr Nobody: The Secret Life of Goronwy Rees* (London: Weidenfeld and Nicolson, 1994), p. 42.

47. John Harris, *Goronwy Rees* (Cardiff: University of Wales Press, 2001), pp. 57–9; Rees, *Looking for Mr Nobody*, chaps. 9, 14; Rees, *Chapter of Accidents*, chaps. 3–5; Morgan, 'Rees', *ODNB* online; HO 345/2, correspondence between Newsam and Wolfenden, Mar.–Apr. 1956; Mort, *Capital Affairs*, pp. 188–92; Higgins, *Heterosexual Dictatorship*, pp. 82–6.

48. *Wolfenden Report*, appendix IV: 'List of Witnesses', pp. 152–5. This list includes 203 individuals representing themselves or their professional and public bodies or government
departments, plus 11 additional organizations that submitted memoranda. There is a
certain amount of double counting, as some individuals represented more than one
organization, but the list does not mention those homosexuals who appeared and/or
submitted memoranda anonymously.

49. HO 345/2, 'Revised list of organisations to which invitations to submit evidence might
be addressed'; Curran to Roberts, 16 Sept. 1954; Linstead to Wolfenden, 23 Sept. 1954;
Roberts to Wolfenden, 30 Sept. 1954.

50. See Deborah Cohen, *Family Secrets: Living with Shame from the Victorians to the Present Day*
(London: Viking, 2013), pp. 154–9; Chris Waters, 'Havelock Ellis, Sigmund Freud and the
State: Discourses of Homosexual Identity in Interwar Britain', in Lucy Bland and Laura
Doan (eds), *Sexology in Culture: Labelling Bodies and Desires* (Chicago, IL: University of
Chicago Press, 1998), pp. 165–76.

51. For the history of homosexual law reform between 1957 and 1967, see Weeks, *Coming
Out*, chap. 15; Stephen Jeffery-Poulter, *Peers, Queers, and Commons: The Struggle for Gay
Law Reform from 1950 to the Present* (London: Routledge, 1991), chaps. 2–5; Antony Grey,
Quest for Justice: Towards Homosexual Emancipation (London: Sinclair Stevenson, 1992),
chaps. 3–9.

52. See, for example, Leslie J. Moran, 'The Homosexualization of English Law', in Didi
Herman and Carl Stychin (eds), *Legal Inversions: Lesbians, Gay Men, and the Politics of Law*
(Philadelphia, PA: Temple University Press, 1995), pp. 7–21; Leslie J. Moran, *The Homo-
sexual(ity) of Law* (London: Routledge, 1996), pp. 115–17; Houlbrook, *Queer London*,
pp. 242–8, 254; Matthew Waites, *The Age of Consent: Young People, Sexuality and Citizenship* (Houndmills: Palgrave Macmillan, 2005), pp. 106–11; Matthew Waites, 'The Fixity
of Sexual Identities in the Public Sphere: Biomedical Knowledge, Liberalism and the Het-
erosexual/Homosexual Binary in Late Modernity', *Sexualities*, 8, 5 (Dec. 2005), 558–9;
Jeffrey Weeks, *Sex, Politics and Society: The Regulation of Sexuality Since 1800* (London:
Pearson; 3rd edn, 2012), p. 314; Weeks, *The World We Have Won*, pp. 52–5; Hornsey,
The Spiv and the Architect, pp. 9–10, chaps. 2–3; Newburn, *Permission and Regulation*,
pp. 61–2; Jeffery-Poulter, *Peers, Queers, and Commons*, p. 263; Frank Mort, 'Mapping
Sexual London: The Wolfenden Committee on Homosexual Offences and Prostitution
1954–1957', *New Formations*, 37 (Spring 1999), 94–5; Derek McGhee, 'Wolfenden and
the Fear of "Homosexual Spread": Permeable Boundaries and Legal Defences', *Studies in
Law, Politics, and Society*, 21 (2000), 71–2.

53. J. Tudor Rees and Harley V. Usill (eds), *They Stand Apart: A Critical Survey of the Problems
of Homosexuality* (London: William Heinemann, 1955), p. xii.

1   Law Enforcers

1. See also HO 345/9: Memorandum of Evidence of the Faculty of Advocates.

2. As did the Association of Chief Police Officers of England and Wales, represented before
the committee by Chief Constables C. C. Martin of Liverpool and C. H. Watkins of
Glamorgan (HO 345/13, 31 Mar. 1955, QQ1405–551).

3. See HO 345/7, Memorandum by Nott-Bower, para. 33. He also pointed to a temporary
steep rise in 1951 because of the number of provincial visitors to the Festival of Britain.

4. The more relaxed attitude of the Scottish police was seen in the testimony of W. Hunter
and James A. Robertson, Assistant Chief Constables of the Edinburgh and Glasgow
police respectively (HO 345/16, 10 Apr. 1956, QQ5141–281). Robertson, for example,
outlined a few cases—of male prostitution, a brothel serviced by youths, gross indecency and associated robberies and blackmail and apparent homosexuals congregating
in a hotel lounge (QQ5141–54)—but asserted: 'We do not seem to have a really big
problem in Glasgow, so far as we know' (Q5141). For a more equivocal stance by the

magistrates of Glasgow, see HO 345/16, 9 Apr. 1956, QQ5010, 5038–9 and *passim*. The lower prosecution rate in Scotland was partly a result of a higher evidential standard: the need of 'corroboration'—that is, more than one witness (see, for example, HO 345/16, Mr. Lionel I. Gordon, O.B.E., Crown Agent, 9 Apr. 1956, QQ4887–97). For more detail, see Davidson and Davis, *Sexual State*, pp. 45–6; *Wolfenden Report*, ss. 9–11, 136–40; Brian Dempsey, 'Piecemeal to Equality: Scottish Gay Law Reform', in Leslie J. Moran, Daniel Monk and Sarah Beresford (eds), *Legal Queeries: Lesbian, Gay and Transgender Legal Studies* (London: Cassell, 1998), pp. 156–7.

5. See also the description by Commander A. Robertson, 'A' Division, Metropolitan Police, HO 345/12, 7 Dec. 1954, QQ582–9.

6. HO 345/12, 7 Dec. 1954, Q602.

7. Wolfenden, *Turning Points*, p. 137. On the policing of West End lavatories since the 1920s, see Mort, *Capital Affairs*, pp. 157–9. For evidence of policemen targeting homosexuals as an easy way to improve their conviction rates, see Wildeblood, *Against the Law*, pp. 174–5 and below, II: 3(a); Harry Daley, *This Small Cloud: A Personal Memoir* (London: Weidenfeld and Nicolson, 1986), pp. 101, 112, 156–8, 213.

8. Home Office statistics revealed that of the 71 cases of blackmail reported to the police in England and Wales between 1950 and 1953, 32 were connected with allegations of homosexual acts (HO 345/3: Note by the Secretary, 22 Apr. 1955; *Wolfenden Report*, s. 110). For a discussion of postwar homosexual blackmail, see Angus McLaren, *Sexual Blackmail: A Modern History* (Cambridge, MA: Harvard University Press, 2002), pp. 222–38; Croft-Cooke, *The Verdict of You All*, pp. 145–8. The *Wolfenden Report*, ss. 109–11, recognized but tended to downplay the significance of the law in inciting blackmail for homosexual acts; but it did recommend that men who reported to the police that they were being blackmailed for homosexual offences should not then be prosecuted for those offences (s. 112).

9. See on this point the evidence of Claud Mullins and his fellow magistrates before the committee, HO 345/15, 14 Dec. 1955, QQ4003–4, 4009.

10. Wildeblood, *A Way of Life*, p. 84. Thirty was the age at which a criminal was assumed to be inveterate and thus liable for preventive detention.

11. Sir Laurence Dunne, Chief Metropolitan Magistrate, asked them to submit individual memoranda (HO 345/14, 4 Oct. 1955, Q2978).

12. HO 345/15, 14 Dec. 1955, Q4005.

13. HO 345/13, 31 Mar. 1855, Q1511.

14. HO 345/12, 5 Jan. 1955, Q804.

15. Elwes' leniency highlights a point made, for example, by C. B. Trusler, Senior Probation Officer for Kent, and Frank Dawtry, General Secretary of the National Association of Probation Officers (HO 345/13, 13 July 1955, QQ2248–51), that the range of judges' attitudes and sentencing of homosexual offences was greater than for almost any other offence. See also HO 345/2, Curran to Roberts, 16 Sept. 1954; *Wolfenden Report*, ss. 173–80.

16. HO 345/15, 1 Nov. 1955, QQ3389, 3395.

17. HO 345/15, 17 Nov. 1955, QQ3870–2.

18. Moran, 'Homosexualization of English Law', p. 19.

19. Section 61: 'Whosoever shall be convicted of the abominable crime of buggery, committed either with mankind or with any animal, shall be liable ... to imprisonment for life'; section 62: 'Whosoever shall attempt to commit the said abominable crime, or shall be guilty of any indecent assault upon any male person, shall be guilty of a misdemeanour, and ... shall be liable ... to imprisonment for any term not exceeding ten years'.

20. 'Any male person who, in public or private, commits, or is a party to the commission of, or procures or attempts to procure the commission by any male person of, any act of gross indecency with another male person, shall be guilty of a misdemeanour, and

being convicted thereof shall be liable, at the discretion of the court, to be imprisoned for any term not exceeding two years.'

21. 'Every male person who...in any public place persistently solicits or importunes for immoral purposes, shall be deemed a rogue and vagabond within the meaning of the Vagrancy Act, 1824, and may be dealt with accordingly.' The *Wolfenden Report*, ss. 117–18, notes that this clause was intended to target men pimping for and living off the earnings of female prostitutes but was used almost exclusively to police homosexual behaviour.

22. This was not strictly true. As the Magistrates' Association pointed out in its memorandum (II: 1, iii(a)), women could be prosecuted for assaults on other females under the provisions of section 52 of the Offences Against the Person Act, 1861—though they could produce no evidence that homosexual acts between consenting adult women had ever actually been prosecuted.

23. *Sexual Offences: A Report of the Cambridge Department of Criminal Science* (London: Macmillan, 1957). A copy of the preface and of 'A Note on Criminal Statistics in so far as they Relate to Sexual Crimes Known to the Police', both by Dr Leon Radzinowicz, the Director of the department, are contained in the Wolfenden archive (HO 345/10). In the latter, Radzinowicz noted (p. 2), '[M]any kinds of homosexual activity practiced with the consent of both parties, will never be reported to the police. The figures of certain types of recorded homosexual crime are often merely the measure of police activity at a particular time in a particular area.'

24. For these common law crimes, penalties of up to three months could be imposed in summary cases, up to two years for proceedings by indictment in the sheriff court and any term of imprisonment in the High Court of Justiciary (where sodomy cases were tried). See also Davidson and Davis, *Sexual State*, p. 42.

25. These provided for summary jurisdiction (up to six months) or indictment (up to two years). The *Wolfenden Report*, ss. 117–18, notes that unlike section 1 of the Vagrancy Act, 1898 (see n. 21), its Scottish counterpart, the Immoral Traffic (Scotland) Act, 1902, was apparently only used for its intended original purpose.

26. Lionel Gordon, Crown Agent, in his testimony (HO 345/16, Q4885) disagreed with this: 'there have been prosecutions for homosexual acts committed by consenting adults in private without the aggravations mentioned in that sentence.'

27. Sir John Reginald Hornby Nott-Bower (1892–1972), Commissioner, Metropolitan Police, 1953–8 (*Who Was Who*, accessed 7 Oct. 2014).

28. Under section 21 of the Criminal Justice Act, 1948, an offender became liable to preventive detention if he was aged at least 30, had three previous convictions since the age of 17 for offences punishable by at least two years of imprisonment and had served sentences on at least two of those occasions. See Leslie T. Wilkins, 'Persistent Offenders and Preventive Detention', *Journal of Criminal Law, Criminology and Police Science*, 57, 3 (1967), 312–17.

29. See *Wolfenden Report*, ss. 125–6, for acknowledgment of the anomaly and a recommendation of how to deal with it.

30. The White Horse, Rupert Street. See Houlbrook, *Queer London*, p. 80.

31. This could be from Thomas Burke, *For Your Convenience: A Learned Dialogue, Instructive to all Londoners and London Visitors, Overheard in the Thélème Club and Taken Down Verbatim by Paul Pry* (London: Routledge, 1937).

32. The statue of the actor Sir Henry Irving behind the National Portrait Gallery.

33. Note that Drs Curran and Whitby regarded this as a highly improper use of entrapment (II: 2, ii(n)).

34. Special Investigation Branch of the Royal Military Police.

35. In April 1951 Arthur Robert Birley, who had worked for the BBC, was arrested with a number of Life Guardsmen in his Curzon Street flat. See Houlbrook, *Queer London*,

pp. 229–31; Higgins, *Heterosexual Dictatorship*, pp. 70–1; HO 345/12, 7 Dec. 1954, Q 610, testimony of T. MacD. Baker, solicitor to the Metropolitan Police.

36. See Michael Evelyn, rev. Mark Pottle, 'Mathew, Sir Theobald (1898–1964)', *ODNB* online, accessed 16 Apr. 2013. He was Director of Public Prosecutions from 1944 until his death.

37. The Law Society, founded in 1825, is the professional association for solicitors.

38. Kenny drew on Jeremy Bentham's *Principles of Morals and Legislation*, chap. xv, and his *Principles of Penal Law*, II, 1.4. See Courtney Stanhope Kenny, *Outlines of Criminal Law: Based on Lectures Delivered in the University of Cambridge*, 15th ed., rev. G. Godfrey Phillips (Cambridge: Cambridge University Press, 1936), pp. 28–30. Given Bentham's (unpublished) advocacy for the abolition of anti-sodomy laws, it is ironic that his principles should be used to argue the opposite case. See Louis Crompton (ed.), 'Jeremy Bentham's Essay on "Paederasty"', Parts 1 and 2, *Journal of Homosexuality*, 3, 4 (Summer 1978), 383–405, and 4, 1 (Fall 1978), 91–107.

39. As the secretary, Conwy Roberts, pointed out (HO 345/3 and /4, 'Note by the Secretary', Sept. 1955), the Law Society did not seem to have dealt with the second proposition in Bentham's first point and—in reference to alleged police improprieties during the Montagu/Wildeblood/Pitt-Rivers trial—'The arguments on Bentham's fourth point may not commend themselves to those who have been at pains to refer to what they allege was an unlawful searching of the Pitt-Rivers flat'.

40. The Bar Council, founded in 1894, is the professional association for barristers.

41. HO 345/16, 20 Feb. 1956, Q4776: N. R. Fox-Andrews, QC (for the Bar Council): The vote by a show of hands on section 11 was roughly three against change for every two in favour.

42. See *Criminal Appeal Reports*, 35 (1951–2), pp. 167–80. Alfred George Hall was found guilty at the 'C.C.C.' (Central Criminal Court of England and Wales, or the Old Bailey) of multiple acts of gross indecency with three young men. He was an instructor at a London institution 'connected with the study of the occult and similar matters' and— on the pretext of *bona fide* medical treatment—applied ointment to and indecently handled their genitals. He had a prior conviction in Canada for gross indecency.

43. This stipulated that a recommendation from two registered medical practitioners, approved by the Board of Control, should accompany any application for a person suffering from mental illness to be received as a temporary patient into a hospital or other institution for treatment.

44. Norman Roy Fox-Andrews, QC (1894–1921), Recorder of Bournemouth, 1945–61 (*Who Was Who*, accessed 2 Oct. 2014).

45. Presumably Roger Ormrod (1911–92), who became a judge of the High Court and was knighted in 1961 (*Who Was Who*, accessed 2 Oct. 2014; Robin Dunn, 'Ormrod, Sir Roger Fray Greenwood', *ODNB* online, accessed 2 Oct. 2014).

46. Reginald Ethelbert Seaton (1899–1978), Recorder of Maidstone, 1951–9 (*Who Was Who*, accessed 2 Oct. 2014). It was Seaton who had sent down Rupert Croft-Cooke for nine months at East Sussex Quarter Sessions in October 1953. See Morley, *John Gielgud*, p. 264.

47. The Church of England Moral Welfare Council's *The Problem of Homosexuality*, 1954.

48. This was the language of section 61 of the Offences Against the Person Act, 1861. The most famous statement of revulsion against *peccatum illud horribile, inter christianos non nominandum*—the 'horrible sin not to be spoken of among Christians'—appears in William Blackstone, *Commentaries on the Laws of England*, IV (1769), p. 215. An earlier formulation had appeared in Sir Edward Coke, *The Third Part of the Institutes of the Laws of England* (1644), chap. 10.

49. The National Prohibition Act or Volstead Act of 1919 was enacted to carry out the intent of the Eighteenth Amendment to the US Constitution, which declared illegal

the production, transportation and sale of alcohol. Prohibition remained in effect until 1933.

50. Alfred Kinsey, Wardell Pomeroy and Clyde Martin, *Sexual Behavior in the Human Male* (Philadelphia, PA: W. B. Saunders, 1948).

51. The phrase is sometimes attributed to the barrister and High Court Judge Sir Travers Humphreys (see, for example, Higgins, *Heterosexual Dictatorship*, p. 98), but in his 'Foreword' to H. Montgomery Hyde, *The Trials of Oscar Wilde* (London: Penguin, 1962), p. 12, Humphreys attributed it to 'a learned Recorder'.

52. The Magistrates' Association, founded in 1920, represented 9,000 magistrates and 100 benches. This memorandum was narrowly adopted by the Association's governing council but rejected at the annual general meeting of the membership (see HO 345/15, 14 Dec. 1955, QQ3984–6).

53. The Wolfenden archive contains a 'Survey of the Law on Homosexuality in Western Europe' submitted by H. A. Hammelmann for the Magistrates' Association (HO 345/8). Of the ten countries surveyed, only two—Germany and Norway—still had legal provisions against consenting adult male sex, and those in Norway were no longer enforced. See also HO 345/9 for a series of letters from the secretary addressed to the foreign ministries and ambassadors of Sweden, the Netherlands and Germany, inquiring about the state of the law, and the responses.

54. The case involved Maggie Hare, who was found guilty at the Central Criminal Court on 20 October 1933 of three counts of indecent assault on (that is, sexual intercourse with) a boy of 12. She appealed on the grounds that the 'whosoever' in various sections of the Offences against the Person Act, 1861, did not apply to females. The Appeal Court judges ruled that, 'A woman may be guilty of indecent assault upon a male person, contrary to section 62 of the Offences against the Person Act, 1861, or upon another female, contrary to section 52 of that Act'. *Criminal Appeal Reports*, 24 (1932–4), pp. 108–12.

55. See Richard Davenport-Hines, 'Dunne, Sir Laurence Rivers (1893–1970)', *ODNB* online, accessed 30 Apr. 2013. Dunne was Chief Metropolitan Magistrate from 1948 to 1960.

56. The memoranda of the War Office (II: 1, v(d)) and of the Lord Chief Justice (II: 1, iii(j)) painted a different picture. For the assertion that prostitution in the Guards continued into the 1960s, see Simon Raven, 'Boys Will Be Boys: The Male Prostitute in London', *Encounter*, xv, 1 (July 1960), 19–20; Matt Houlbrook, 'Soldier Heroes and Rent Boys: Homosex, Masculinities, and Britishness in the Brigade of Guards, circa 1900–1960', *Journal of British Studies*, 42, 3 (July 2003), 353.

57. Dunne is confused here. The familiar injunctions against a man lying with a man 'as with a woman' appear in the Book of Leviticus, 18:22 and 20:13.

58. The Gateways club in Chelsea was the most famous such establishment. For the lesbian pub and club scene, see Rebecca Jennings, *A Lesbian History of Britain: Love and Sex between Women since 1500* (Oxford: Greenwood, 2007), chap. 8; Rebecca Jennings, *Tomboys and Bachelor Girls: A Lesbian History of Post-War Britain 1945–71* (Manchester: Manchester University Press, 2007), chap. 4.

59. Captain Eugene Paul Bennett (1892–1970) was a Metropolitan Police Magistrate from 1935 to 1961 (*Who Was Who*, accessed 7 Oct. 2014).

60. It is not clear which BBC television programme Bennett was watching. Possible candidates, which screened in the weeks prior to his interview, included *In the News*, *Panorama* and *Press Conference* (BBC Genome Project, http://genome.ch.bbc.co.uk, accessed 29 Oct. 2014). Homosexuality had earlier been discussed on BBC radio—on the Light Programme's *Any Questions?* (without the word being mentioned), 30 October 1953 (Kynaston, *Family Britain*, pp. 333–4) and on the Midland Home Service's *Behind the News*, 11 November 1953, a discussion chaired by Graham Hutton between Wolfenden, the Oxford historian Alan Bullock and the scientist and author Jacob Bronowski (BBC, TX 11/11/1953).

61. See S. M. Cretney, 'Mullins, Claud William (1887–1968)', *ODNB* online, accessed 30 Apr. 2013.

62. Sir William Clarke Hall (1866–1932), Metropolitan Magistrate, Thames (1913–14), Old Street (1914 ff); Chairman of the Magistrates' Association and of the National Association of Probation Officers (*Who Was Who*, accessed 7 Oct. 2014).
63. See Malcolm Pines, 'Glover, Edward George' (1888–1972), *ODNB* online, accessed 7 Oct. 2014. Glover was a noted psychoanalyst, a founder of the Institute for Scientific Treatment of Delinquency and a pioneer of the psychoanalytic investigation and treatment of delinquency.
64. Claud Mullins, *Crime and Psychology* (London: Methuen, 1943).
65. See Malcolm Pines, 'Rees, John Rawlings (1890–1969)', *ODNB* online, accessed 7 Oct. 2014. Rees was a psychiatrist, the director of the Tavistock Clinic from 1932 to 1947 and subsequently president and director of the World Federation of Mental Health.
66. HM Prison Grendon, which opened in 1962 to treat prisoners with antisocial personality disorders. See n. 103.
67. Geoffrey Keith Rose (1889–1959), Metropolitan Magistrate from 1934 (*Who Was Who*, accessed 7 Oct. 2014).
68. Harold Francis Ralph Sturge (1902–93), Metropolitan Magistrate 1947–68 (*Who Was Who*, accessed 7 Oct. 2014).
69. Sturge framed much of his memorandum in response to points raised by the Church of England Moral Welfare Council's report *The Problem of Homosexuality*.
70. The particular distaste for homosexual behaviour that transgressed class boundaries is discussed below, p. 205.
71. The Wolfenden Report did not specify, but the Sexual Offences Act of 1967 did: any number over two.
72. Allan Grierson Walker (1907–94), Sheriff Substitute of Lanarkshire at Glasgow, 1950–63; knighted 1968 (*Who Was Who*, accessed 29 Jan. 2015).
73. Wolfenden marked this sentence with a large exclamation mark.
74. See also HO 345/16, 9 Apr. 1956, QQ5077–140: The negative opinions of Sheriffs Substitute F. Middleton (Dunfermline and Kinross) and A. M. Prain (Perth and Dunblane), speaking for the Association of Sheriffs Substitute.
75. Stuart Grace Kermack (1888–1981), Sheriff Substitute of Lanarkshire at Glasgow, 1936–55, of Renfrew and Argyll at Oban, 1955–62 (*Who Was Who*, accessed 29 Jan. 2015).
76. Archibald Hamilton (1895–1974), Sheriff Substitute of Aberdeen, Kincardine and Banff, 1952–71 (*Who Was Who*, accessed 29 Jan. 2015).
77. Richard Everard Augustine Elwes (1901–68), Recorder of Northampton from 1946 to 1958 and Chairman of the Derbyshire Quarter Sessions from 1954 to 1958, became a High Court Judge and was knighted in 1958 (*Who Was Who*, accessed 9 Oct. 2014). He was also a member of the Roman Catholic Advisory Committee that presented a report to Wolfenden (II: 4(d)).
78. This is incorrect. As H. G. Cocks makes clear (*Nameless Offences: Homosexual Desire in the 19th Century* (London: I. B. Tauris, 2003), pp. 17–18), all types of homosexual act could be and were prosecuted under the existing anti-sodomy laws.
79. For details of the passage of the Labouchère Amendment, see F. B. Smith, 'Labouchère's Amendment to the Criminal Law Amendment Bill', *Historical Studies (Australia)*, 17, 67 (1976), 165–75; Matt Cook, *London and the Culture of Homosexuality, 1885–1914* (Cambridge: Cambridge University Press, 2003), pp. 42–5.
80. See HO 345/2, Roberts to Wolfenden, 8 Aug. 1955: Roberts considered the police actions in this case to be 'pretty shocking', but thought that he should get the police's own perspective, hence the solicitation of these reports (annexes IV and V).
81. John Scott Henderson (1895–1964), Recorder of Portsmouth, 1945–62 (*Who Was Who*, accessed 9 Oct. 2014).
82. See K. J. M. Smith, 'Goddard, Rayner (1877–1971)', *ODNB* online, accessed 30 Apr. 2013. He was Lord Chief Justice from 1946 to 1958.
83. Annex A consists of a series of questions posed by the secretary.

84. In a 'Note by the Chairman', HO 345/4, Sept. 1955, Wolfenden commented that this 'authoritative statement' was influential to his reaching the conclusion that the law should be changed.

85. Presumably a reference to the Montagu trial (Michael Pitt-Rivers had 'a highly distinguished war record' (Wildeblood, *Against the Law*, p. 41) and was promoted to captain in 1946) or to the trial of Gilbert Nixon et al. (see p. 3).

86. Sentences of whipping were abolished by the Criminal Justice Act, 1948—A reform that Goddard opposed at the time (see *Times*, 24 May 1948, p. 4).

87. Of the judges surveyed, Devlin, Donovan, Finnemore, Hallett and Oliver were broadly in favour of reform, and Sellers to a limited extent; Barry, Byrne, Cassels, Gerrard, Gorman, McNair, Parker, Pilcher, Slade, Stable and Streatfield were opposed.

88. Sir Patrick Redmond Barry (1898–1972), High Court Judge 1950–66 (*Who Was Who*, accessed 29 Jan. 2015).

89. Hon. Sir Laurence Austin Byrne (1896–1965), High Court Judge 1947–60 (*Who Was Who*, accessed 29 Jan. 2015).

90. Sir Patrick Arthur Devlin (1905–92), High Court Judge 1948–60, life peer 1961 (*Who Was Who*, accessed 29 Jan. 2015).

91. Sir Terence Norbert Donovan (1898–1971), High Court Judge 1950–60, life peer 1964 (*Who Was Who*, accessed 29 Jan. 2015).

92. Sir Donald Leslie Finnemore (1889–1974), High Court Judge 1948–64 (*Who Was Who*, accessed 29 Jan. 2015).

93. Sir Hugh Imbert Periam Hallett (1886–1967), High Court Judge 1939–57 (*Who Was Who*, accessed 29 Jan. 2015).

94. Sir William Lennox McNair (1892–1979), High Court Judge 1950–66 (*Who Was Who*, accessed 29 Jan. 2015).

95. Rt Hon. Sir Frederic Aked Sellers (1893–1979), High Court Judge 1946–57 (*Who Was Who*, accessed 29 Jan. 2015).

96. 'Approved schools' for juvenile delinquents, which superseded industrial schools and reformatories, were created by the Children and Young Persons Act 1933. See Steven Schlossman, 'Delinquent Children: The Juvenile Reform School', in Norval Morris and David J. Rothman (eds), *The Oxford History of the Prison: The Practice of Punishment in Western Society* (Oxford: Oxford University Press, 1995), p. 343.

97. HO 345/8, Memorandum of the Association of Headmasters, Headmistresses and Matrons of Approved Schools (endorsed by the Association of Managers of Approved Schools).

98. The Prison Commissioners also submitted detailed data on the 1069 prisoners and Borstal inmates serving time for homosexual offences on 15 November 1954 or who were incarcerated up to 14 February 1955 (HO 345/9: Summary of information available to date from the Prison Commissioners' Questionnaire). The information—32 items about each inmate—was coded and punched on cards for machine analysis. 42% of the prisoners or Borstal boys had been convicted of buggery, 5.9% of attempted buggery, 32.6% of assault with intent, 13.7% of gross indecency, 0.7% of attempted indecency, 4.2% of importuning and 1.5% of procuring an act of gross indecency.

99. 'Star' prisoners were those gaoled for the first time. Wildeblood, *Against the Law*, p. 96, describes the prison uniform 'with a scarlet star on each shoulder to indicate that we were First Offenders'.

100. For the history of Borstals—institutions for young offenders aged 16 to 21, first introduced in 1902—see Sean McGonville, 'The Victorian Prison: England, 1865–1965', in Morris and Rothman (eds), *The Oxford History of the Prison*, pp. 141–4.

101. Harvie Kennard Snell (1898–1969), Director of Medical Services, Prison Commission, 1951–63 (*Who Was Who*, accessed 11 Feb. 2015).

102. Snell appears to be unaware of F. J. Kallmann's study (see II: 2, n. 6).

103. William Norwood East and William Henry de Bargue Hubert, *Report on the Psychological Treatment of Crime* (London: HMSO, 1939), which recommended the setting up of a special prison to provide psychological treatment for certain categories of prisoner—and resulted in the opening of HM Prison Grendon in 1962. See also Waters, 'Havelock Ellis, Sigmund Freud and the State', pp. 172–3.

104. Psalms 127: 3–5.

105. For Wildeblood's impressions of Landers, 'an agreeable Irish psychiatrist', see II: 3(a) and *Against the Law*, pp. 143–5, 187.

106. For the perceived threat of venereal disease emanating from 'morally lax' women during the War, see Sonya O. Rose, *Which People's War? National Identity and Citizenship in Wartime Britain 1939–1945* (Oxford: Oxford University Press, 2003), pp. 80–1.

107. Wolfenden's pencil note in the margin against this paragraph: 'it does sometimes'.

108. A reference to a statement by Lord Jowitt in his Maudsley Lecture to the Royal Medico-Psychological Association, 1953, about the period when he became Attorney-General in 1929: 'A very large percentage of the blackmail cases—nearly 90 per cent. of them—were cases in which the person blackmailed had been guilty of homosexual practices with an adult person' (Jowitt, 'The Twenty-Eighth Maudsley Lecture: Medicine and the Law', *Journal of Mental Science*, 100, 419 (Apr. 1954), 353). This figure was inaccurate (see n. 8), but in a speech to parliament the following year, Jowitt inflated it even further, suggesting that, 'at least 95 per cent. of the cases of blackmail which came to my knowledge arose out of homosexuality' (*Hansard's Parliamentary Debates*, Lords, 5th ser., 187 (1954), c. 745).

109. The ATS—Auxiliary Territorial Service—was the women's branch of the army during the War.

110. The memorandum is presumably 'A Special Problem', by Letitia Fairfield, the retired chief medical officer of the ATS, but Emma Vickers puts the correct date of this undated study as 1943. See London Metropolitan Archives, PH/GEN/3/19, Papers of Letitia Fairfield, quoted by Vickers, *Queen and Country: Same-Sex Desire in the British Armed Forces, 1939–45* (Manchester: Manchester University Press, 2013), pp. 12, 121–4, 130, 140. My thanks to Dr Vickers for clarification on this point.

111. General Officer Commanding.

112. An apparent reference to the case of 14 Coldstream Guards drummer boys (aged 16–18) from Windsor barracks who had sex with three men. See HO 345/12, 7 Dec. 1954, Q503, testimony of Sir Theobald Mathew; HO 345/2, Wolfenden to Roberts, 14 Jan. 1955; Higgins, *Heterosexual Dictatorship*, p. 71.

2 Medical Practitioners and Scientists

1. For the rise of sexology, see Weeks, *Sex, Politics and Society*, chaps. 6, 8; Weeks, *Coming Out*, part 2; Lucy Bland and Laura Doan (eds), *Sexology in Culture: Labelling Bodies and Desires* and *Sexology Uncensored: The Documents of Sexual Science* (Chicago, IL: University of Chicago Press, 1998); Chris Waters, 'Sexology', in Matt Houlbrook and Harry Cocks (eds), *The Modern History of Sexuality* (Houndmills, Basingstoke: Palgrave Macmillan, 2006); Ivan Crozier, 'Introduction: Havelock Ellis, John Addington Symonds and the Construction of *Sexual Inversion*', in Ellis and Symonds, *Sexual Inversion: A Critical Edition*, ed. Crozier (Houndmills, Basingstoke: Palgrave Macmillan, 2008); Kenneth Plummer (ed.), *The Making of the Modern Homosexual* (London: Hutchinson, 1981); Anomaly, *The Invert and His Social Adjustment* (London: Baillière, Tindall and Cox, 1927).

2. Waters, 'Havelock Ellis, Sigmund Freud and the State', 165–76; Cohen, *Family Secrets*, pp. 152–3.

3. Kinsey et al., *Sexual Behavior in the Human Male*; Alfred C. Kinsey, Wardell B. Pomeroy, Clyde E. Martin and Paul H. Gebhard, *Sexual Behavior in the Human Female* (Philadelphia: W. B. Saunders, 1953).

4. For a classic discussion of 'minoritizing' and 'universalizing', see Eve Kosofsky Sedgwick, *Epistemology of the Closet* (Berkeley and Los Angeles, CA: University of California Press, 1990).
5. The BMA published its memorandum in a booklet, *Homosexuality and Prostitution*, in December 1955. The *Observer* (18 Dec. 1955, p. 6) commented that 'it might have been better if the committee had kept to medical aspects, and had not discussed moral and legal issues on which doctors are not specially qualified to give verdicts'. Wildeblood, *A Way of Life*, p. 156, ridiculed the BMA's moralizing: ' "as though anybody would think of consulting a doctor about a question of morals"'. On the enduring strength of religious faith and of 'discursive Christianity' until the 1960s, see Callum Brown, *The Death of Christian Britain: Understanding Secularization 1800–2000* (London: Routledge, 2001).
6. Franz Josef Kallmann (1897–1965) was a German-born psychiatrist who spent much of his career in the Department of Medical Genetics of the New York Psychiatric Institute. See his obituary in the *American Journal of Psychiatry*, 123, 1 (July 1966), 105–6.
7. HO 345/15, 15 Dec. 1955, Q4222.
8. For example, Edward Glover, 'Introduction' to Westwood, *Society and the Homosexual*, p. 16; Eustace Chesser, *Live and Let Live: The Moral of the Wolfenden Report* (London: Heinemann, 1958), pp. 25, 37.
9. Dr M. Grünhut, Reader in Criminology at Oxford, provided a report on the relatively small number dealt with under section 4 of the Criminal Justice Act, 1948, which gave courts the power to place offenders on probation with a requirement that they submit to mental health treatment for up to twelve months (HO 345/9, Memorandum on the Mental Treatment of Probationers Found Guilty of Homosexual Offences).
10. See the testimony of Dr T. D. Inch, Medical Adviser, Scottish Prison and Borstal Services (HO 345/15, 1 Nov. 1955, QQ3544–5), but also the doubts because of the alarming side effects (possible atrophy of the testicles and the development of breasts) expressed by Drs Matheson and Roper, Senior Medical Officers of Brixton and Wakefield Prisons respectively (HO 345/15, Q3454). See also the extended discussion by Drs Leonard Simpson, Ronald Gibson, T. C. N. Gibbens and Dennis Carroll on behalf of the BMA (HO 345/15, 15 Dec. 1955, QQ4261–71). For the history of hormone treatment, see Andrew Hodges, *Alan Turing: The Enigma* (Princeton, NJ: Princeton University Press, 2012; 1st edn 1983), pp. 467–73. Turing, its most famous victim, opted for this kind of treatment to avoid prison after pleading guilty to acts of gross indecency. This followed the logic of section 4 of the Criminal Justice Act, 1948, and an emphasis on the duty of the community to provide treatment for habitual sexual offenders (see F. L. Golla and R. Sessions Hodge, 'Hormone Treatment of the Sexual Offender', *The Lancet*, CCLVI, I (11 June 1949), 1006–7). For other forms of treatment of homosexuality of the period, including aversion therapy, see Michael King, Glenn Smith and Annie Bartlett, 'Treatments of Homosexuality in Britain Since the 1950s—An Oral History', *British Medical Journal*, 328, 7437 (21 Feb. 2004), 427–32; Tommy Dickinson, *'Curing Queers': Mental Nurses and Their Patients, 1935–74* (Manchester: Manchester University Press, 2015); Hugh David, *On Queer Street: A Social History of British Homosexuality 1895–1995* (London: HarperCollins, 1997).
11. HO 345/2, Newsam to Wolfenden, 31 Jan. 1955; Wolfenden to Newsam, 10 Feb. 1955.
12. *Wolfenden Report*, ss. 209–11, which recommended the lifting of the ban on oestrogen treatment in English and Welsh prisons if the prisoner desired it and if the prison medical officer believed that it would be beneficial.
13. Comprising Ronald Gibson, GP, Winchester (chairman); T. H. Blench, Lecturer in Forensic Medicine, University of Manchester; Dennis C. Carroll, Consultant Psychiatrist, Portman Clinic; T. C. N. Gibbens, Lecturer in Forensic Psychiatry, Institute of Psychiatry, London; J. Glaister, Regius Professor of Forensic Medicine, University of Glasgow; Ambrose King, Senior Physician, London Hospital Clinic for Venereal

Diseases; J. A. Moody, GP, London; Doris Odlum, Senior Physician for Psychological Medicine, Elizabeth Garrett Anderson Hospital (Royal Free Hospital); T. P. Rees, Medical Superintendent, Warlingham Park Hospital; S. Leonard Simpson, Consultant Endocrinologist, St Mary's Hospital; R. D. Summers, Divisional Surgeon, Metropolitan Police; E. E. Claxton, Assistant Secretary of the BMA (secretary).

14. This paragraph in the booklet version of the memorandum, *Homosexuality and Prostitution*, raised the ire of Lt-Col. Marcus Lipton, Labour MP for Lambeth Brixton. He claimed in the House of Commons that 'any assertion of the existence of practising homosexuals in Parliament must be regarded as serious', since 'the charging of Members with conduct rendering them unworthy to sit in Parliament is a breach of Privilege'. *Hansard's Parliamentary Debates*, Commons, 5th ser., 547 (19 Dec. 1955), cc. 1660–2. See also HO 345/2, Roberts to Wolfenden, 21 Dec. 1955.

15. Comprising Doris Odlum (chairman); Josephine Barnes, Assistant Obstetrician and Gynaecologist, Charing Cross Hospital and Elizabeth Garrett Anderson Hospital; Mary Esslemont, GP; Letitia Fairfield, barrister, formerly Senior Medical Officer, LCC; Dorothy Fenwick, Assistant Medical Officer, LCC; Annis Gillie, GP; Mona Macnaughton, GP; Joan Malleson, Medical Officer in charge of Dyspareunia Clinic, Obstetric Unit, University College Hospital; Marjorie Murrell, Consultant, Venereal Diseases Department, Edinburgh Royal Infirmary; Gladys M. Sandes, Consultant Venereologist and Gynaecologist, Queen Mary's Hospital for Children, Carshalton; Patricia Shaw, barrister, Senior Medical Officer, Nottingham; Albertine Winner, Senior Medical Officer, Ministry of Health.

16. Cf above, II: 1, v(d).

17. The Bible contains only one apparent reference: Romans 1:26: 'God gave them up unto vile passions: for their women changed the natural use into that which is against nature.'

18. As the *Wolfenden Report* makes clear (ss. 37–40), this was guesswork based on practically no reliable data. It was probably derived from the unsupported supposition of Havelock Ellis, *Studies in the Psychology of Sex*, vol. 2: 'Sexual Inversion' (3rd edn, 1927), pp. 51–2. For Kinsey's critique of Ellis' statistics, see Kinsey et al., *Sexual Behavior in the Human Male*, pp. 618–19.

19. This understates the case. The figures for homosexual offences known to the police published in the *Wolfenden Report*, p. 130, point to a 1,068% increase between 1931 and 1955.

20. See A. M. Cooke, 'Penrose, Lionel Sharples (1898–1972), physician', and Hugh Series, 'Slater, Eliot Trevor Oakeshott (1904–1983), psychiatrist', *ODNB* online, accessed 25 Feb. 2015. Penrose was Galton Professor of Eugenics at University College London and President of the Genetical Society of Great Britain. Slater was Senior Lecturer at the Institute of Psychiatry at the Maudsley Hospital and Physician in Psychological Medicine at the National Hospital for Nervous Diseases, London.

21. Samuel Leonard Simpson (1900–83) was the author of a standard work in the field, *Major Endocrine Disorders* (1938) (see M. Ginsburg, 'Simpson family', *ODNB* online, accessed 25 Feb. 2015).

22. Electroconvulsive therapy.

23. A term in psychoanalysis for (in the words of the Oxford English Dictionary) 'the relief of anxiety by the expression and release of a previously expressed emotion, through reliving the experience that caused it'.

24. Stilboestrol and ethinyl oestradiol were both first synthesized in 1938, in Oxford and Berlin, respectively. The former was written up in *Nature*, 141 (5 Feb. 1938), 247–8, the latter in *Naturwissenschaften*, 26, 6 (11 Feb. 1938), 96.

25. Boyd was Consultant Venereologist and Director VD Depts, Essex County Hospitals, and Consultant in Infertility, Wanstead Hospital.

26. On Jefferiss and St Mary's, see Elsbeth Heaman, *St Mary's: The History of a London Teaching Hospital* (Liverpool: Liverpool University Press, 2003).

27. Franz J. Kallmann, *Heredity in Health and Mental Disorder: Principles of Psychiatric Genetics in the Light of Comparative Twin Studies* (New York: W. W. Norton, 1953), pp. 116–19.
28. In his evidence to the committee (HO 345/14, 31 Oct. 1955, QQ3189–98), Sessions Hodge commented that he had treated 250–300 men with oestrogen, the longest for ten years; he had been able to relax the treatment in only two cases; but he had experienced only two failures. In some instances he had to reduce the dose because of the degree of gynaecomastia.
29. The memorandum was written by D. W. Winnicott, FRCP, who was in charge of the Psychology Department at the hospital. See Clifford Yorke, 'Winnicott, Donald Woods (1896–1971), paediatrician and psychoanalyst', *ODNB* online, accessed 26 Feb. 2015; Brett Kahr, *D. W. Winnicott: A Biographical Portrait* (London: H. Karnac, 1996).
30. The memorandum was prepared by a committee comprising Drs Noel G. Harris (President; Physician for Psychological Medicine, Middlesex Hospital), M. J. Brookes (Medical Superintendent, Shelton Hospital, Shrewsbury), J. M. Crawford (Physician Superintendent, Botleys Park Hospital), J. A. Hobson (Assistant Physician in Psychological Medicine, Middlesex Hospital), J. D. W. Pearce (Physician-in-Charge, Dept of Psychiatry, St Mary's Hospital; Psychiatrist, Queen Elizabeth Hospital for Children), J. S. I. Skottowe (Consultant Psychiatrist, the Warneford Hospital, Oxford) and A. Walk (Physician Superintendent, Cane Hill Hospital, Coulsdon, Surrey). In his evidence before the Wolfenden committee (HO 345/14, 31 Oct. 1955, Q3241), Dr Harris gave some background about the Royal Medico-Psychological Association: it was founded [in 1841] as an association of the medical officers of England's mental hospitals; it was the biggest and oldest association associated with mental diseases; and it had approximately 1,000 members (working in mental hospitals or specializing in psychological medicine in teaching hospitals, neurosis centres and mental deficiency centres), representing practically the whole of the psychiatric profession in the UK. (It became the Royal College of Psychiatrists in 1971.)
31. But see above, II: 1, i(a), n. 22.
32. Published as Lennox Ross Broster and Walter Langdon-Brown, *The Adrenal Cortex and Intersexuality* (London: Chapman and Hall, 1938).
33. It was published in 1949. Allen submitted a copy of it to the committee.
34. Perhaps the most emphatic iteration of this long-standing claim can be found in Xavier Mayne [Edward Irenaeus Prime Stevenson], *The Intersexes: A History of Similisexualism as a Problem in Social Life* (privately printed, 1908), chap. 12.
35. See Lesley A. Hall, 'Chesser, Eustace (1902–1973), psychiatrist and social reformer', *ODNB* online, accessed 30 Apr. 2013. A prolific writer and broadcaster, Chesser was perhaps best known for his first book, *Love Without Fear: A Plain Guide to Sex Technique for Every Married Adult* (London: Rich and Cowan, 1941).
36. Edward Carpenter, *The Intermediate Sex: A Study of Some Transitional Types of Men and Women* (New York and London: Mitchell Kennerley, 1912).
37. The Tavistock Clinic, founded in Tavistock Square, London, in 1920, was the brainchild of Dr Hugh Crichton-Miller. He and his collaborators wished to extend pioneering psychotherapeutic techniques for shell-shocked soldiers to civilians of modest means. It quickly became established as a leader in the analysis and treatment of nervous disorders. See H. V. Dicks, *Fifty Years of the Tavistock Clinic* (London: Routledge and Kegan Paul, 1970), chap. 2; K. Loughlin, 'Miller, Hugh Crichton- (1877–1959), psychotherapist', *ODNB* online, accessed 26 Feb. 2015.
38. Wolfenden was sceptical about the use of this label. He double-underlined 'an illness' and pencilled in on his copy, '??? -?symptoms?'
39. The distinction between buggery and other homosexual offences was going to divide the Wolfenden Committee. Wolfenden's opinion was that, 'We cannot overlook the fact that there are a great many people who believe (however much we may disagree with them) that buggery, as distinct from other forms of homosexual behaviour, has

demoralizing effects not only on individuals but on nations and empires' (HO 345/10, 'Note by the Chairman', n.d.). In the end, the *Wolfenden Report* recommended that the distinction be largely scrapped, but retained it in one respect: s. 91 suggested a maximum penalty of life imprisonment for buggery with a boy under sixteen but ten years for gross indecency with such a boy. Cohen, Curran, Stopford and Whitby all expressed reservations (*Wolfenden Report*, pp. 123–7), wanting to see the distinction abolished entirely. For a discussion of the debate, see *Wolfenden Report*, ss. 78–94; HO 345/3, Curran and Whitby, 'Buggery: Points for abolitionism as a separate offence'; Moran, *The Homosexual(ity) of Law*, chap. 2.

40. The Maudsley Hospital Medical School, established in 1924, was renamed the Institute of Psychiatry in 1948. The memorandum was put together by a sub-committee in consultation with the senior medical staff of the Institute and of two associated mental hospitals, the Bethlem Royal and the Maudsley.
41. Children and Young Persons Act 1933, s. 107: ' "Young person" means a person who has attained the age of fourteen years and is under the age of seventeen years'.
42. Probation of Offenders Act 1907, s. 4: 'It shall be the duty of a probation officer ... to advise, assist, and befriend [the offender], and, when necessary, to endeavour to find him suitable employment.'
43. See above, II: 1, n. 103.
44. The provision of the Mental Deficiency Act 1913 that allowed a court to send an offender, if proven to be a mental defective, to a mental institution in lieu of imprisonment.
45. The Institute was founded in 1924. For details, see Shapira, *The War Inside*, p. 7.
46. Sigmund Freud, *Three Essays on the Theory of Sexuality* (trans. James Strachey, New York: Basic Books, 1962; 1st edn 1905), essay II, 'Infantile Sexuality'.
47. The BPS was founded as the Psychological Society at University College London in 1901 and renamed the British Psychological Society in 1906. The memorandum was prepared by a special committee comprising: Michael Balint, Psychiatrist, Tavistock Clinic; G. M. Carstairs, Deputy Director of the Unit for Research in Occupational Adaptation, Institute of Psychiatry; M. A. Davidson, Consultant Clinical Psychologist, Warneford and Park Hospitals, Oxford; Michael Fordham, Director of the Clinic of the Society of Analytical Psychology, London; T. H. Pear, Emeritus Professor of Psychology, University of Manchester; Erwin Popper, Consultant Psychiatrist, Tavistock Clinic; Seymour Spencer, First Assistant, Dept. of Psychological Medicine, University of Durham; C. Anthony Storr, Consultant Psychiatrist, Tilbury and Riverside Hospital, London; E. B. Strauss, Physician of Psychological Medicine, St Bartholomew's Hospital, London; P. E. Vernon, Professor of Educational Psychology, University of London.
48. This was Kinsey's estimate of the number of 'Kinsey 6s' in the US, as the memorandum notes in point (6), but 'most investigators' in Britain guessed at a lower figure for the UK.
49. The references listed for this appendix are: Robert Briffault, *The Mothers: A Study of the Origins of Sentiments and Institutions* (1927); Clellan S. Ford and Frank A. Beach, *Patterns of Sexual Behaviour* (1951); James George Frazer, *The Golden Bough* (1890 and subsequent, extended editions); Melville J. Herskovits, *Man and His Works: The Science of Cultural Anthropology* (1948); Bronislaw Malinowski, *The Sexual Life of Savages in North-Western Melanesia* (1929) and *Sex and Repression in Savage Society* (1927); Margaret Mead, *From the South Seas: Studies of Adolescence and Sex in Primitive Societies* (1939) and *Male and Female* (1949); and Havelock Ellis, *Studies in the Psychology of Sex* (1897–1928).
50. A reference to the Human Relations Area Files, Inc., founded at Yale in 1949. This interuniversity consortium, under the direction of Clellan S. Ford, grew out of the Cross-Cultural Survey at Yale's Institute of Human Relations—an information-gathering and indexing system designed to retrieve and share comparative information about a broad range of societies—and was intended to make it more widely available (see http://

hraf.yale.edu/about/history-and-development/, accessed 3 Mar. 2015). Since Ford and Beach, *Patterns of Sexual Behavior*, drew much of their data from the HRAF, presumably it is this study to which the memorandum refers.

51. The Siwa Oasis is in the Libyan Desert in western Egypt.
52. The Chukchi of far eastern Siberia.
53. An island off the north coast of New Guinea.
54. The group was convened in the autumn of 1954 because of a concern that 'certain more general social aspects' of homosexuality and prostitution might be overlooked (HO 345/14, 28 July 1955, Q2626, Professor Norman Haycocks). The National Council of Social Service, an umbrella for various voluntary organizations, was founded in 1919 (and renamed the National Council of Voluntary Organisations in 1980), the National Association for Mental Health in 1946 (and renamed MIND in 1972).
55. In other words, the 'care or protection' clauses of the Act (ss. 61–3) should be extended above the age of 16 (see above, n. 41). Wolfenden pencilled in against this 'but they can go to Korea'—A reminder that young men below the age of 21 could be sent to war, and that it made little sense to extend to the 17–21 age group provisions intended for children and adolescents. The *Wolfenden Report*, s. 74, ruled out such an extension.
56. See above, II: 1, n. 103.
57. The Davidson Clinic was founded in Edinburgh in 1940 to provide 'psychotherapeutic treatment for psychoneuroses and psychosomatic disorders in patients who are unable to afford the fees of private psychotherapists'. The fees were met by voluntary subscriptions and donations. Both Freudian and Jungian techniques were used. See also HO 345/16, 10 Apr. 1956, Dr Rushforth, QQ5282–3: in 1956 the clinic had five medically qualified workers, six lay therapists and a children's department.
58. In her interview with the committee (HO 345/16, 10 Apr. 1956, QQ5350, 5314), Rushforth suggested that 'attachment to the mother is almost diagnostic' and that mother-fixation and paedophilia were closely related: 'I often wonder if . . . the mother fixator is really looking for the mother's nipple with the one with the small penis.'
59. It was signed by O. Cargill, Dennis Carroll, T. C. N. Gibbens, Edward Glover, John Mackwood, W. Paterson Brown, L. H. Rubenstein and P. D. Scott. In his evidence to the committee on behalf of the ISTD, Dr Dennis Carroll claimed that the memorandum was unanimously accepted by the clinic's staff of 23 or 24—that is, from every school of psychiatry dealing with homosexuality—Freudians, Jungians, Adlerians, general psychiatrists and so on (HO 345/16, 1 Feb. 1956, Q4566). For Glover, the chairman of the ISTD, who drafted the memorandum, see Shapira, *The War Inside*, pp. 146–7, 190–3; Waters, 'Havelock Ellis, Sigmund Freud and the State', pp. 174–6; Malcolm Pines, 'Glover, Edward George (1888–1972), psychoanalyst', *ODNB* online, accessed 11 Mar. 2015.
60. For details see Shapira, *The War Inside*, chap. 5.
61. Prepared by Mary Woodward, Research Fellow of the Institute.
62. This undated memorandum by the medical practitioners on the committee gave a useful summary of the scientific and medical evidence and suggestions for the writing of the report. The report did not, in the end, go far enough for the two doctors, so they added a 'Note by Dr. Curran and Dr. Whitby' (*Wolfenden Report*, pp. 72–6).
63. Kinsey et al., 'Concepts of Normality and Abnormality in Sexual Behavior', in Paul H. Hoch and Joseph Zubin (eds), *Psychosexual Development in Health and Disease* (Proceedings of the 38th Annual Meeting of the American Psychopathological Association, New York, June 1948) (New York: Grune and Stratton, 1949), p. 12.
64. Institute of Psycho-Analysis, Tavistock Clinic and ISTD/Portman Clinic respectively.
65. British Medical Association.
66. Royal Medico-Psychological Association.
67. P.36 is presumably a typo for M.36 = Winifred Rushforth of the Davidson Clinic.
68. Rushforth, Sir Laurence Dunne (Chief Metropolitan Magistrate), Institute of Psycho-Analysis.

69. ISTD/Portman, Tavistock Clinic, Institute of Psycho-Analysis, Institute of Psychiatry, Tavistock Clinic.
70. Institute of Psychiatry.
71. Ethical Union.
72. BMA.
73. Kinsey et al., 'Concepts of Normality', p. 27.
74. ISTD.
75. Kinsey et al., 'Concepts of Normality', p. 21; interview with Kinsey.
76. Prison Commissioners of England and Wales.
77. British Psychological Society.
78. Royal Medico-Psychological Association.
79. For accounts of the suicide in 1822 of Lord Castlereagh, the Foreign Secretary, see H. Montgomery Hyde, *The Strange Death of Lord Castlereagh* (London: Heinemann, 1959); John Bew, *Castlereagh: A Life* (Oxford: Oxford University Press, 2012), chaps. 20–1.
80. The expression, indicating a progressive deterioration, is taken from William Hogarth's series of eight paintings (1735) depicting the decline and fall of Tom Rakewell.
81. See above, II: 1, iv(b), appx II.
82. Senior Medical Officers Matheson (Brixton), Landers (Wormwood Scrubs), Roper (Wakefield) and Brisby (Liverpool) and Dr Boyd (Consultant Psychiatrist to the Scottish Prison and Borstal Services).
83. Drs Clifford Allen, Eustace Chesser and R. Sessions Hodge.
84. See above, II: 1, i(d).
85. See above, II: 1, i(b).
86. Kinsey et al., *Sexual Behavior in the Human Male*, pp. 650–1.
87. British Psychological Society.
88. See above, n. 18.
89. BMA.
90. HO 345/8, 'Survey of the Law on Homosexuality in Western Europe'.
91. See Sir Samuel Romilly, *Observations on the Criminal Law of England, as It Relates to Capital Punishments, and on the Mode in Which It Is Administered* (3rd edn, London: Cadell and Davis, 1813), pp. 21–2.
92. Case studies supplied by Sessions Hodge, HO 345/9.
93. Clifford Allen.
94. The Institute of Biology was founded in 1950 as a professional body for Britain's biologists.
95. Crew was Buchanan Professor of Animal Genetics (1928–44) and Professor of Public Health and Social Medicine (1945–55) at the University of Edinburgh. See D. S. Falconer, 'Crew, Francis Albert Eley (1886–1973), animal geneticist', *ODNB* online, accessed 19 Mar. 2015.
96. Darlington was Sherardian Professor of Botany at Oxford and Fisher was Arthur Balfour Professor of Genetics at Cambridge; both founded and co-edited the journal *Heredity*. See Vassiliki Betty Smocovitis, 'Darlington, Cyril Dean (1903–1981), cytogeneticist and evolutionist' and Hamish G. Spencer, 'Fisher, Sir Ronald Aylmer (1890–1962), statistician and geneticist', *ODNB* online, accessed 30 Apr. 2013. For Huxley's prolific career as a scientist and public intellectual, see Robert Olby, 'Huxley, Sir Julian Sorell (1887–1975), zoologist and philosopher', *ODNB* online, accessed 30 Apr. 2013.
97. The memorandum footnoted Kinsey's studies.
98. The memorandum footnoted William II, Richard I, Edward II, Richard II and James I as the five English kings. Henry III of France and four of the English kings (the exception being James I) met violent ends, but scarcely 'on account of' their 'symptoms of homosexuality'.
99. Memorandum footnote: 'James I and Oscar Wilde'.

100. Memorandum footnote: 'Oscar Wilde and Lord Montagu'. This is inaccurate regarding Montagu. He identified as bisexual, and although his fiancée, Anne Gage, broke off their engagement before his release from prison, he later married twice, in 1959 and 1974. See Montagu, *Wheels within Wheels*, pp. 100, 129 and *passim*.
101. See Francis Galton, 'The History of Twins, as a Criterion of the Relative Powers of Nature and Nurture', *Fraser's Magazine*, 12 (Nov. 1875), 566–76. Galton's twin study method involved the tracking of life-changes in similar and dissimilar twins to see whether they diverged or converged if raised in similar or dissimilar environments.
102. Franz J. Kallmann, 'Twin and Sibship Study of Overt Male Homosexuality', *American Journal of Human Genetics* 4, 2 (June 1952), 136–46.
103. Curran, Lovibond, Scott, Whitby and Roberts were in attendance. For Kinsey (1894–1956), see James H. Jones, *Alfred C. Kinsey: A Public/Private Life* (New York: Norton, 1997); Jonathan Gathorne-Hardy, *Alfred C. Kinsey: Sex the Measure of All Things* (London: Chatto and Windus, 1998); Donna J. Drucker, *The Classification of Sex: Alfred Kinsey and the Organization of Knowledge* (Pittsburgh, PA: University of Pittsburgh Press, 2014).
104. A 'Note by the Secretary', May 1956 (HO 345/9) concluded—after consultation between Mishcon and a justice of the Supreme Court of New York State—that Kinsey was in error here. The law had indeed changed five years previously, but to provide for three degrees of sodomy: in the first degree (amounting to rape), liable to a punishment of up to life imprisonment; in the second degree (by persons over 21 against persons under 18)—up to ten years' imprisonment; and all other forms of sodomy (including oral-genital intercourse), classified as misdemeanours, maximum punishment unstated.
105. Kinsey was not impressed by Kallmann's study. See HO 345/14, 31 Oct. 1955, Q3313: Dr Curran: 'Dr. Kinsey was telling us the other day that we could throw away all this study.'
106. Theo Lang, 'Studies on the Genetic Determination of Homosexuality', *Journal of Nervous and Mental Disease*, 92, 1 (July 1940), 55–64. Lang worked at the Research Institute for Psychiatry in Munich. His studies during the Nazi era, drawing on lists of homosexuals supplied by the Munich and Hamburg police, posited that the apparent tendency for male homosexuals to have a higher than average number of brothers rather than sisters could be explained by the fact that a proportion of homosexuals were 'at least to a certain extent defined as intersexes' (p. 55).

3 Homosexuals

1. See HO 345/6, minutes of the fifth meeting, 4 and 5 Jan. 1955.
2. For example, HO 345/2, Roberts to Wolfenden, 30 Sept. 1954; Wolfenden to Roberts, 14 Jan. 1955.
3. HO 345/2, Roberts to Wolfenden, 11 Oct. 1954.
4. HO 345/2, Roberts to Wolfenden, 15 Dec. 1954. This unnamed individual was not, in fact, interviewed.
5. HO 345/2, Wolfenden to Roberts, 25 June 1955; Roberts to Wolfenden, 28 June 1955; Wolfenden to Roberts, 29 June 1955.
6. Wildeblood, *Against the Law*, pp. 173–4.
7. HO 345/2, 23 May 1955.
8. British Library, National Sound Archive, Hall-Carpenter Oral History Project, C456/089, Patrick Trevor-Roper interviewed by Margot Farnham, 1 Aug. 1990. Trevor-Roper provided a slightly different recollection of events in an interview with Frank Mort in 1999. See Mort, *Capital Affairs*, pp. 173–4.
9. HO 345/2, Wolfenden to Roberts, 22 Dec. 1954.
10. HO 345/2, Wolfenden to Roberts, 8 May 1955.

11. NSA, C456/089, Patrick Trevor-Roper.
12. Houlbrook, *Queer London*, pp. 242–8, 254; Houlbrook and Waters, '*The Heart in Exile*', 145–55. For the queer scene, see: Houlbrook, *Queer London*, chap. 3; Alkarim Jivani, *It's Not Unusual: A History of Lesbian and Gay Britain in the Twentieth Century* (Bloomington: Indiana University Press, 1997), pp. 128–33; Daniel Farson, *Soho in the Fifties* (London: Michael Joseph, 1987), pp. 71–81; Garland, *Heart in Exile*, pp. 56–70.
13. Even though Trevor-Roper and Wildeblood had had a brief affair (NSA, C456/089, Patrick Trevor-Roper).
14. 'Purported' since he had picked up his lover, RAF corporal Edward McNally, in a public place, Piccadilly Circus. Trevor-Roper too had breached the public/private divide that they were all preaching: he was lucky to escape merely with a warning from a policeman who caught him cottaging in St James's Park during the war. Their testimonies relied on a good deal of calculated amnesia. See Wildeblood, *Against the Law*, p. 37; NSA, C456/089, Patrick Trevor-Roper; Houlbrook, *Queer London*, p. 260.
15. For example, in the anti-Wildeblood camp, Higgins, *Heterosexual Dictatorship*, pp. 40–1, and in the pro-Wildeblood camp, Matthew Parris, 'Preface to the 1999 edition', in Wildeblood, *Against the Law*, pp. v–viii. See also Mort, *Capital Affairs*, p. 178; Weeks, *Sex, Politics and Society*, p. 310.
16. His hostility towards the 'pathetically flamboyant pansy with the flapping wrists' is also evident in *Against the Law*, p. 7. But in *A Way of Life*, for example, he associates with a colourful array of mainstream and marginal queer characters in public and private settings and—although he makes clear his preference for monogamy (for example, pp. 86–7)—he allows a voice to those in the 'community' who denounce his priggishness (for example, p. 160).
17. For testimonies to his courtesy, patience and sympathy, see HO 345/2, Trevor Roper to Wolfenden, 21 June 1955; Winter to Wolfenden, 1 Aug. 1955.
18. HO 345/2, Wolfenden to Roberts, 20 July 1955; HO 345/3, Roberts to Wolfenden, 25 July 1955.
19. Gordon Westwood [Michael Schofield], *A Minority: A Report on the Life of the Male Homosexual in Great Britain* (London: Longmans, 1960), p. 131, to some extent corroborated such a suggestion. In his sample of 127 homosexual men, only 32 per cent regularly had anal sex—though many others had tried it and found it unpleasant or unsatisfactory.
20. Associating with 'social inferiors' also helped seal the fate of Montagu, Pitt-Rivers and Wildeblood (*Against the Law*, pp. 26–7, 77; *A Way of Life*, p. 150).
21. But P. Allen of the Home Office was alive to the dangers (HO 345/12, Q71). In a reference to the Croft-Cooke case, he talked about a wealthy gentleman and his servant picking up two sailors in a London pub and taking them to his country house, and asked 'whether it is desirable that people of this kind, wealthy gentlemen, should be able to go out and pick up adults to gratify their desires ... It is not only the case of the two consenting adults of the same sort of social status.' Croft-Cooke and Wilde joined a long list of others—John Addington Symonds, Edward Carpenter, J. R. Ackerley, E. M. Forster (and his creation Maurice Hall) come to mind—who searched for lower-class lovers. On class transgression and its history, see Jeffrey Weeks, 'Inverts, Perverts, and Mary-Annes: Male Prostitution and the Regulation of Homosexuality in England in the Nineteenth and Early Twentieth Centuries', *Journal of Homosexuality*, 6, 1/2 (Fall/Winter 1980–1), 121–2; Hornsey, *The Spiv and the Architect*, pp. 96–8.
22. See Matthew Parris, 'Wildeblood, Peter (1923–1999), journalist and campaigner for homosexual law reform', *ODNB* online, accessed 30 Apr. 2013. Points made, stories recounted and language used in this memorandum parallel very closely *Against the Law*.
23. Wolfenden's pencil note in the margin: 'Nothing in all of this that questions man <u>was</u> guilty.'
24. Wildeblood presumably had in mind the Buggery Act, which actually dates to 1533; it replaced ecclesiastical jurisdiction with the country's first civil buggery law.

25. Scott was Metropolitan Police Commissioner from 1945 to 1953. See Kenneth Parker, 'Scott, Sir Harold Richard (1887–1969), civil servant and commissioner of police', *ODNB* online, accessed 23 Mar. 2015.
26. Wildeblood was repeating a common error here. The Code Napoleon of 1804 referred only to civil law whereas sexual offences came under criminal law. It was the French Revolutionaries in the Penal Code of 1791 who abrogated the anti-sodomy laws, and this was perpetuated in Napoleon's Penal Code of 1810. See Michael David Sibalis, 'The Regulation of Male Homosexuality in Revolutionary and Napoleonic France, 1789–1815', in Jeffrey Merrick and Bryant T. Ragan, Jr (eds), *Homosexuality in Modern France* (Oxford: Oxford University Press, 1996), pp. 80–101.
27. This allegation is consistent with the 'sexual McCarthyism' thesis, but the arrest and prosecution numbers do not support an abrupt shift in policing methods c.1952 or between the Scott and Nott-Bower regimes at Scotland Yard.
28. Wolfenden's note: 'As you believe everything anybody in prison tells you?'
29. In *Against the Law*, pp. 101–2, Wildeblood made plain his lack of respect for the prejudices of this 'hard-bitten little Scot'.
30. In *Against the Law*, pp. 107, 187, Wildeblood wrote that there were in fact very few opportunities for physical contact, and that homosexual expression was almost exclusively emotional.
31. Wolfenden's note: 'No: you broke the law'.
32. See, for example, *Studies in the Psychology of Sex*, vol. 2, pp. 322–5.
33. Wolfenden's note: 'Yes, if c-b results in them driving cars across traffic lights'.
34. Wolfenden underlined and put an exclamation mark and a tick beside this last phrase.
35. This witness was Patrick Trevor-Roper. For his biography, see Stephen Lock, 'Roper, Patrick Dacre Trevor- (1916–2004), ophthalmic surgeon and author', *ODNB* online, accessed 17 Apr. 2013.
36. The case involved two off-duty police constables, John Warhurst and Thomas Collister. They were sentenced at the Old Bailey to two years' imprisonment for attempting to pick up Kenneth Jeffries, a commercial traveller, in Piccadilly Circus and demanding money in exchange for their silence (*Manchester Guardian*, 29 Mar. 1955, p. 3; 30 Mar., p. 3).
37. Carl Winter (1906–66). For biographical details, see his obituary, *Times*, 23 May 1966, p. 14.
38. HM Prison Holloway, North London, had been a female-only prison since 1903.
39. That is, Trevor-Roper and Winter respectively.
40. In his 1990 interview (NSA, C456/089), Trevor-Roper recalled with gratitude this initial 'very warm, encouraging' intervention by Rees.
41. A reference to E. M. Forster, who began a long-term relationship with Hammersmith policeman Bob Buckingham in 1930. See Wendy Moffat, *A Great Unrecorded History: A New Life of E. M. Forster* (New York: Farrar, Straus and Giroux, 2010), chap. 10 and *passim*.
42. A reference to the Labouchère Amendment. See above, II: 1, i(a), n. 20.
43. But see Kinsey et al., *Sexual Behavior in the Human Male*, p. 579 ('anal activity in the heterosexual is not frequent enough to make it possible to determine the incidence of individuals who are specifically responsive to such stimulation') and *Sexual Behavior in the Human Female*, p. 585 ('good incidence data are not available').
44. A note by the secretary simply comments that the author was a professional man, known to the chairman.
45. A footnote reads, 'The writer's full name, address and qualifications have been supplied in a covering letter to the Secretary.' But, since this letter is missing, the writer retains his anonymity.
46. As examples of this 'ignorance' and 'intolerance', the writer footnotes the article of a general practitioner, Dr G. C. Learoyd of Beckley, Sussex, 'The Problem of

Homosexuality', in the *Practitioner*, 172 (Apr. 1954), 355–63 (p. 362: the idea that adult homosexuality in private should be legalized was 'dangerously decadent' and would 'lower moral standards incalculably'), and Learoyd's letters to the editor in the *British Medical Journal*, 5 June 1954, p. 1326 and the *Lancet*, 5 June 1954, pp. 1186–7;18 Sept. 1954, pp. 604–5.

47. Kinsey et al., *Sexual Behavior in the Human Female*, p. 680: 'There are some psychiatrists who consider all transvestism homosexual, but this is incorrect. Transvestism and homosexuality are totally independent phenomena, and it is only a small portion of transvestites who are homosexual in their physical relationships.'

48. Roberta (born Robert) Cowell (1918–2011), a racing driver and wartime RAF pilot who had been married with children and who did not identify as homosexual, was among the first British male-to-female transsexuals to undergo—in the late 1940s and early 1950s—sex reassignment surgery. Her story came out in 1954, first in a serialization in the *Picture Post* (13 Mar.–24 Apr. 1954), then in *Roberta Cowell's Story by Herself: Her Autobiography* (New York: British Book Centre, 1954). See also her obituary in the *Independent*, 27 Oct. 2013; Joanne Meyerowitz, *How Sex Changed: A History of Transsexuality in the United States* (Cambridge, MA: Harvard University Press, 2002), pp. 83–4; Alison Oram, 'Cross-Dressing and Transgender', in Cocks and Houlbrook (eds), *Modern History of Sexuality*, pp. 276–80; Pagan Kennedy, *The First Man-Made Man: The Story of Two Sex Changes, One Love Affair, and a Twentieth-Century Medical Revolution* (New York: Bloomsbury, 2007).

49. Samuel Butler, *Hudibras* (1663), canto I, ll. 215–16.

50. The writer footnotes the Church of England Moral Welfare Council's *The Problem of Homosexuality* (1954).

51. Kinsey et al., *Sexual Behavior in the Human Male*, fig. 161.

4 Christians, Moralists and Reformers

1. Wolfenden wrote in a cautionary note (HO 345/10, 'Note by the Chairman', n.d.): 'The views of the Church of England Moral Welfare Council and the Roman Catholic Advisory Committee, while representing the views of theologians and churchmen who have studied the matter dispassionately, may be far from representing the views of the churches as a whole.' For example, in two letters to Wolfenden (HO 345/2, 1 and 8 Aug. 1956), the Bishop of Rochester wanted to make it clear that the memorandum and its liberal views reflected only the opinion of the Moral Welfare Council and not of the Church more broadly. In the second letter he was particularly concerned 'that Sodomy Societies and Sodomy Week-end House parties must not be made legal' and expressed 'more sympathy with a Curate or Scout-Master who has offended with a boy (horrible though this is: and possibly because I have had to deal with such cases) than with two grown men misbehaving together'. For the bishop's more detailed views, see his article 'The Church and Sex', *Practitioner*, 172 (Apr. 1954), 350–4.

2. Church of England Moral Welfare Council, *The Problem of Homosexuality*; Jones, *Sexual Politics in the Church of England*, pp. 176–82; Timothy W. Jones, 'Moral Welfare and Social Well-Being: The Church of England and the Emergence of Modern Homosexuality', in Lucy Delap and Sue Morgan (eds), *Men, Masculinities and Religious Change in Twentieth-Century Britain* (Houndmills: Palgrave Macmillan, 2013), pp. 206–8; Grimley, 'Law, Morality and Secularisation', pp. 728–9; Grimley, 'Bailey, Derrick Sherwin (1910–1984), Church of England priest and sexual ethicist', *ODNB* online, accessed 30 Apr. 2013; HO 345/7, 'The Homosexual, the Law, and Society', preamble.

3. The memorandum was based on the chapter ('Homosexuality and Society') he had written for the collection edited by Tudor Rees and Usill, *They Stand Apart*. See HO 345/16, 31 Jan. 1956, Viscount Hailsham, Q4496.

4. In a letter to Roberts (HO 345/2, 13 Jan. 1956), Wolfenden hinted at his disdain for Hailsham's opinions.

5. The Public Morality Council began life in 1899 as the London Council for the Promotion of Public Morality, with the aim of fighting vice and indecency. It largely represented the churches and an evangelical crusading zeal against sexual nonconformity, variously targeting over the decades street prostitution, unwholesome plays and films, queer spaces and morally deficient commercial venues in general. See G. I. T. Machin, *Churches and Social Issues in Twentieth-Century Britain* (Oxford: Clarendon Press, 1998), p. 81; Houlbrook, *Queer London*, pp. 25, 78–9.

6. The sub-committee consisted of Kenneth Walker, MD, FRCS, MRCS, LRCP, John Mackwood, MRCS, LRCP, MC, and Mrs S. Neville-Rolfe, OBE (Convenor). A memorandum of the Public Morality Council (HO 345/8) stated that it accepted the findings of the sub-committee, with two caveats: '(1) There are some members of our widely representative body who are anxious lest it should be assumed that this is a problem in which no blameworthiness rests upon any offender'; and '(2) There are many others who are anxious that in the treatment of offenders the remedial and redemptive power of religion should not be ignored'. But Fr T. Holland, DD, one of the witnesses testifying on behalf of the Council (HO 345/14, 27 July 1955, QQ2392-428), made it clear that he did not favour decriminalization. His testimony was cited by Wolfenden as evidence that the Roman Catholic Advisory Committee did not speak for the entire church (HO 345/10, 'Note by the Chairman', n.d.).

7. The first British ethical society, the South Place Ethical Society, was founded in London in 1888 at the instigation of Stanton Coit, a disciple of fellow American Felix Adler. Other societies swiftly followed, and by the mid-1890s the four London societies joined together in the Ethical Union. See Edward Royle, *Radicals, Secularists, and Republicans: Popular Freethought in Britain, 1866–1915* (Manchester: Manchester University Press, 1980), p. 42; Stephen Law, *Humanism: A Very Short Introduction* (Oxford: Oxford University Press, 2011), chap. 1.

8. This was not quite right. See *Wolfenden Report*, appendix III: 'Homosexual Offences in European Countries', p. 151: Swedish law prohibited homosexual acts 'committed with young persons under 18, if the offender had himself reached that age at the time of the offence', and homosexual acts 'committed with persons under 21, if the offender is 18 or over and commits the act by abusing the other person's inexperience or dependence on him'.

9. This organization had its roots in late-Victorian moral reform societies. The White Cross Army, established by Ellice Hopkins in 1883 to recruit men for the cause of social purity, amalgamated with the Church of England Purity Society in 1891 to form the White Cross League. In 1939 this in turn amalgamated with the Archbishops' Advisory Board for Moral Welfare Work to form the Church of England Moral Welfare Council. See Frank K. Prochaska, *Women and Philanthropy in Nineteenth-Century England* (Oxford: Oxford University Press, 1980), pp. 215–16; Jones, *Sexual Politics in the Church of England*, p. 177.

10. Reprinted in 'Homosexuality, Prostitution and the Law', *Dublin Review*, 230, 471 (Summer 1956), 60–5; the copy in the Wolfenden archive (HO 345/8) is missing. The advisory committee, set up by the Cardinal Archbishop of Westminster, Bernard Griffin, comprised: G. A. Tomlinson (Chairman), Catholic Students Chaplain, University of London; J. McDonald, Professor of Moral Theology, St Edmund's College, Ware; John Preedy, Parish Priest, Englefield Green, Surrey; B. S. McFie, Psychiatric Social Worker; C. M. Jenner, Probation Officer; E. B. Strauss, Psychiatrist; and Richard Elwes, QC, Recorder of Northampton and Chairman of Derbyshire Quarter Sessions.

11. See above, II: 1, i(c), n. 49.

12. The followers of Austrian psychotherapist Alfred Adler (1870–1937).

13. Denmark was the first European country to introduce a surgical castration law for sex offenders, in 1929, and the other Scandinavian countries, Germany, Estonia, Latvia

and Iceland followed suit in the 1930s and 1940s. Switzerland, the Netherlands and Greenland used castration without legislation. Britain and the Catholic countries did not practise surgical castration. See Louis Le Maire, 'Danish Experiences Regarding the Castration of Sexual Offenders', *Journal of Criminal Law and Criminology*, 47, 3 (1956), 294; Nikolaus Heim and Carolyn J. Hursch, 'Castration for Sex Offenders: Treatment or Punishment? A Review and Critique of Recent European Literature', *Archives of Sexual Behavior*, 8, 3 (May 1979), 282–3.

14. It was originally named the Federation of Progressive Societies and Individuals and became the Progressive League in 1940. The first president was the philosopher and public intellectual C. E. M. Joad, and the vice-presidents included H. G. Wells, Bertrand Russell, Barbara Wootton, Vera Brittain, Aldous Huxley, Kingsley Martin and Leonard Woolf. See Joad (ed.), *Manifesto: Being the Book of the Federation of Progressive Societies and Individuals* (London: George Allen and Unwin, 1934); Tony Judge, *Radio Philosopher: The Radical Life of Cyril Joad* (London: Alpha House Books, 2012), pp. 81–2.

15. The Montagu/Pitt-Rivers/Wildeblood case.

16. See above, II: 4(b).

17. Inspired by the ideas of Sir Francis Galton, this was established in 1907 as the Eugenics Education Society, and became the Eugenics Society in 1926. See Richard A. Soloway, *Demography and Degeneration: Eugenics and the Declining Birthrate in Twentieth-Century Britain* (Chapel Hill, NC: University of North Carolina Press, 1990), pp. 26–7.

18. Founded in 1945 by Gerald Gardiner QC (later Lord Chancellor, 1964–70) as a professional association of lawyers on the left of politics who supported the Labour Party. See Norman S. Marsh, 'Gardiner, Gerald Austin, Baron Gardiner (1900–1990)', *ODNB* online, accessed 2 Apr. 2015. Gardiner and other members of the society submitted a memorandum to and appeared as witnesses before the Wolfenden Committee (HO 345/14, 27 July 1955, QQ 2472–2587).

19. Founded by freethinkers in 1899 to publish secular and humanist literature deemed too controversial by mainstream publishers. See Bill Cooke, *The Gathering of Infidels: A Hundred Years of the Rationalist Press Association* (Amherst, NY: Prometheus Books, 2004), pp. 27–9.

20. An obscure organization led by humanist philosopher J. B. Coates. See Emmanuel Mounier, *Personalism* (London: Routledge and Kegan Paul, 1952), p. xx.

21. For a similar comment see the opinion of Dr W. P. Kreamer, Medical Director, Davidson Clinic, Edinburgh: 'it is quite obviously very nearly a prerequisite to be homosexual to make a career as an actor in London' (HO 345/16, 10 Apr. 1956, Q5344).

22. Wolfenden pencilled in '?evidence' next to this paragraph and '!' next to this last sentence.

23. The homosexual writer and Nobel laureate André Gide (1869–1951).

24. In 1942 Marshal Philippe Pétain's Vichy regime had reintroduced a distinction (abolished during the French Revolution, see above, II: 3(a), n. 26) between 'natural' and 'unnatural' sex by criminalizing sexual relations with anyone of the same sex under the age of 21. The age of heterosexual consent remained 15. Charles de Gaulle's provisional government after the liberation perpetuated this distinction: by the decree of 8 February 1945 (article 331) it reaffirmed a sentence of from six months to three years for this offence. See Michael D. Sibalis, 'Homophobia, Vichy France, and the "Crime of Homosexuality": The Origins of the Ordinance of 6 August 1942', *GLQ: A Journal of Lesbian and Gay Studies*, 8, 3 (2002), 302–3; Guy Hocquenghem, *Homosexual Desire* (Durham, NC: Duke University Press, 1993; 1st edn 1972), p. 64.

25. The Howard Association (established in 1866) and the Penal Reform League (1907) merged to form the Howard League for Penal Reform in 1921.

26. The National Council of Social Service was created in 1919, in part with a legacy from Edward Birchall, who had been killed in action in France in 1916. The Standing Conference of Juvenile Organisations was created under its auspices in 1936; it acquired

the name heading this memorandum in 1943. See Margaret E. Brasnett, *Voluntary Social Action: A History of the National Council of Social Service, 1919–1969* (London: National Council of Social Service, 1969); Keith Laybourn, 'Birchall, Edward Vivian Dearman (1884–1916), philanthropist', *ODNB* online, accessed 3 Apr. 2015.

27. These included the Army Cadet Force Association, the Boys' Brigade, the Boy Scouts Association, the British Red Cross Society, the Church Lads' Brigade, the Co-operative Youth Movement, the Girls' Life Brigade, the Methodist Association of Youth Clubs, the National Association of Boys' Clubs, the National Association of Mixed Clubs and Girls' Clubs, the National Federation of Young Farmers' Clubs, the St John Ambulance Brigade, the Salvation Army, the Young Men's Christian Association, Toc H, the Youth Hostels Association, the Covenanter Union and the Christian Alliance of Women and Girls (345/8: Prof. Norman Haycocks, Chairman of the Standing Conference, to Wolfenden, 15 June 1955).

28. See S. M. Cretney, 'Hogg, Quintin McGarel, second Viscount Hailsham and Baron Hailsham of St Marylebone (1907–2001), lawyer and politician', *ODNB* online, accessed 30 Apr. 2013.

29. Wolfenden's pencil note in margin: 'exc[ept]. that the h[omosexual]. goes to prison if he "falls"'.

30. Wolfenden note: 'many aren't'.

31. Wolfenden note: 'dogma'.

32. Wolfenden note: 'dogma'.

Conclusion

1. Wolfenden, *Turning Points*, p. 138. The assistant secretary, E. J. Freeman of the Scottish Home Department, had sufficient reservations about the report that he was reluctant to sign—until whipped into line by his superiors. See Michael McManus, *Tory Pride and Prejudice: The Conservative Party and Homosexual Law Reform* (London: Biteback Publishing, 2011), p. 26.

2. HO 345/16, 1 Feb. 1956 (Magistrates' Association), Q4690.

3. Trudeau, when Canadian Minister of Justice, defended the decriminalization of homosexual acts in the Omnibus Bill by telling reporters that 'there's no place for the state in the bedrooms of the nation' and that 'what's done in private between adults doesn't concern the Criminal Code'—two impeccably Wolfendenian statements (CBC News, 21 Dec. 1967).

4. 'Reservation by Mr. Adair', *Wolfenden Report*, pp. 117–23. See also the discussion on the BBC Home Service, 19 September 1957, 'The Wolfenden Report, 2: Homosexuality and the Law'. Adair argued, '[T]here has been a vast increase in the adultery and in the fornication since these were taken out of the criminal code, and I think just as it occurred then, so it will occur today if we take this away from the criminal code even to the limited extent that is proposed by the Committee' (BBC, TX 19/09/1957, p. 4).

5. Patrick Devlin, *The Enforcement of Morals* (London: Oxford University Press, 1965), pp. 8–25. For a similar argument see Lord Denning's address to the Law Society's conference (*Times*, 27 Sept. 1957, p. 7): 'I would say most emphatically that standards and morals are the concern of the law, and that whether done in private or in public...I would say that without religion there can be no morality and without morality there can be no law'.

6. Stephen's *Liberty, Equality, Fraternity* (1873) was a reply to Mill's *On Liberty* (1859).

7. H. L. A. Hart, *Law, Liberty and Morality* (Stanford, CA: Stanford University Press, 1963), chaps. 1–3; quotation p. 81. See also a similarly themed talk by Hart, 'Immorality and Treason', broadcast on the BBC Third Programme, 14 July 1959 (BBC, TX 14/07/1959).

8. Hall, 'Reformism and the Legislation of Content', 1, 11–12; Weeks, *The World We Have Won*, pp. 54–5, 88; Mark Jarvis, *Conservative Governments, Morality and Social Change in*

Affluent Britain, 1957–64 (Manchester: Manchester University Press, 2005), pp. 6, 94–9. Hall's list (p. 1) includes the Homicide Act (1957), the Street Offences Act (1959), the Obscene Publications Acts (1959 and 1964), the Suicide Act (1961), the Murder (Abolition) Act (1965), the Sexual Offences Act (1967), the Family Planning Act (1967) and the Abortion Act (1967), plus legislation concerning licensing, gambling, theatre censorship, Sunday entertainments and divorce.

9. Canon Demant, in a discussion on 12 Sept. 1957 on the BBC's General Overseas Service (BBC, 'The Wolfenden Report', TX 12/09/1957, p. 2). See also Wolfenden on *At Home and Abroad*, the Home Service, 6 Sept. 1957 (BBC, TX 6/09/1957, p. 1), and on *Woman's Hour*, the Light Programme, 11 Sept. 1957 (BBC, TX 11/09/1957, p. 2).

10. *Hansard's Parliamentary Debates*, Commons, 5th ser., 596 (1958), cc. 369–70. See also Newburn, *Permission and Regulation*, p. 56.

11. *Hansard's Parliamentary Debates*, Lords, 5th ser., 266 (1965), c. 136.

12. Ibid., c. 138. See also Higgins, *Heterosexual Dictatorship*, pp. 120–1.

13. HO 345/4: notes by committee members concerning their recommendations, n.d. There is no note from Mishcon and Adair was against any age of consent.

14. HO 345/4, 'Note by the Secretary', Sept. 1955.

15. HO 345/15, 14 Dec. 1955 (British Psychological Society), Q4100.

16. HO 345/4: 'Note by the Chairman; Comments on the Secretary's Note of September, 1955'.

17. See Lucy Robinson, *Gay Men and the Left in Post-War Britain: How the Personal Got Political* (Manchester: Manchester University Press, 2007), pp. 38–40; Stephen Brooke, *Sexual Politics, Family Planning, and the British Left from the 1880s to the Present Day* (Oxford: Oxford University Press, 2011), pp. 154–5, 177–9; Harold Nicolson, *Diaries and Letters*, vol. III: *The Later Years 1945–1962* (New York: Atheneum, 1968), p. 355.

18. Wolfenden, *Turning Points*, pp. 140–1; Higgins, *Heterosexual Dictatorship*, pp. 116–17; Davidson and Davis, *The Sexual State*, pp. 57–8. The *Daily Mirror*, in its support for reform, was one of the few exceptions among the popular press. See Bingham, *Family Newspapers*, pp. 188–91.

19. *Daily Mail*, 5 Sept. 1957, p. 1.

20. *Times*, 15 Nov. 1957, p. 7. The Church of Scotland's General Assembly voted against reform (Davidson and Davis, *Sexual State*, pp. 55–7).

21. *Economist*, 7 Dec. 1957, p. 844.

22. Brooke, *Sexual Politics*, pp. 155–6, 181.

23. Moran, 'The Homosexualization of English Law', pp. 21–3; Moran, *The Homosexual(ity) of Law*, pp. 115–17; Waites, *The Age of Consent*, pp. 110–13.

24. See Lisa Power, *No Bath but Plenty of Bubbles: An Oral History of the Gay Liberation Front 1970–73* (London: Cassell, 1995).

25. At the Conservative Party Conference in Manchester in 2011 (*Guardian*, 5 Oct. 2011).

26. Barbara Wootton, 'Sickness or Sin?' *The Twentieth Century*, 159 (May 1956), 433–42.

27. HO 345/9, correspondence between Roberts and the Swedish authorities, annex III: interim reply from the Swedish Ministry of Foreign Affairs, 19 Jan. 1956: '[A] recent investigation in Gothenburg showed a frequency of about 1% of all men primarily homosexual, that is sexually exclusively interested in other men, and 4% secondarily homosexual, that is with both homosexual and heterosexual impulses. Other information available seems to indicate, however, that these figures are too low.'

Bibliography

Primary Sources

BBC Written Archives Centre, Caversham, Reading
   Transcripts of radio broadcasts.
British Library, National Sound Archive, London
   Hall-Carpenter Oral History Project: C456/089, Patrick Trevor-Roper, 1 Aug. 1990.
Hansard's Parliamentary Debates, Commons, 5th ser.; Lords, 5th ser.
The National Archives, Kew
   Cabinet Office Papers: CAB 128–9.
   Home Office Papers [the Wolfenden Papers]: HO 345/1–18.

Newspapers and Journals

British Medical Journal, Daily Mail, Economist, Glasgow Herald, Lancet, Manchester Guardian, Observer, Picture Post, Sunday Times, The Times.

Published Sources

Anomaly. *The Invert and His Social Adjustment*. London: Baillière, Tindall and Cox, 1927.
Bailey, Derrick Sherwin. 'The Problem of Sexual Inversion', *Theology*, 55, 380 (1952), 47–52.
Bengry, Justin. 'Queer Profits: Homosexual Scandal and the Origins of Legal Reform in Britain', in Heike Bauer and Matt Cook (eds), *Queer 1950s: Rethinking Sexuality in the Postwar Years*. Houndmills: Palgrave Macmillan, 2012.
Berg, Charles. *Fear, Punishment, Anxiety and the Wolfenden Report*. London: George Allen and Unwin, 1959.
Bew, John. *Castlereagh: A Life*. Oxford: Oxford University Press, 2012.
Bingham, Adrian. *Family Newspapers? Sex, Private Life, and the British Popular Press, 1918–1978*. Oxford: Oxford University Press, 2009.
———. 'The "K-Bomb": Social Surveys, the Popular Press, and British Sexual Culture in the 1940s and 1950s', *Journal of British Studies*, 50, 1 (January 2011), 156–79.
Bland, Lucy and Laura Doan (eds). *Sexology in Culture: Labelling Bodies and Desires*. Chicago, IL: University of Chicago Press, 1998.
———. *Sexology Uncensored: The Documents of Sexual Science*. Chicago, IL: University of Chicago Press, 1998.
Bowlt, Eileen M. *Justice in Middlesex: A Brief History of the Uxbridge Magistrates' Court*. Winchester: Waterside Press, 2007.
Brasnett, Margaret E. *Voluntary Social Action: A History of the National Council of Social Service, 1919–1969*. London: National Council of Social Service, 1969.
Brooke, Stephen. *Sexual Politics, Family Planning, and the British Left from the 1880s to the Present Day*. Oxford: Oxford University Press, 2011.
Brown, Callum. *The Death of Christian Britain: Understanding Secularization, 1800–2000*. London: Routledge, 2001.
Burke, Thomas. *For Your Convenience: A Learned Dialogue, Instructive to all Londoners and London Visitors, Overheard in the Thélème Club and Taken Down Verbatim by Paul Pry*. London: Routledge, 1937.
Cambridge Department of Criminal Science. *Sexual Offences: A Report*. London: Macmillan, 1957.

Carpenter, Edward. *The Intermediate Sex: A Study of Some Transitional Types of Men and Women.* New York and London: Mitchell Kennerley, 1912.

Chesser, Eustace. *Live and Let Live: The Moral of the Wolfenden Report.* London: Heinemann, 1958.

Church of England Moral Welfare Council. *The Problem of Homosexuality: An Interim Report.* Oxford: Church Information Board, 1954.

Cocks, H. G. *Nameless Offences: Homosexual Desire in the 19th Century.* London: I. B. Tauris, 2003.

Cohen, Deborah. *Family Secrets: Living with Shame from the Victorians to the Present Day.* London: Viking, 2013.

Cook, Matt. *London and the Culture of Homosexuality, 1885–1914.* Cambridge: Cambridge University Press, 2003.

———. 'Queer Conflicts: Love, Sex and War, 1914–1967', in Matt Cook (ed.), *A Gay History of Britain: Love and Sex Between Men Since the Middle Ages.* Oxford: Greenwood, 2007.

Cooke, Bill. *The Gathering of Infidels: A Hundred Years of the Rationalist Press Association.* Amherst, NY: Prometheus Books, 2004.

Cowell, Roberta. *Roberta Cowell's Story by Herself: Her Autobiography.* New York: British Book Centre, 1954.

Croft-Cooke, Rupert. *The Verdict of You All.* London: Secker and Warburg, 1955.

Crompton, Louis (ed.). 'Jeremy Bentham's Essay on "Paederasty"', Parts 1 and 2, *Journal of Homosexuality*, 3, 4 (Summer 1978), 383–405, and 4, 1 (Fall 1978), 91–107.

Crozier, Ivan. 'Introduction: Havelock Ellis, John Addington Symonds and the Construction of *Sexual Inversion*', in Ellis and Symonds, *Sexual Inversion: A Critical Edition.* Houndmills, Basingstoke: Palgrave Macmillan, 2008.

Daley, Harry. *This Small Cloud: A Personal Memoir.* London: Weidenfeld and Nicolson, 1986.

David, Hugh. *On Queer Street: A Social History of British Homosexuality, 1895–1995.* London: HarperCollins, 1997.

Davidson, Roger and Gayle Davis. *The Sexual State: Sexuality and Scottish Governance, 1950–80.* Edinburgh: Edinburgh University Press, 2012.

Dempsey, Brian. 'Piecemeal to Equality: Scottish Gay Law Reform', in Leslie J. Moran, Daniel Monk and Sarah Beresford (eds), *Legal Queeries: Lesbian, Gay and Transgender Legal Studies.* London: Cassell, 1998.

Devlin, Patrick. *The Enforcement of Morals.* London: Oxford University Press, 1965.

Dickinson, Tommy. *'Curing Queers': Mental Nurses and Their Patients, 1935–74.* Manchester: Manchester University Press, 2015.

Dicks, H. V. *Fifty Years of the Tavistock Clinic.* London: Routledge and Kegan Paul, 1970.

Drucker, Donna J. *The Classification of Sex: Alfred Kinsey and the Organization of Knowledge.* Pittsburgh, PA: University of Pittsburgh Press, 2014.

Ellis, Havelock. *Studies in the Psychology of Sex*, vol. 2: 'Sexual Inversion'. 3rd edn, 1927.

Farson, Daniel. *Soho in the Fifties.* London: Michael Joseph, 1987.

Faulks, Sebastian. *The Fatal Englishman: Three Short Lives.* London: Hutchinson, 1996.

Ferris, Paul. *Sex and the British: A Twentieth-Century History.* London: Michael Joseph, 1993.

Freud, Sigmund. *Three Essays on the Theory of Sexuality.* Trans. James Strachey. New York: Basic Books, 1962; 1st edn 1905.

Galton, Francis. 'The History of Twins, as a Criterion of the Relative Powers of Nature and Nurture', *Fraser's Magazine,* 12 (November 1875), 566–76.

Garland, Rodney [Adam de Hegedus]. *The Heart in Exile.* London: W. H. Allen, 1953.

Gathorne-Hardy, Jonathan. *Alfred C. Kinsey: Sex the Measure of All Things.* London: Chatto and Windus, 1998.

Golla, F. L. and R. Sessions Hodge. 'Hormone Treatment of the Sexual Offender', *The Lancet,* CCLVI, I (11 June 1949), 1006–7.

Grey, Antony. *Quest for Justice: Towards Homosexual Emancipation.* London: Sinclair Stevenson, 1992.

Grimley, Matthew. 'Law, Morality and Secularisation: The Church of England and the Wolfenden Report, 1954–1967', *Journal of Ecclesiastical History*, 60, 4 (October 2009), 725–41.

Hall, Lesley. *Sex, Gender and Social Change in Britain since 1880*. London: Palgrave Macmillan, 2nd edn, 2013.

Hall, Stuart. 'Reformism and the Legislation of Consent', in National Deviancy Conference (ed.), *Permissiveness and Control: The Fate of the Sixties Legislation*. New York: Barnes and Noble, 1980.

Harris, John. *Goronwy Rees*. Cardiff: University of Wales Press, 2001.

Harrison, Brian. *Seeking a Role: The United Kingdom, 1951–1970*. Oxford: Clarendon Press, 2009.

Hart, H. L. A. *Law, Liberty and Morality*. Stanford, CA: Stanford University Press, 1963.

Heaman, Elsbeth. *St Mary's: The History of a London Teaching Hospital*. Liverpool: Liverpool University Press, 2003.

Heim, Nikolaus and Carolyn J. Hursch. 'Castration for Sex Offenders: Treatment or Punishment? A Review and Critique of Recent European Literature', *Archives of Sexual Behavior*, 8, 3 (May 1979), 281–304.

Higgins, Patrick. *Heterosexual Dictatorship: Male Homosexuality in Postwar Britain*. London: Fourth Estate, 1996.

Hocquenghem, Guy. *Homosexual Desire*. Durham, NC: Duke University Press, 1993; 1st edn 1972.

Hodges, Andrew. *Alan Turing: The Enigma*. Princeton, NJ: Princeton University Press, 2012; 1st edn 1983.

Holden, Andrew. *Makers and Manners: Politics and Morality in Postwar Britain*. London: Politico's, 2004.

Home Office and Scottish Home Department. *Report of the Committee on Homosexual Offences and Prostitution* [*The Wolfenden Report*]. London: HMSO, 1957.

Hornsey, Richard. *The Spiv and the Architect: Unruly Life in Postwar London*. Minneapolis, MN: University of Minnesota Press, 2010.

Houlbrook, Matt. 'Soldier Heroes and Rent Boys: Homosex, Masculinities, and Britishness in the Brigade of Guards, circa 1900–1960', *Journal of British Studies*, 42, 3 (July 2003), 351–88.

———. *Queer London: Perils and Pleasures in the Sexual Metropolis, 1918–1957*. Chicago, IL: University of Chicago Press, 2005.

——— and Chris Waters. 'The Heart in Exile: Detachment and Desire in 1950s London', *History Workshop Journal*, 62 (Autumn 2006), 142–65.

Hyde, H. Montgomery.*The Strange Death of Lord Castlereagh*. London: Heinemann, 1959.

———. *The Trials of Oscar Wilde*. London: Penguin, 1962.

———. *The Love That Dared Not Speak Its Name*. Boston, MA: Little, Brown and Company, 1970

Jarvis, Mark. *Conservative Governments, Morality and Social Change in Affluent Britain, 1957–64*. Manchester: Manchester University Press, 2005.

Jeffery-Poulter, Stephen. *Peers, Queers, and Commons: The Struggle for Gay Law Reform from 1950 to the Present*. London: Routledge, 1991.

Jennings, Rebecca. *A Lesbian History of Britain: Love and Sex between Women since 1500*. Oxford: Greenwood, 2007.

———. *Tomboys and Bachelor Girls: A Lesbian History of Post-War Britain, 1945–71*. Manchester: Manchester University Press, 2007.

Jivani, Alkarim. *It's Not Unusual: A History of Lesbian and Gay Britain in the Twentieth Century*. Bloomington, IN: Indiana University Press, 1997.

Joad, C. E. M. (ed.). *Manifesto: Being the Book of the Federation of Progressive Societies and Individuals*. London: George Allen and Unwin, 1934.

Jones, James H. *Alfred C. Kinsey: A Public/Private Life*. New York: Norton, 1997.

Jones, Timothy W. 'Moral Welfare and Social Well-Being: The Church of England and the Emergence of Modern Homosexuality', in Lucy Delap and Sue Morgan (eds), *Men,*

Masculinities and Religious Change in Twentieth-Century Britain. Houndmills: Palgrave Macmillan, 2013.

———. *Sexual Politics in the Church of England, 1857–1957.* Oxford: Oxford University Press, 2013.

Jowitt, Lord. 'The Twenty-Eighth Maudsley Lecture: Medicine and the Law', *Journal of Mental Science,* 100, 419 (April 1954), 351–9.

Judge, Tony. *Radio Philosopher: The Radical Life of Cyril Joad.* London: Alpha House Books, 2012.

Kahr, Brett. *D. W. Winnicott: A Biographical Portrait.* London: H. Karnac, 1996.

Kallmann, Franz J. 'Twin and Sibship Study of Overt Male Homosexuality', *American Journal of Human Genetics* 4, 2 (June 1952), 136–46.

———. *Heredity in Health and Mental Disorder: Principles of Psychiatric Genetics in the Light of Comparative Twin Studies.* New York: W. W. Norton, 1953.

Kennedy, Pagan. *The First Man-Made Man: The Story of Two Sex Changes, One Love Affair, and a Twentieth-Century Medical Revolution.* New York: Bloomsbury, 2007.

Kenny, Courtney Stanhope. *Outlines of Criminal Law: Based on Lectures Delivered in the University of Cambridge,* 15th edn, rev. G. Godfrey Phillips. Cambridge: Cambridge University Press, 1936.

King, Michael, Glenn Smith and Annie Bartlett, 'Treatments of Homosexuality in Britain since the 1950s—An Oral History', *British Medical Journal,* 328, 7437 (21 February 2004), 427–32.

Kinsey, Alfred et al. *Sexual Behavior in the Human Male.* Philadelphia, PA: W. B. Saunders, 1948.

Kinsey, Alfred et al. 'Concepts of Normality and Abnormality in Sexual Behavior', in Paul H. Hoch and Joseph Zubin (eds), *Psychosexual Development in Health and Disease* (Proceedings of the 38th Annual Meeting of the American Psychopathological Association, New York, June 1948). New York: Grune and Stratton, 1949.

———. *Sexual Behavior in the Human Female.* Philadelphia, PA: W. B. Saunders, 1953.

Kynaston, David. *Family Britain, 1951–57.* London: Bloomsbury, 2009.

Lang, Theo. 'Studies on the Genetic Determination of Homosexuality', *Journal of Nervous and Mental Disease,* 92, 1 (July 1940), 55–64.

Law, Stephen. *Humanism: A Very Short Introduction.* Oxford: Oxford University Press, 2011.

Le Maire, Louis. 'Danish Experiences Regarding the Castration of Sexual Offenders', *Journal of Criminal Law and Criminology,* 47, 3 (1956), 294–310.

Learoyd, G. C. 'The Problem of Homosexuality', *Practitioner,* 172 (April 1954), 355–63.

Lewis, Brian (ed.). *British Queer History.* Manchester: Manchester University Press, 2013.

McGhee, Derek. 'Wolfenden and the Fear of "Homosexual Spread": Permeable Boundaries and Legal Defences', *Studies in Law, Politics, and Society,* 21 (2000), 65–97.

McGonville, Sean. 'The Victorian Prison: England, 1865–1965', in Norval Morris and David J. Rothman (eds), *The Oxford History of the Prison: The Practice of Punishment in Western Society.* Oxford: Oxford University Press, 1995.

Machin, G. I. T. *Churches and Social Issues in Twentieth-Century Britain.* Oxford: Clarendon Press, 1998.

McLaren, Angus. *Sexual Blackmail: A Modern History.* Cambridge, MA: Harvard University Press, 2002.

McManus, Michael. *Tory Pride and Prejudice: The Conservative Party and Homosexual Law Reform.* London: Biteback Publishing, 2011.

Mayne, Xavier [Edward Irenaeus Prime Stevenson]. *The Intersexes: A History of Similisexualism as a Problem in Social Life.* Privately printed, 1908.

Meek, Jeff. 'Scottish Churches, Morality and Homosexual Law Reform, 1957–1980', *Journal of Ecclesiastical History,* 66, 3 (July 2015), 596–613.

Meyerowitz, Joanne. *How Sex Changed: A History of Transsexuality in the United States.* Cambridge, MA: Harvard University Press, 2002.

Moffat, Wendy. *A Great Unrecorded History: A New Life of E. M. Forster.* New York: Farrar, Straus and Giroux, 2010.

Montagu of Beaulieu. *Wheels within Wheels: An Unconventional Life.* London: Weidenfeld and Nicolson, 2000.

Moran, Leslie J. 'The Homosexualization of English Law', in Didi Herman and Carl Stychin (eds), *Legal Inversions: Lesbians, Gay Men, and the Politics of Law.* Philadelphia, PA: Temple University Press, 1995.

———. *The Homosexual(ity) of Law.* London: Routledge, 1996.

Morley, Sheridan. *John Gielgud: The Authorized Biography.* New York: Simon and Schuster, 2010.

Mort, Frank. 'Mapping Sexual London: The Wolfenden Committee on Homosexual Offences and Prostitution 1954–57', *New Formations*, 37 (Spring 1999), 92–113.

———. *Capital Affairs: London and the Making of the Permissive Society.* New Haven, CT: Yale University Press, 2010.

Mounier, Emmanuel. *Personalism.* London: Routledge and Kegan Paul, 1952.

Newburn, Tim. *Permission and Regulation: Law and Morals in Post-War Britain.* London: Routledge, 1992.

Nicolson, Harold. *Diaries and Letters*, vol. III: *The Later Years, 1945–1962.* New York: Atheneum, 1968.

Norwood East, William and William Henry de Bargue Hubert. *Report on the Psychological Treatment of Crime.* London: HMSO, 1939.

Oram, Alison. 'Cross-Dressing and Transgender', in Matt Houlbrook and Harry Cocks (eds), *The Modern History of Sexuality.* Houndmills, Basingstoke: Palgrave Macmillan, 2006.

Oxford Dictionary of National Biography [ODNB online]. Oxford: Oxford University Press, 2004.

Plummer, Kenneth (ed.). *The Making of the Modern Homosexual.* London: Hutchinson, 1981.

Porter, Kevin and Jeffrey Weeks (eds). *Between the Acts: Lives of Homosexual Men, 1885–1967.* London: Routledge, 1991.

Power, Lisa. *No Bath but Plenty of Bubbles: An Oral History of the Gay Liberation Front, 1970–73.* London: Cassell, 1995.

Prochaska, Frank K. *Women and Philanthropy in Nineteenth-Century England.* Oxford: Oxford University Press, 1980.

Raven, Simon. 'Boys will be Boys: The Male Prostitute in London', *Encounter*, xv, 1 (July 1960), 19–24.

Rayside, David M. 'Homophobia, Class and Party in England', *Canadian Journal of Political Science*, 25, 1 (March 1992), 121–49.

Rees, Goronwy. *A Bundle of Sensations: Sketches in Autobiography.* London: Chatto and Windus, 1960.

———. *A Chapter of Accidents.* London: Chatto and Windus, 1972.

Rees, Jenny. *Looking for Mr Nobody: The Secret Life of Goronwy Rees.* London: Weidenfeld and Nicolson, 1994.

Renault, Mary. *The Charioteer.* London: Longman, 1953.

Rhodes James, Robert. *Robert Boothby: A Portrait of Churchill's Ally.* New York: Viking, 1991.

Robinson, Lucy. *Gay Men and the Left in Post-War Britain: How the Personal Got Political.* Manchester: Manchester University Press, 2007.

Rochester, Bishop of [Christopher Chavasse]. 'The Church and Sex', *Practitioner*, 172 (April 1954), 350–4.

Roman Catholic Advisory Committee. 'Homosexuality, Prostitution and the Law', *Dublin Review*, 230, 471 (Summer 1956), 60–5.

Rose, Sonya O. *Which People's War? National Identity and Citizenship in Wartime Britain, 1939–1945.* Oxford: Oxford University Press, 2003.

Royle, Edward. *Radicals, Secularists, and Republicans: Popular Freethought in Britain, 1866–1915.* Manchester: Manchester University Press, 1980.

Schlossman, Steven. 'Delinquent Children: The Juvenile Reform School', in Norval Morris and David J. Rothman (eds), *The Oxford History of the Prison: The Practice of Punishment in Western Society*. Oxford: Oxford University Press, 1995.

Sedgwick, Eve Kosofsky. *Epistemology of the Closet*. Berkeley and Los Angeles, CA: University of California Press, 1990.

Shapira, Michal. *The War Inside: Psychoanalysis, Total War, and the Making of the Democratic Self in Postwar Britain*. Cambridge: Cambridge University Press, 2013.

Sibalis, Michael D. 'The Regulation of Male Homosexuality in Revolutionary and Napoleonic France, 1789–1815', in Jeffrey Merrick and Bryant T. Ragan, Jr (eds), *Homosexuality in Modern France*. Oxford: Oxford University Press, 1996.

———. 'Homophobia, Vichy France, and the "Crime of Homosexuality": The Origins of the Ordinance of 6 August 1942', *GLQ: A Journal of Lesbian and Gay Studies*, 8, 3 (2002), 301–18.

Sinfield, Alan. *Out on Stage: Lesbian and Gay Theatre in the Twentieth Century*. New Haven, CT: Yale University Press, 1999.

Smith, F. B. 'Labouchère's Amendment to the Criminal Law Amendment Bill', *Historical Studies (Australia)*, 17, 67 (1976), 165–75.

Soloway, Richard A. *Demography and Degeneration: Eugenics and the Declining Birthrate in Twentieth-Century Britain*. Chapel Hill, NC: University of North Carolina Press, 1990.

Stanley, Liz. *Sex Surveyed, 1949–1994: From Mass-Observation's 'Little Kinsey' to the National Survey and the Hite Reports*. London: Taylor and Francis, 1995.

Szreter, Simon and Kate Fisher. *Sex before the Sexual Revolution: Intimate Life in England, 1918–1963*. Cambridge: Cambridge University Press, 2010.

Tudor Rees, J. and Harley V. Usill (eds). *They Stand Apart: A Critical Survey of the Problems of Homosexuality*. London: William Heinemann, 1955.

Vickers, Emma. *Queen and Country: Same-Sex Desire in the British Armed Forces, 1939–45*. Manchester: Manchester University Press, 2013.

Waites, Matthew. *The Age of Consent: Young People, Sexuality and Citizenship*. Houndmills: Palgrave Macmillan, 2005.

———. 'The Fixity of Sexual Identities in the Public Sphere: Biomedical Knowledge, Liberalism and the Heterosexual/Homosexual Binary in Late Modernity', *Sexualities*, 8, 5 (December 2005), 539–69.

Waters, Chris. 'Havelock Ellis, Sigmund Freud and the State: Discourses of Homosexual Identity in Interwar Britain', in Lucy Bland and Laura Doan (eds), *Sexology in Culture: Labelling Bodies and Desires*. Chicago, IL: University of Chicago Press, 1998.

———. 'Disorders of the Mind, Disorders of the Body Social: Peter Wildeblood and the Making of the Modern Homosexual', in Becky Conekin, Frank Mort and Chris Waters (eds), *Moments of Modernity: Reconstructing Britain, 1945–1964*. London: Rivers Oram Press, 1999.

———. 'Sexology', in Matt Houlbrook and Harry Cocks (eds), *The Modern History of Sexuality*. Houndmills, Basingstoke: Palgrave Macmillan, 2006.

———. 'The Homosexual as a Social Being in Britain, 1945–1968', in Brian Lewis (ed.), *British Queer History*. Manchester: Manchester University Press, 2013.

Weeks, Jeffrey. *Coming Out: Homosexual Politics in Britain, from the Nineteenth Century to the Present*. London: Quartet Books, 1977.

———. 'Inverts, Perverts, and Mary-Annes: Male Prostitution and the Regulation of Homosexuality in England in the Nineteenth and Early Twentieth Centuries', *Journal of Homosexuality*, 6, 1/2 (Fall/Winter 1980–1), 113–34.

———. *The World We Have Won: The Remaking of Erotic and Intimate Life*. New York: Routledge, 2007.

———. *Sex, Politics and Society: The Regulation of Sexuality Since 1800*. London: Pearson, 3rd edn, 2012.

West, D. J. *Homosexuality*. London: Duckworth, 1955.

Westwood, Gordon [Michael Schofield]. *Society and the Homosexual*. London: Victor Gollancz, 1952.

———. *A Minority: A Report on the Life of the Male Homosexual in Great Britain.* London: Longmans, 1960.

Who Was Who [online edn]. Oxford: Oxford University Press, 2014.

Wildeblood, Peter. *Against the Law.* London: Phoenix, 2000; 1st edn 1955.

———. *A Way of Life.* London: Weidenfeld and Nicolson, 1956.

Wilkins, Leslie T. 'Persistent Offenders and Preventive Detention', *Journal of Criminal Law, Criminology and Police Science*, 57, 3 (1967), 312–17.

Wills, Abigail. 'Delinquency, Masculinity and Citizenship in England, 1950–1970', *Past and Present*, 187 (May 2005), 157–85.

Wolfenden, John. *Turning Points: The Memoirs of Lord Wolfenden.* London: Bodley Head, 1976.

Index

Note: The letter 'n' following locators refers to notes.

Lightning Source UK Ltd.
Milton Keynes UK
UKOW06f2207310316

271253UK00003BA/87/P